THE GEOPOLITICS READER
Second edition

This extensively revised second edition of *The Geopolitics Reader* draws together the most influential and significant geopolitical readings from the last hundred years.

This Reader considers five different domains of geopolitical thought and practice: imperialist, Cold War and twenty-first century geopolitics as well as special sections that address global dangers and resistances to the practices of geopolitics. The editors provide comprehensive introductions and critical analysis for each of the five parts. Political cartoons are integrated throughout. Divergent viewpoints on geopolitics are represented in the selection of readings, which have been updated to incorporate the best critical scholarship on the geopolitics of 9/11, the Iraq War and global terrorism.

Including articles by Halford Mackinder, Theodore Roosevelt, Adolf Hitler, George Kennan, Samuel Huntington, Edward Said, Osama Bin Laden and American neoconservatives, this volume is great for classroom instruction and debates, provoking lively discussion of how questions of discourse and power are at the centre of the critical study of geopolitics.

Gearóid Ó Tuathail is Professor in Government and International Affairs at Virginia Tech's campus in Washington, DC; **Simon Dalby** is Professor in the Department of Geography and Environmental Studies at Carleton University in Ottawa; **Paul Routledge** is Reader in Human Geography in the Department of Geographical and Earth Sciences at the University of Glasgow, Scotland.

The Geopolitics Reader

Second edition

Edited by

Gearóid Ó Tuathail,

Simon Dalby and

Paul Routledge

Routledge
Taylor & Francis Group

LONDON AND NEW YORK

First edition published 1998
by Routledge

Second edition published 2006
by Routledge
2 Park Square, Milton Park, Abingdon, Oxon OX14 4RN

Simultaneously published in the USA and Canada
by Routledge
270 Madison Ave, New York, NY 10016

Routledge is an imprint of the Taylor & Francis Group

Transferred to Digital Printing 2006

Typeset in Amasis and Akzidenz Grotesk by Keystroke, Jacaranda Lodge, Wolverhampton

British Library Cataloguing in Publication Data
A catalogue record for this book is available from the British Library

Library of Congress Cataloging in Publication Data
The geopolitics reader / edited by Gearóid Ó Tuathail, Simon Dalby, and
Paul Routledge. — 2nd ed.
p. cm.
Includes bibliographical references and index.
1. Geopolitics. I. Ó Tuathail, Gearóid. II. Dalby, Simon.
III. Routledge, Paul, 1956– .
JC319.G646 2006
320.1'2—dc22   2005019590

ISBN10: 0–415–34147–7 (hbk)
        0–415–34148–5 (pbk)

ISBN13: 9–78–0–415–34147–9 (hbk)
        9–78–0–415–34148–6 (pbk)

Printed and bound by CPI Antony Rowe, Eastbourne

untrenched solidarity movement
Irish
British AID worker.

To Rachel Corrie (1979–2003), Margaret Hassan (1945–2004), Daniel Pearl (1963–2002) and Edward Said (1936–2003).

American
journalist
killed by
Al Quaeda.

Contents

Illustrations

Acknowledgements

The editors would like to thank the numerous instructors and students who provided us with valuable comments on the first edition of *The Geopolitics Reader*. We have incorporated this feedback to create this improved, updated and slightly expanded second edition. We would like to thank our publisher Routledge for their commitment to the second edition and patience with the many delays in its production. Gearóid Ó Tuathail would like to thank David Tomblin for his assistance in tracking down copyright permissions and in doing some scanning when needed. The Alexandria Center of the Virginia Tech campus in Northern Virginia provided computer facilities and other equipment which made this project possible. John Agnew, Matt Coleman, Carl Dahlman, Michael Dear, Tim Luke, David Newman, John O'Loughlin, Anna Secor, Matt Sparke and Lasha Tchantouridze provided encouragement. Sabine Durier provided inspiration with her dedication to the global struggle against AIDS.

Simon Dalby once again thanks Susan Tudin for her excellent work in the Carleton University Library, Dale Armstrong for his patience in tracking down materials and getting the details right, and Cara Stewart for her forbearance through the preparation of another edition. Paul Routledge would like to thank the Department of Geographical and Earth Sciences, University of Glasgow for support and technical assistance, and Teresa Flavin and all rebel clowns for inspiration, encouragement and good counsel.

All three of us would like to thank the institutions, publishing houses, agents and authors who kindly provided us with the copyright permissions needed to compose this edition. As before, we have decided to dedicate the volume to an inspiring intellectual who has passed the scene, and to inspiring professionals and activists who lost their lives because they demonstrated the courage to make a difference in the face of danger and hatred.

GENERAL INTRODUCTION

Thinking Critically about Geopolitics

Gearóid Ó Tuathail

All words have histories and geographies and the term "geopolitics" is no exception. Coined in 1899 by a Swedish political scientist named Rudolf Kjellen, the word "geopolitics" had a twentieth century history that was intimately connected with the belligerent dramas of that century. For Kjellen at the beginning of the century, geopolitics was a useful word to describe the geographical base of the state, its natural endowment and resources, which many claimed determined its power potential (Holdar, 1992). Kjellen's term was taken up in Germany after World War I, with a former general named Karl Haushofer founding a journal called *Zeitschrift für Geopolitik* (*Journal of Geopolitics*) in 1924 to promote conservative nationalist thinking. Haushofer's aide-de-camp during World War I was Rudolf Hess, who introduced General Haushofer to another veteran and aspirant politician, Adolf Hitler. As Hitler rose to power and launched wars of aggression against Germany's neighbors, "geopolitics" entered the English language as a translation of *Geopolitik* and developed an association with expansionist Nazi foreign policy. As this policy culminated in the horrors of World War II, the word "geopolitics" became taboo and fell out of favor with many writers and commentators. During the Cold War, however, the word returned as a description of the global contest between the Soviet Union and the United States for influence over the states and strategic resources of the world. The reason for its revival was the rise of a persecuted Jewish refugee from Nazi Germany to exalted positions of power in the US state. As National Security advisor and later Secretary of State, Henry Kissinger almost single-handedly revived the term in the 1970s, using it as a synonym for the balance-of-power politics he envisioned himself performing heroically across the globe (Hanhimäki, 2004; Hepple, 1986; Isaacson, 1992).

Since then the term geopolitics has enjoyed wide circulation. Journalists, foreign policy makers and strategic analysts employ it frequently, with its meaning defined by the particular context of its use. But, irrespective of whether the word geopolitics is used or not, the conventional understanding today is that geopolitics is discourse about world politics, with a particular emphasis on state competition and the geographical dimensions of power. Thus to study geopolitics we must study discourse, which can be defined as the representational practices by which cultures creatively constitute meaningful worlds (what Spivak (1988) terms "worlding"). As we will see, most cultures do this by means of stories (narratives) and images. Since geopolitics is a discourse with distinctive "world" constitutive ambitions – it seeks to make "world politics" meaningful – we must be attentive to the ways in which global space is labeled, metaphors are deployed, and visual images are used in this process of making stories and constructing images of world politics.

We can begin to think critically about geopolitics by considering why it is an attractive discourse to journalists, politicians and strategic thinkers. First, geopolitical discourse deals with compelling questions of power and danger in world affairs. Where are the axes of power and conflict in the world? What are the dangers and threats that face the world? These are important questions for political elites and educated segments of the general public. While these queries can be scholarly, there is also a self-interested agenda behind many geopolitical questions. Ruling elites and an educated general public usually want to know what the distribution of power and danger in world affairs mean for their state and its role in world affairs. What are the emergent threats *we* face? How should *our* state conduct its foreign policy in a world of dangers and enemies? What resources and friends do *we* need in order to protect *ourselves* from *them*? Obviously, these questions are not neutral inquiries but bound up with varied political agendas and nationalist identity formations – particular construction of our, we and them – within states (Campbell, 1998). The critical point to grasp at the outset is that geopolitics is already involved in world politics; it is not separate neutral commentary on it.

Second, geopolitics is attractive because it purports to explain a great deal in simple terms. Geopolitical discourse offer busy policy makers, journalists and citizens a seemingly comprehensive vision of world politics. It provides a framework within which local events in one place can be related to a large global picture. A hotel bomb in Jakarta, an assassination in Beirut or an embassy attack in Nairobi, for example, can be pulled together into a single framework: "the global struggle against Islamic terrorism" (irrespective of whether it is true or not). We need to take the metaphors "vision," "framework" and "picture" seriously here because one of the seductive qualities of geopolitical discourse is how it transforms the opaqueness of world affairs into an apparently clear picture (Agnew, 2003). Geopolitics involves "framework," the work of creating frames for interpreting events and making them meaningful. Many geopolitical narratives are enframed by essentialized oppositions between "us" and "them," the "civilized" versus the "fanatical." Whole regions of the world are divided into oppositional zones, a frameworking we can call "earth labeling." For example, geopoliticians use grand spatial abstractions like the Islamic World, the Non-Integrating Gap, the Global South or the Civilized World. Other spatial metaphors like heartlands, faultlines, and axes are popular. All these expressions draw rhetorical force from their ability to reduce the complexity of world politics to a simplified framework.

Finally, geopolitics is popular because it promises insight into the future direction of world affairs. What is the coming shape of the world political map? Where will the wars of the future be fought and over what? Geopolitics has a certain magical appeal because it aspires to be prophetic discourse. Journalists and politicians look to geopolitics for crystal ball visions of the future, visions that get beyond the fog of everyday events to reveal the supposed permanent clashing faultlines or primordial struggles between essentialized differences (freedom versus tyranny, the West versus the Rest, Christianity versus Islam, etc.). The publishing industry caters to this desire by promoting variations on simplified overstatements as the next "big idea" in geopolitics. An inane example from a few years ago is the declaration that "on major strategic and international questions today, Americans are from Mars and Europeans are from Venus" (Kagan, 2002: 3). Because those most interested in international affairs live in a globalizing world characterized by information saturation, the desire for simplified nostrums packaged as "strategic insight" is strong.

The goal of *The Geopolitics Reader* is to foster critical thinking about the history and contemporary forms of geopolitical discourse. The first edition of *The Geopolitics Reader* was put together in the mid to late 1990s and reflected debate about the end of the Cold War and the Gulf War of 1991. Since then the dramas of world politics have been characterized by continuity and change. There is continuity in that global environmental trends, like global warming and attendant rising sea levels, remain daunting challenges with potentially significant implications for the future of human habitation on the planet. In February 2005, the Kyoto Protocol regulating the emission of greenhouse gases became international law after 141 states endorsed it, including all states in the enlarged European Union and the Russian

Federation. The United States, the world's largest polluter, withdrew from the Protocol in 2001 in deference to corporate interests, particularly the fossil fuel industrial complex.

Hostility to collective action against the long term degradation of the planet by the occupants of the White House is not new (we documented this in the first *Geopolitics Reader*). What is new, from their point of view, is the global war against terrorism that began when terrorists attacked the World Trade Center and Pentagon on September 11, 2001. For most Americans, that date marks a decisive moment of rupture with the past, a new post-September 11 era that marked the end of the post-Cold War era. The US president declared the United States at war and the phrase "global war on terror" became so ubiquitous within the US government that it earned a bureaucratic acronym: GWOT.

Amidst this trumpeting of a new war, certain continuities in how US state leaders conceptualized the role of their state in world affairs are apparent. As during the Cold War, US leaders view their state as leading the "free world" in a crusade against "tyranny" and "evil" (Daalder and Lindsey, 2003; Frum and Pearle, 2003). Instead of an "evil empire," Ronald Reagan's simplifying description of the Soviet Union, the United States is battling an "axis of evil," George W. Bush's shorthand for the official enemies of the United States in 2002: Iraq, Iran and North Korea. Yet, unlike the Cold War era and the strategy of his father during the first Gulf War, the Bush administration subsequently authorized the invasion of Iraq without the support of the United Nations and the backing of only a few allies. This illegal action and the general unilateralism of the Bush administration produced a significant rift in transatlantic relations. The subsequent exposure of the Bush administration's reasons for going to war – that Iraq constituted an immanent threat to the United States because it had weapons of mass destruction – as empirically false deepened the rift with international opinion generally. But GWOT and the Iraq war has been good for certain groups within the United States. The US Department of Defense budget is at a record level and it remains the most powerful bureaucracy within the US state. US defense contractors, some with strong ties to the White House, are cashing in on the swelling appropriations. And, despite dangerously low popularity ratings, George W. Bush was able to use his self-appointed status as a "wartime president" to win a close re-election battle in November 2004. Bush's Republican Party also made electoral gains, leaving it in control of both the Congress and White House. GWOT, in short, has been very good for the GOP (the Grand Old Party, the nickname for the US Republican Party).

It is in the context of these events, and with the prospect that the Bush administration may square up against another "axis of evil" state, that we have compiled the second edition of *The Geopolitics Reader*. Like the first edition, we have organized the Reader into five parts, two of which address geopolitics historically and two of which deal with geopolitics at the present time, while the final part addresses resistance to geopolitics both historically and today. While retaining the original structure, we have revised and updated our reading selections. As before, we have made difficult decisions to leave out certain readings, perspectives and aspects of geopolitics that we would have liked to address more systematically. The introduction to Part Three discusses some readings not included in this volume but available on the internet. We encourage instructors to supplement the Reader with these readings and other scholarly articles as changing conditions warrant.

Part One of the Reader is the shortest in terms of readings. It addresses the imperialist origins of geopolitical discourse, documenting the entwining of imperialist strategy and racist thought in the period leading up to World War II. While all the imperial powers of this time had geopolitical narratives marked by racist attitudes and beliefs, we have chosen to concentrate on the key rivalry between the British empire and the German state in the early twentieth century, a rivalry at the heart of World Wars I and II. In this edition, we have also sought to address more explicitly the geopolitical narratives found in the United States on the eve of the Cold War by adding an article by Isaiah Bowman and expanding the introduction to Part One.

Part Two addresses Cold War geopolitics, documenting the origins, consequences and eventual passing of the Cold War as a structure of world order and as a complex of geopolitical discourses and practices. Again, we have chosen to focus on the key rivalry, this time between the United States and the Soviet Union. We have added an essay on Robert McNamara and Vietnam. Because of its influence as a triumphalist summation of this period, we have incorporated Francis Fuyukama's essay "The End of History" into this part.

Part Three of the Reader is completely new and addresses the geopolitical dramas and debates of the early twenty-first century. While there are compelling intellectual reasons to question the centrality of September 11, 2001 to this period, the fact is that this event happened on the territory of the largest military power in the world. Those holding power in the US state and military apparatus at the time chose to interpret the attacks, as indeed the perpetrators wanted, as a "declaration of war." The geopolitical events that followed, which included the overthrow of the governments of Afghanistan and Iraq, are marked by the shadow not of 9/11 itself but of *the interpretation of 9/11 by the Bush administration*. This era is defined by debate over how the United States is exercising its power and influence in world affairs. Given its repudiation of international treaties, institutions and law, has the United States become a "rogue superpower"? To some, our central focus on the United States may seem too Amerocentric. To this we plead guilty but argue in our defense that, since the United States is an "elephant" among states in the world community, what drives the elephant inevitably affects everyone else in the community.

Part Four on "global dangers" deals with a number of other geopolitical themes that have preoccupied the "elephant" since the early 1990s, threats and foreign policy concerns that have technological and environmental dimensions and which involve disease, either as a destructive factor in populations in many places or, in the case of post-September 11 fears of bioterrorism, as a weapon. The readings in this part emphasise the importance of environmental formulations in geopolitical thinking, both because of the explicit geographical representations of danger and also because they are tied into the networks of power in the United States that control both technological innovations and energy supplies. Bioweapons are a matter of technological surveillance at a global scale; petroleum supplies are also very much a matter of global politics, especially so in the region that is still frequently simply designated by the geopolitical term "the Middle East." The daily toll of preventable deaths from AIDS (an average of 8493 people *every day* by 2004: see http://www.unaids.org), nearly three times as many as were killed in the September 11 attacks, suggests that this is also a significant trend in world affairs. AIDS too is now discussed in geopolitical discourses of security.

Part Five is devoted to the theme of resistance and geopolitics. Although we stress the essentially contested nature of geopolitical discourses throughout other parts of the book, we also wanted to document anti-geopolitical discourses and practices. Geopolitics in the past was concerned with imperialist expansion and ideological struggles between competing territorial states. But debate was not only between the different state creeds or different strategies of imperialism. Many also questioned the fundamental foundations of imperialist geopolitical discourses: state-centric reasoning, racism, ethnocentrism and the economic exploitation of dominated regions and peoples. Since geopolitics has a long history of association with militarists, nationalists and conservative "wise men," it is important that we broaden the debate and consider many different voices – anti-nationalists, feminists, post-colonial critics, trade unionists, indigenous peoples and, nowadays, so-called "Islamic fundamentalists" – opposing the dominant understanding and practice of geopolitics. Resistance to geopolitics is not the monopoly of the political left or carried out only by admirable dissidents. Historically and at the present time, xenophobic nationalists, power-hungry indigenous elites, religious fundamentalists and terrorists have also resisted dominant and dominating geopolitical discourses.

As in the previous edition, each part of the Reader has a comprehensive introduction to the readings that follow. These introductions place the readings within their historical and geographical contexts, and discuss their significance within the history of international politics and world order. Whenever possible, we have tried to include readings that directly comment on and/or critique each other. In this way, readers will be able to appreciate the essentially contested nature of geopolitical discourses. To further this goal, we have also chosen to illustrate the Reader with maps and political cartoons. As we have already noted, cartographic images and spatial metaphors are central to the operation of geopolitical discourse. These visual texts help illustrate some of the themes we wish to examine and provide critical commentary on questions of power. The latter function is admirably performed by the cartoons of the independent illustrator Matt Wuerker (http://www.mwuerker.com) and we have used his work extensively throughout this volume.

Some may charge that this Reader is biased and unbalanced. Charges of bias, however, are relative to that which is accepted as objective. The Fox News channel in the United States, for example, presents itself as a "fair and balanced" media network yet has its coverage of world affairs shaped by a GOP media powerbroker, a former political consultant to President Reagan and Bush senior (Auletta, 2003). Fox is controlled by Rupert Murdoch's News Corp, which also owns Sky broadcasting in the United Kingdom and a series of chauvinistic tabloid newspapers like *The New York Post* and the British *Sun* (Shawcross, 1997). In the Murdoch media empire, notions of what is "fair and balanced" are shaped by the normalization of national chauvinism and right of center politics. Anything that does not confirm to this worldview can be labeled "biased." The problem with this labeling is not merely its hypocrisy, with ideologues charging bias because their worldview is not confirmed, but its anti-intellectualism. Critical inquiry is eclipsed by the daily affirmation of popular prejudices and myths. Learning is not valued but seen as threatening to comfortable chauvinistic preconceptions.

Our goal is to foster learning through critical thinking. We have organized this Reader around a set of intellectual arguments about geopolitics which we contend have analytical insight and explanatory power. They help us understand what geopolitics is, and how it works. Like all social scientific arguments, these arguments are constantly being challenged and refined by academic debate and scholarly research (to see this process in action, consult academic journals like *Geopolitics*, *Political Geography* or *Millennium*). Together, these arguments constitute a distinctive perspective called *critical geopolitics* (Dalby, 1990; Ó Tuathail, 1996). In the remainder of this introduction, we will examine the key arguments that critical geopolitics advances (Dodds and Sidaway, 1994; Ó Tuathail and Dalby, 1998). Readers can then judge from themselves whether these arguments are analytically powerful and sufficiently robust to answer criticisms they may have. Through the process of debate and critical thinking, we can advance our understandings of the history and contemporary workings of geopolitics.

CRITICAL GEOPOLITICS

Critical geopolitics is a perspective within political geography and international relations that has developed since the early 1980s within international academia. A key article in the development of the approach (Reading 12) is included in this volume. To help readers better grasp what critical geopolitics is all about, we have broken it down into three sets of intellectual arguments. Each of these arguments is characterized by criticism of a conventional conception of geopolitics and a move beyond this conception to richer and more sophisticated sets understandings. The first of these "intellectual moves" addresses how we think about how geopolitics works, the second what we define as geopolitics, and the third the structures of power that promotes certain geopolitical discourses over others.

Beyond political realism: geopolitics as a discourse

Because of the predominant influence of Henry Kissinger on our conventional conception of geopolitics, many people associate geopolitical thinking with a political realist approach to international politics. Political realism, or the power politics school of thought, holds that international relations is characterized by a struggle for power between competing sovereign states (Morgenthau, 1985). The interstate system is an anarchic realm with no overarching authority. Because of this structural condition, states must compete to survive (Waltz, 1979). The purpose of their statecraft is national survival. The means to this is the accumulation of power by states: that, it is argued, is their transcendent "national interest." For stability in the system as a whole, it is best if there is a "balance of power" between the major states in the interstate system. Kissinger defined geopolitics as "an approach that pays attention to the requirements of equilibrium" in international politics (Kissinger, 1979: 914).

This dominant state-centric discourse has a number of assumptions that are worth being clear about. Whether the emphasis is on the structure of the interstate system itself or human nature (with bad leaders wanting to accumulate power), the struggle for power is taken to be a *natural condition* in international politics (Waltz, 2001). Indeed, realists tend to think of international relations as a *primordial state of nature* where, in Hobbes's famous phrase, "life is nasty, brutish and short" (Hobbes, 1982: book 1, ch. 13). This is the way it is, they claim, irrespective of what we think about it. This is why proponents of this school call themselves realists. They "tell it like it is," and those who believe other things are "naive" and "idealist."

There are many problems with this approach but we will focus on just three (see George, 1994; Weber, 2001). First, political realism provides a very poor guide to the empirical history of international politics. While states do indeed compete and go to war, they also have long histories of cooperation and friendship. Many develop joint institutions together. We cannot explain the European Union, for example, within the terms of political realism (this doesn't stop some realists from trying!). Second, political realism is a discourse that thinks it is not a discourse; rather it is "the real." It alone clearly sees the dark primordial state of nature and how things really are, not what we would like to believe. The problem, however, is that *it does not see that it itself is a set of beliefs*. It is merely a constructed story with a simplified and reductionist account of the complex historical processes and structures that have produced modern states (Wendt, 1999). It is blind to the conditions of its own possibility, to the cultural and political assumption that make it possible to tell such an intellectually impoverished story.

This brings us to the third problem that is not specific to political realist storylines but is common to most approaches to world politics: divine methodology. Most geopolitical discourse assumes that "the world" is independent of our beliefs and understandings about it. The visual, theatrical and geological metaphors geopoliticians use imply that world politics is "out there" as a separate reality. Geopoliticians are like gods who stand above it all, detached observers who view the globe as if they were not on it! These assumptions have been termed "god-tricks" by critical intellectuals and they are demonstratably flawed (Haraway, 1991). Because we can know the world only through the conceptual schemas provided by our cultures and languages, we cannot ever assume that the world is independent of the representation conventions we use to describe it. Human beings, after all, are embedded in cultures, places and histories; they are not all-seeing gods. We cannot ever get outside our representational conventions to access "how things really are" unmediated by culture. Our seeing is already a writing (i.e. a cultural frame working) of the world.

As critical thinkers, however, what we can do is become aware that we are all embedded in cultural ways of seeing and constructing the world. We can develop a critical understanding of how these ways of seeing/writing operate. We can become aware that studying international politics requires that we confront questions of cultural meanings. We can grasp that discourse is not a neutral tool that describes objects already existing in the world but is involved in the very recognition and constitution of those objects (in "worlding"). Our world is constituted through discursive languages and practices. In sum, we can recognize that political realist and other geopolitical storylines are mere discourses, particular ways of making sense of and telling stories about international politics.

Critical geopolitics is an intellectual move beyond political realism, and the god-tricks that characterize uncritical geopolitical narratives generally. It rejects state-centric and cognitively miserly stories about how the interstate system works. Most importantly, it recognizes that how people know, categorize and make sense of world politics is an interpretative cultural practice. To understand this process requires studying geopolitics as discourse and the cultural context that gives it meaning.

Beyond "wise men": the cultural embeddedness of geopolitics

Another legacy of Henry Kissinger is the popular notion that geopolitics is all about wise men at the center of state power (Isaacson and Thomas, 1986). Besides its outdated patriarchy, this notion is misleading for another reason (Enloe, 1990). Geopolitics is more than an elite activity. While geopolitical discourses may be articulated by those at the center of state power, they emerge from an historical cannon of narratives about state formation and identity. American presidents generally do not write their own speeches but have teams of professional speechwriters to draw from this cannon to make geopolitical discourse. These speechwriters draw upon already existing images, metaphors and storylines from a state's historical and geographical experiences to produce the required affect. They construct geopolitical storylines that are embedded within a much broader set of cultural practices marked by boundaries between good and evil, friend and enemy, self and other, and "our" space and "their" space (Agnew, 1983).

Thus, critical geopolitics' second intellectual move is to broaden understandings of geopolitics beyond elitist conceptions. Critical geopolitics does this by introducing the general concept of geopolitical culture which it connects to a series of other concepts (see Figure 1). All states, as recognized territorial institutions within an international system of states, have a geopolitical culture, namely a culture of conceptualizing their state and its unique identity, position and role in the world. Geopolitical culture emerges from a state's encounter with the world. It is conditioned by a series of factors: a state's geographical situation, historical formation and bureaucratic organization, discourses of national identity and traditions of theorizing its relationship to the wider world, and the networks of power that operate within the state. We will consider the importance of the last factor on its own after we have briefly reviewed the significance of the former.

As classic geopolitics suggests, the physical geographic inheritance and location of a state inevitably influences its geopolitical culture. Germany, for example, has a long history of geopolitical thought on its relationship to *Mitteleuropa* (Middle Europe) and to Poland and Russia (see Reading 4). As a group of islands off the coast of continental Europe, Great Britain has a long history of thought about its relationship to "the continent" and "Europe" (Reading 1). States that are mountainous, landlocked and without significant petroleum resources will have a geopolitical culture that addresses these qualities. In critical geopolitics, however, physical geography is not destiny or in any way determinate. The *cultural interpretation* of a state's geographical situation and resource endowments is what is significant.

In addition, how states came into being and what particular state apparatus and juridical structures were constructed in the process of state and nation-building has a great conditioning impact on geopolitical culture. What particular type of bureaucratic apparatus has developed to manage foreign

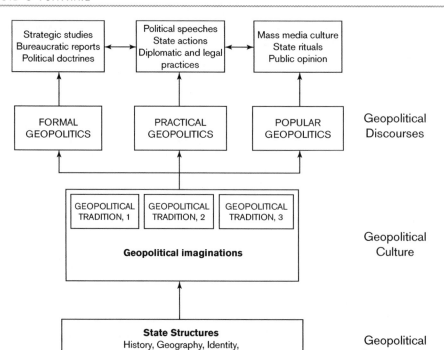

Figure 1   Geopolitics: a critical geopolitics conceptualization

policy and the ways in which constitutional laws are made and implemented shape how foreign policy is conceptualized and practiced across the territory of the state (Coleman, 2005).

The geopolitical culture of a state is also shaped by the primary forms of identification and boundary-formation that characterize its social, cultural and political life. How do the various groups within a state draw boundaries between themselves and others, between insiders and outsiders, us and them? What worldwide communities do they imagine themselves part of and what communities do they define themselves in opposition to? What states are seen as friendly and what ones are considered potential or actual enemies? What collective national identity predominates within the state and how does this shape the state's relationship to the international community. Edward Said (Readings 18 and 29) and others have written extensively on the power of fluid *geopolitical imaginations* (Said uses the term "imaginary geographies") in shaping how states behave culturally in the world.

Geopolitical cultures are characterized by particular schools of thought on foreign policy and the state. We can call these *geopolitical traditions*. Coming together as distinctive approaches in the process of debate with each other, these foreign policy philosophies and geopolitical orientations are the building blocks from which intellectuals of statecraft produce particular geopolitical discourses. Graham Smith (1999), for example, argues that there are three distinct geopolitical traditions in Russia. The first places Russia within Europe and is associated with the long historical tradition of Westernization exemplified by Peter the Great and the city of St Petersburg. Westernizers (*zapadniki*) argue that Russia needs to adopt Western models and institutions in order to achieve modernity and economic development. An opposing tradition is the notion that Russia is a distinctive Eurasian territory and state. Slavophilism in nineteenth century Russia and Eurasianism nowadays are examples. The third geopolitical tradition sees Russia as a bridge between East and West, between Europe and Asia. There are many other examples though they may not use the term geopolitical tradition. Mead (2002) defines four distinct schools of American foreign policy which he names after presidents: Wilsonian, Hamiltonian, Jeffersonian and Jacksonian. Ash (2004) identifies four different faces of contemporary Britain: an island world (little England), a world island (cosmopolitan Britain), a European Britain (an active member of the European Union) and an American Britain (a Britain with a "special relationship" with the United States). Tony Blair's vision could be a fifth in that he envisages Britain as a bridge between America and Europe (which Ash wryly terms the "Blair Bridge Project").

Geopolitics finds concrete expression in the form of particular discourses or narratives of world politics. These discourses are not only produced by "wise men" in strategic studies institutes. Rather, geopolitics is produced throughout state-centered society at multiple sites. Critical geopolitics distinguishes between three different types of geopolitical discourses. *Formal geopolitics* refers to the advanced geopolitical theories and visions produced by intellectuals of statecraft (what we commonly associate with geopolitics). *Practical geopolitics* refers to the narratives used by policy makers and politicians in the actual practice of foreign policy. The public forms of these storylines – in speeches and public addresses – are the raw material of practical geopolitics. *Popular geopolitics* refers to the narratives of world politics that find expression in the popular culture of a state, in its cinema, magazines, novels and even cartoons (Dodds, 1996). Sharp (2000), for example, has studied the popular geopolitics found in *Reader's Digest*. Others have explored the popular geopolitics of films (Power and Crampton, 2005). This informal geopolitics circulates geopolitical understandings beyond the political class to ordinary people. All these genres of geopolitical narration – formal, practical and popular – are the products of prevailing imaginations, cultures and traditions so the lines between them can blur. Discourses of resistance to dominant geopolitical understandings and practices also have formal, practical and popular expressions.

The layered understanding of geopolitics found in the critical geopolitical literature makes it a much more complex phenomenon that simply the study of the "great ideas of great men" (Halberstam, 1972). Let us now turn to the third intellectual move that characterizes critical geopolitics, the contextualization of geopolitical discourses within networks of power.

Beyond the absence of power: power networks and geopolitical discourses

Conventional conceptions of geopolitics are characterized by an interesting paradox: on the one hand, they address power struggles between states very explicitly yet, on the other hand, they usually contain little reflection on the social structures of power within states, and how these shape geopolitical discourse itself. We need to acknowledge that not all geopolitical discourses are created or treated equally. Some are produced by state institutions, like foreign policy bureaucracies, military academies or state universities, and are central to the political life of the state. Others are the product of civil society. Certain geopolitical discourses at the present time emerge from private research and policy institutions, often called think tanks or idea factories dedicated to producing and distributing certain types of ideas. Intellectuals of statecraft willing to espouse them are hired and financed to do research. For example, some think tanks promote geopolitical discourses that imply large defense expenditures, military exports, and the aiding of favored states no matter what their human rights record. Defense corporations and lobbying organizations are likely to richly endow such think tanks. Journals and newspapers often promote particular geopolitical discourses to serve political agendas. Commercial publishers will try to hype certain geopolitical discourses to stimulate the sale of the book. Equally insightful ideas or important discourses may attract little financial support because they threaten vested interests and powerful political constituencies. Think tanks questioning levels of defense expenditure or the small arms market are usually not wealthy institutions. Any critical perspective on geopolitics must acknowledge the workings of power networks in shaping the prevailing forms of geopolitical discourse one finds in states (Foucault, 1980; Agnew and Corbridge, 1995).

To grasp the operation of power, we need to think about the sources of social power. The sociologist Michael Mann (1993) offers us a useful framework which identifies four bases of social power:

- *Political power* refers to the centralized territorial regulation of social life by a state bureaucratic complex. This complex is a source of great structural power. The army and police exercise coercive power. The apparatus of governance has the ability to formulate and enforce the law. Those in control can determine levels of taxation and the rules under which economic accumulation occurs. Political power is the subject of fierce struggle between different social groups because the rewards of capturing the state apparatus are so great.
- *Ideological power* is the power to shape, formulate and mobilize the values, norms and rituals that characterize human social life. It is the power to steer culture in certain directions, to shape what values are predominant in a society, what norms are to be considered legitimate and what rituals characterize the life of the community. Religious institutions, mass media organizations, political parties and think tanks are some of the many sources of ideological power in a society.
- *Economic power* is the power that resides in the circuits of production, distribution, exchange and consumption in a society. The relationship of people to the mode of production and consumption in a society generates a class system which has vectors of power that vary from state to state. Power can be concentrated in the hands of a small asset owning class or widely dispersed throughout society depending on the rules governing economic accumulation.
- *Military power* is the power that comes from the organization of security and defense in a particular state. It resides with the military elites who run the various state bureaucracies that are charged with providing "national security." Because this role is considered so vital to the existence of the state, these bureaucracies are able to command a considerable chunk of the state's resources. In some instances, military elites can take over the whole state apparatus or become future state leaders because of their perceived leadership qualities.

Mann's (1993) model is useful because it is neither unidimensional (there are multiple sources of power) nor monocausal (economic power, for example, does not determine in the last instance as some Marxists claim). Rather, the sources of social power *interact with each other and form entwined structural networks of power*. For example, in the United States and the United Kingdom, business corporations have tremendous economic power which they are able to leverage into other sources of power: political (through influencing the political system with campaign contributions) and ideological (through owning or influencing media organizations or sponsoring think tanks to produce ideas they like). Political parties can increase their chances of political power by aligning themselves with the most powerful economic, ideological and military interests in a state. Ideological groups can augment their power by working closely with a powerful political party and associated law and lobbying firms. Military bureaucracies can expand their power by building linkages to powerful industrial groups and entrenched political interests. Military leaders and veterans organizatons often endorse certain political candidates. In short, what develops is a structural affinity of interests between particular economic interests, ideological groups, political parties and military bureaucracies in the battle to control the state apparatus.

A good example of this structural affinity of interests, particularly relevant to understanding present day geopolitics, is the operation of a so-called "iron triangle" in US political life (H. Smith, 1988). The first side of this triangle is made up of powerful US defense contractors who have historically accumulated profits by securing lucrative state defense expenditure contracts. Their corporate strategy is to get the US state to spend ever larger amounts of money on the hardware products they make (whether these products are really needed or not). The second side of the triangle comprises powerful politicians in Congress who have close ties to these defense corporations through campaign donations. These politicians are interested in allocating defense expenditures to and keeping military bases open in their districts so they can claim credit and get re-elected. The third side of the triangle consists of the leaders of the US Department of Defense who seek to augment their own power and that of their agency by securing ever greater levels of state appropriations. This triangle does not always operate smoothly. The Pentagon, for example, may not want a weapons system that contractors want to sell and politicians want to fund (for example, the Comanche helicopter, the Crusader artillery system and the C-130J aircraft made by Lockheed Martin Corporation; see Wayne, 2005). It may want more funds for personnel whereas Congressional politicians may want to spend taxpayer money on defense contractor hardware built in their districts. Defense contractors and politicians also compete among themselves. Occasionally, evidence of flagrant corruption by politicians, politically appointed government officials and lobbyists also emerges. Yet, despite tensions within its operation, the iron triangle is an enduring structure of power within the US state that helps explain why US defense spending remains at astoundingly high levels despite the relative absence of an existential threat to the country since the end of the Cold War. World military expenditure exceeded $1 trillion in 2004, with the United States accounting for 47 percent of this spending total. In other words, *United States military spending alone is almost as much as the rest of the world combined* (SIPRI, 2005). President George W. Bush requested $420.7 billion for the US military in fiscal year 2005 ($401.7 billion for the Defense Department and $19.0 billion for the nuclear weapons functions of the Department of Energy). This is an increase of 7.9 percent above 2004. This figure did not include the cost of the Afghanistan and Iraqi wars, which were funded by emergency "supplementary" Congressional appropriations. The supplementary appropriations allocated to the Department of Defense for financial years 2003–2005 amounted to approximately $238 billion and exceeded the combined military spending of the entire developing world (SIPRI, 2005). In all, the US plans to spend $2.2 trillion on its military between 2005 and 2010, all at a time when the US federal deficit is estimated at almost half a trillion dollars per year.

The iron triangle is a congealment of military, economic and political power but it requires ideological power to rationalize this enormous level of expenditure. Politicians, defense contractors and generals need to justify the contracts they sign, the taxpayer money they spend, and the profits they accumulate. The justification is provided by the prevailing geopolitical discourse in the United States, the dominant storylines about America's mission in the world (usually "indispensable") and the threats and dangers it faces (usually pervasive and immediate). The most powerful interest groups in the state work hard to shape these discourses of danger to serve their interests (Weldes, 1999). The ideological power structure in Washington DC is conditioned by commercial media organizations and corporate endowed think tanks. Some are well established like the *Washington Post*, the traditional three US television networks (NBC, ABC and CBS) and research centers like the Brookings Institution and the Carnegie Endowment for International Peace.

The media organizations and think tanks that have developed in the city since the mid-1970s are much more partisan and ideologically conservative. The *Washington Times*, for example, is the city's second daily newspaper and is owned by the Reverend Sun Myung Moon. It is an agenda setting instrument for his evangelical and conservative worldview. The *Weekly Standard* is a vehicle of influence for Rupert Murdoch and the voice of American neoconservatives (see Parts Two and Three). Powerful right wing think tanks like the American Enterprise Institute, the Heritage Foundation, the Hudson Institute and the Cato Institute employ professional "policy wonks" whose job it is to produce books, reports and opinion editorials advocating the position of their organization and its corporate backers. Media organizations select and promote "analysts" from these and other institutes to serve as "experts" on their news shows (even though their intellectual qualifications may be thin, and their expertise more in self-promotion!). These "experts" may later move into government itself and serve in various positions within the policy bureaucracy and then return to a think tank or to the lucrative world of post-government service consulting. This is the revolving door of power networks, with most celebrity policy leaders forming their own consulting firms upon exit from government service. Henry Kissinger, the founder of Kissinger Associates, has perhaps the most powerful geopolitical network in Washington of all, with every US national security advisor since the mid-1970s associated with him in one way or another (Rothkopf, 2005). In sum, geopolitics is not a contest of ideas produced by free floating intellectuals in which truthful and intellectually superior arguments triumph over mendacious, flawed and weak ones. Geopolitical discourse operates within networks of power and the discourses that emerge as the prevailing ones in any state reflect the influence of that power structure; indeed, these discourses are part of the very operation of this power structure.

The three intellectual moves we have just described are part of the critical geopolitical perspective found throughout this *Geopolitics Reader*. They inform how we put the volume together, what readings we selected, and how we tell the story of these readings in our introductions to the five parts. This is a book to help you think critically about geopolitics, to ask good questions and to push for better answers than our anti-intellectual media provide. Throughout, we emphasize the essentially contested nature of geopolitical discourse by presenting readings that directly comment and critique each other. As academics, our goal is to advance learning through promoting better debate and argumentation, not to promote any state over any other state. Geopolitics, as this volume makes clear, is not a domain of objective stories about world politics. It is world politics itself, about states, cultures, identities, discourses and power. Geopolitics comes in a variety of forms; it is not just formal theories but also presidential speeches, think tank manifestos and the operation of geopolitical cultures more generally. This edition incorporates more instances of practical geopolitics than the first. The goal is the same: to render the relations of power embedded in geopolitical discourses visible and manifest. We all live in a world shaped by the prevailing hegemony of certain geopolitical discourses over others. In pulling back the curtain on how geopolitics works, we hope this volume contributes to helping you think critically about what geopolitical discourses deserve support and what ones deserve opposition.

REFERENCES

Agnew, J. (1983) "An Excess of 'National Exceptionalism'," *Political Geography Quarterly*, 1: 151–166.
—— (2003) *Geopolitics: Re-Visioning World Politics*, 2nd edn. New York: Routledge.
Agnew, J. and Corbridge, S. (1995) *Mastering Space: Hegemony, Territory and International Political Economy*. London: Routledge.
Ash, T.G. (2004) *Free World: America, Europe and the Surprising Future of the West*. New York: Random House.
Auletta, K. (2003) "Vox Fox: How Roger Ailes and Fox News are Changing Cable News," *The New Yorker*, May 26.

Campbell, D. (1998) *Writing Security: United States Foreign Policy and the Politics of Identity*, revised edn. Minneapolis, MN: University of Minnesota.

Coleman, M. (2005) "US Statecraft and the US–Mexico Border," *Political Geography*, 24: 185–210.

Daalder, I. and Lindsey, J. (2003) *America Unbound: The Bush Revolution in Foreign Policy*. Washington DC: Brookings Institution Press.

Dalby, S. (1990) *Creating the Second Cold War: The Discourse of Politics*. London: Pinter.

Dodds, K. (1996) "The 1982 Falklands War and a Critical Geopolitical Eye: Steve Bell and the If . . . Cartoons," *Political Geography*, 15: 571–591.

Dodds, K. and Sidaway, J. (1994) "Locating Critical Geopolitics," *Society and Space*, 12: 515–524.

Enloe, C. (1990) *Bananas, Beaches, and Bases: Making Feminist Sense of International Politics*. Berkeley, CA: University of California Press.

Foucault, M. (1980) *Power/Knowledge*, New York: Pantheon.

Frum, D. and Pearle, R. (2003) *An End to Evil: How to Win the War on Terror*. New York: Random House.

George, J. (1994) *Discourses of Global Politics: A Critical (Re)Introduction*. Boulder, CO: Lynne Rienner.

Halberstam, D. (1972) *The Best and the Brightest*. New York: Penguin.

Hanhimäki, J. (2004) *The Flawed Architect: Henry Kissinger and American Foreign Policy*. Oxford: Oxford University Press.

Haraway, D. (1991) *Simians, Cyborgs, and Women: The Reinvention of Nature*. New York: Routledge.

Hepple, L. (1986) "The Revival of Geopolitics," *Political Geography Quarterly*, 5 (supplement): S21–S36.

Hobbes, T. (1982) *Leviathan*, new edn. New York: Penguin Classics.

Holdar, S. (1992) "The Ideal State and the Power of Geography: The Life-Work of Rudolph Kjellen," *Political Geography*, 11: 307–324.

Isaacson, W. (1992) *Kissinger: A Biography*. New York: Simon and Schuster.

Isaacson, W. and Thomas, E. (1986) *The Wise Men: Six Friends and the World They Made: Acheson, Bohlen, Harriman, Kennan, Lovett, McCloy*. New York: Simon and Schuster.

Kagan, R. (2002) "Power and Weakness," *Policy Review*, 113 (June–July): 3–28.

Kissinger, H. (1979) *White House Years*. Boston, MA: Little, Brown.

Mann, M. (1993) *The Sources of Social Power*, Volume II. Cambridge: Cambridge University Press.

Mattern, J.B. (2005) *Ordering International Politics: Identity, Crisis, and Representational Force*. New York: Routledge.

Mead, W.R. (2002) *Special Providence: American Foreign Policy and How it Changed the World*. New York: Routledge.

Morgenthau, H. (1985) *Politics Among Nations*, 6th edn. New York: McGraw-Hill.

Ó Tuathail, G. (1996) *Critical Geopolitics*. London: Routledge.

Ó Tuathail, G. and Dalby, S. (1998) *Rethinking Geopolitics*. London: Routledge.

Power, M. and Crampton, A. (eds) (2005) *Geopolitics and Cinema*. London: Routledge.

Rothkopf, D. (2005) *Running the World: The Inside Story of the National Security Council and the Architects of American Power*. New York: Public Affairs.

Sharp, J. (2000) *Condensing the Cold War: Reader's Digest and American Identity*. Minneapolis, MN: University of Minnesota.

Shawcross, W. (1997) *Murdoch: The Making of a Media Empire*, revised and updated. New York: Simon and Schuster.

SIPRI (Stockholm Peace Research Institute) (2005) *SIPRI Yearbook 2005*. Oxford: Oxford University Press.

Smith, G. (1999) *Post-Soviet States: The Politics of Transition*. London: Arnold.

Smith, H. (1988) *The Power Game: How Washington Works*. New York: Random House.

Spivak, G.C. (1988) *In Other Worlds: Essays in Cultural Politics*. London: Routledge.

Waltz, K. (1979) *Theory of International Relations*. New York: McGraw-Hill.

—— (2001) *Man, the State and War*, revised edn. New York: Columbia University Press.

Wayne, L. (2005) "The Flawed Plane Congress Loves," *New York Times*, March 24, C1–2.

Weber, C. (2001) *International Relations Theory: A Critical Introduction*. London: Routledge.

Weldes, J. (ed.) (1999) *Cultures of Insecurity: States, Communities and the Production of Danger*. Minneapolis, MN: University of Minnesota Press.

Wendt, A. (1999) *Social Theory of International Politics*. Cambridge: Cambridge University Press.

PART ONE

Imperialist Geopolitics

INTRODUCTION TO PART ONE

Gearóid Ó Tuathail

Modern geopolitical discourse was born in the era of imperialist rivalry between the decades from 1870s to 1945, when competing empires clashed and fought numerous wars, all the time drawing and redrawing the borders of the world political map. An era characterized by colonial expansionism abroad and industrial modernization at home, it was a time of tremendous technological achievement, social upheaval and cultural transformation. The dominant imperialist structure of the age was the British empire, which, despite its increasing territorial size over the decades, was poorly adjusting to the transforming conditions of world power, particularly those in the early twentieth century. The other "great" imperial powers of the time – Russia, France, Italy, the United States, Germany and later Japan – were its rivals and sought gain from its difficulties and relative decline. Each of these imperialist states produced their own leading intellectuals of statecraft and developed from their own distinctive geopolitical cultures and traditions various imperialist geopolitical discourses.

The most historically and geographically fated imperialist rivalry of the period was that between the British empire and the rising imperial aspirations of the German state in central Europe, a rivalry that was at the crux of the two world wars that destroyed millions of lives in the twentieth century. It is this rivalry that we examine here through an investigation of the geopolitical writings of the British geographer Halford Mackinder, the German general turned geopolitician Karl Haushofer, and the political agitator who became the German *Führer*, Adolf Hitler. To remind us that geopolitics was not a European monopoly, we also examine US President Theodore Roosevelt's 1905 corollary to the Monroe Doctrine, and the geopolitical discourse articulated by a figure described as the American Haushofer, the geographer and prominent presidential advisor, Isaiah Bowman.

HALFORD MACKINDER'S GEOPOLITICS AND BRITISH IMPERIAL GEOPOLITICAL CULTURE

Halford Mackinder began his career teaching geography in 1887 at Oxford University thanks to the influence and sponsorship of the Royal Geographical Society (RGS). Mackinder had impressed a number of fellows of the RGS earlier that year when, at the young age of 25, he addressed the society and made the case for a "new geography" of academic synthesis to supersede the "old geography" of exploration and discovery that largely defined geography in the nineteenth century. Not everyone was impressed, however. One crusty old admiral sat in the front row muttering "damn cheek, damn cheek" as he spoke (Blouet, 1987: 40). To those traditionalists who saw geography as a manly science of military adventuring and "lion hunting," Mackinder must have appeared as a young bookish upstart. Many on the leadership council of the society, however, were sympathetic to his arguments and subsequently championed him for a position at Oxford, agreeing to pay half his salary for five years.

Because geography was seen by many within the RGS as a manly outdoorish science, Mackinder felt over the subsequent years the need to demonstrate exploratory prowess to the traditionalists in the all-male RGS. Consequently, Mackinder and some social acquaintances undertook in 1899 an expedition to climb Mount Kenya in what was then "British East Africa" and is now the independent state of Kenya. From Mackinder's point of view, the expedition was a ripping success – he successfully climbed the mountain, gave a triumphant address at the RGS upon his return, and garnered the esteem of traditionalists – though it was hardly that for those native Africans who were shot by Mackinder's party during the course of the African expedition (Kearns, 1997).

As a new "geographical expert" who had proved his worth in the colonies, Halford Mackinder felt strongly about the role geographical knowledge could play in addressing the relative decline of the British empire, a relative decline dramatically illustrated by the difficulties the British army had in winning the Boer War (1899–1902). Mackinder supported the imperial reform movement of Joseph Chamberlain, the former Colonial Secretary who sought to modernize the British empire by imposing a common external tariff against the products of other "Great Powers" at this time (Ó Tuathail, 1992). Like many of his compatriots, Mackinder worried about the rising power of the German empire on the European continent. Geographical education, for him, was an important weapon in the struggle for "relative efficiency" between the Great Powers, particularly between Great Britain and the German state of Kaiser Wilhelm II. Geography, he argued, was a necessary subject in educating "the children of an Imperial race" (Mackinder, 1907: 36, original capitalization). Most of the British masses were of "limited intelligence" so it was the duty of an elite of experts to educate them to think like the rulers of a vast overseas empire. It is essential, he argued, "that the ruling citizens of the worldwide Empire should be able to visualize distant geographical conditions . . . Our aim must be to make our whole people think Imperially – think that is to say in spaces that are world wide – and to this end our geographical teaching should be directed" (Mackinder, 1907: 37–38, original capitalization).

The idea of educating the "children of an Imperial race" sounds alien to us today but the conceptualization provides insight into Mackinder's worldview. The biological inheritances of different races, the conditioning influence of the environment and the "relative efficiency" of states, especially in their conduct of strategy, are the three determining factors shaping international affairs in Mackinder's writings (Kearns, 1985, 2003). As a formal disciplined way of thinking that studies the interaction of the first two, geography, for Mackinder, can help influence the third by educating Britain's children to think in terms of empire. It can also help educate the political leaders of the empire about race and the geographical factors conditioning human history, so they can pursue clearer grand strategy. On a cold January evening in 1904, Mackinder gave an address to the RGS on precisely this theme. Mackinder's talk on "The Geographical Pivot of History" (Reading 1) created little stir at the time – few political leaders heard him speak – but it was destined to make him famous decades later when, during World War II, the American and British public discovered German geopolitics and the reverence it accorded the ideas of Mackinder. Mackinder's address is important in the history of geopolitics for three reasons: for its god's eye global view, for its division of the globe into vast swaths of territory, and for its sweeping story of geography's conditioning influence on the course of history and politics. These three "innovations" helped define geopolitics as a particular narrative form and account for its subsequent influence and appeal.

First, though he does not use the word geopolitics, Mackinder's essay "invents" geopolitics as a new detached perspective that surveys the globe as "closed" political space. Geopolitics is a new way of seeing the competition for power between states as a unified worldwide scene. Mackinder adopts a god's eye view which looks down on what he calls "the stage of the whole world":

For the first time we can perceive something of the real proportion of features and events on the stage of the whole world and may seek a formula which shall express certain aspects, at any rate, of geographical causation in history.

(Mackinder 1904: 421)

This sentence is extremely important. The "we" it invokes is the community of geographical or geopolitical experts, educated and privileged white men like Mackinder who, by virtue of these social privileges, can adopt an Olympian perspective on the world, perceive "the real proportion of features and events," and seek formulae or laws to explain history. All the elements of formal imperialist geopolitics are in this one sentence: the divine gaze upon the world, the claim to perceive "the real" and the reduction of human history to a formula of geographical causation. Over and over again, these elements are found in imperialist and more modern geopolitical discourse. What they reveal is the operation of a god's eye view that is blind to its own situatedness and partiality.

For critical geopolitics, geopolitical experts are not divine but embedded in economic, political, ideological and military networks of power. They do not see objectively but within the structures of meaning provided by their privileged identities (in Mackinder's case as a white gentleman who had proved himself to the powerful) and socialization within certain intellectual circles, state institutions and geopolitical cultures. They do not see "the real" but see that which their geopolitical culture *interprets and constructs* as "the real." Their so-called "laws" of strategy are often no more than self-justifications for their own political ideology and that of those in power within their state. Their storytelling about international politics, in other words, is a form of power which they wield to serve their own political ends.

Second, Mackinder's text is remembered for its map of "The Natural Seats of Power" and its invention of the game of labeling huge swaths of the world's territory with a singular identity. Like all maps, this Mercator projection map is an *interpretation of the earth* and not a true representation of it. Mercator projections radically distort the size of the northern latitudes, enlarging Greenland and Russia, for example, and shrinking the Australian and African continents. The centering of the map on Eurasia inevitably renders that region pivotal and North and South America marginal. To illustrate his thesis graphically, Mackinder labels enormous tracts of territory with simple identities like "pivot area," "inner or marginal crescent" and "lands of the outer or insular crescent." The great irony of this god-like labeling of the earth is that, in so doing, Mackinder eliminates the tremendous geographical diversity and particularity of places on the surface of the earth. Difference becomes sameness. Geographical heterogeneity becomes geopolitical homogeneity. This "loss of geography" is, as we shall see, a recurrent feature in geopolitical discourses that play the game of earth labeling.

Third, Mackinder's address is remembered because of the sweeping story he tells about "the geographical causation of history." At the center of this story is the relationship between physical geography and transportation technology. Mackinder claims that there are three epochs of history (represented in Table 1) which he names after the explorer Christopher Columbus. Each epoch is defined by dominant dramas – remember his stage metaphor! – and "mobilities of power." With the era of geographical exploration and discovery at an end, Mackinder suggests that history is now entering the post-Columbian epoch, an epoch of closed space where events in one part of the globe will have ripple effects across the globe. More significantly, from a British imperial point of view, "trans-continental railways are now transmuting the conditions of land-power, and nowhere can they have such effect as in the closed heart-land of Euro-Asia." This is alarming to Mackinder because it threatens to change the balance of power between landpower (continental Europe, particularly Germany) and seapower (the British empire) in Eurasia.

Epoch	Dominant drama	Dominant mobility of power type	Ascendant region and power
Pre-Columbian	Asiatic invasions of Europe	Horse and camel	The landpower of the Asian steppes
Columbian	European overseas expansionism	Sailing vessels and sea transportation	The seapower of the European colonial empires
Post-Columbian	Closed space and the struggle for relative efficiency	Railways	The landpower of those who control the heartland

Table 1 Halford Mackinder's geopolitical story

> The oversetting of the balance of power in favour of the pivot state . . . would permit of the use of vast continental resources for fleet-building, and the empire of the world would then be in sight. This might happen if Germany were to ally herself with Russia.
>
> (Mackinder 1904: 436)

This latter scenario is Mackinder's greatest fear. The political leaders of the British Empire must do everything in their power to prevent an alliance between Germany and the "heartland" of world power then ruled by the Romanov dynasty. It is worth asking if Mackinder's reasoning is "realist" here. An alliance between imperial Germany and imperial Russia was conceivable but the phrase from the previous sentence, "empire of the world would then be in sight," indicates paranoia not realism. Human history has never seen an "empire of the world." The very idea has more affinity with the science fiction of Mackinder's contemporary, H.G. Wells, the author of *War of the Worlds* (1898), than with the empirical record of human history. Yet this paranoid fantasy is at the core of Mackinder's geopolitical vision. Indeed, fear and paranoia play a big role in geopolitical discourse.

In a subsequent book called *Democratic Ideals and Reality* (1919), written immediately after World War I with a view to influencing the Versailles Peace negotiations, Mackinder's paranoid fantasy is more explicit in his strategic recommendation to the victorious political leaders. Renaming what he called Euro-Asia the "World-Island" and the "pivot area" the "Heartland," he declared:

> Who rules East Europe commands the Heartland; Who rules the Heartland commands the World-Island; Who rules the World-Island commands the World.
>
> (Mackinder 1919: 150)

Behind this sloganistic strategy is a simple recommendation. What must be prevented is German expansionism in Eastern Europe and a German alliance with what had become the Soviet Union.

Mackinder's formal geopolitics had little impact on British foreign policy during his lifetime though they did, as we shall see, earn the admiration of a school of German militarist geographers led by Karl Haushofer. One reason for their lack of impact at home is that his arguments had many flaws. His thesis was too sweeping, his interpretation of human history too simplistic and geographically deterministic, and his claims about the importance of mobility in the development of power one-sided. Mackinder neglected the importance of social organization in the development of power, he missed the revolutionary implications of airpower for the twentieth century (a point noted by one of his questioners in 1904, Leo Amery) and, most significantly, he underestimated the emergent power of the United States (which he strangely describes as an eastern power!) while overestimating the strategic significance of the vast spaces of the Russian "heartland."

Yet, while flawed, Mackinder's ideas emerged from the geopolitical culture of the British empire. His writings, particularly the textbooks he wrote for the "children of an Imperial race," provide insight into the practical and popular geopolitics of that culture. The foundational categorization scheme in the geopolitical culture of the British empire is an imagined "natural" hierarchy of races. It divided the world into stereotypical racial "types" – often personified in the cartoons of the time as "John Bull" (the British race), "the Hun" (Germans) and "the Slav" (Eastern European peoples) – and organized these types according to imputed levels of civilization. At the pinnacle of this imagined hierarchy of racial qua national types was the "white" "Anglo-Saxon" "race" (we need to use inverted commas because all of these discursive categories are arbitrary social constructions which have no truth separate from belief in them; the essence of racism is the *naturalization of that which is wholly social*; see Gilroy, 2000). At the bottom of the hierarchy are so-called "black" and "colored" races, including Africans and Chinese. Slavs and Irish were also considered "inferior races." One feature of this elaborate racial hierarchy was to justify the right of the so-called "civilized" and "white" races to dominate and rule "barbarous," non-white others. Mackinder voiced racist thinking in passages throughout his writings and work (Ryan, 1998). In *Britain and the British Seas* he described "John Bull" as a "genus" of the human species distinguished by his commitment to "freedom" and "civilized values" (at the same time as Britain was the oppressive imperial overlord of millions of colonial subjects!) (Mackinder, 1902). He expanded this theme later describing "the English race" and "English blood" as "valuable" because it supposedly carries both desirable physical "character" and a tradition of "responsible government" (Mackinder, 1925). Read "The Geographical Pivot" essay closely and you will find him declaring Professor Edward Augustus Freeman's (1823–1892) notion of history as "Mediterranean and European race history" as "in a sense, of course, true" (i.e. an established genetic given prior to the effects of environmental influences on history). Freeman, a Professor of History at Oxford University, acknowledged race as a problematic category but retained the term, nevertheless, in his writings to designate an "original stock of blood" (see Freeman, 1879). Racist beliefs were not particular to Mackinder but found throughout the popular culture of the time, including advertisements for soap, and in the practical geopolitical discourse of some of Britain's most famous politicians (e.g. Winston Churchill) (McClintock, 1995). Indeed, the racist assumptions of British geopolitical culture at this time were shared by other imperial geopolitical cultures.

THEODORE ROOSEVELT AND AMERICAN GEOPOLITICS

By 1904, the year of Mackinder's "Pivot" address, the United States had emerged as a significant player on the world's stage (as his favorite metaphor would describe it). Consequent to its humiliating defeat of the Spanish empire in 1898, the United States acquired the Philippines as a colony and became the imperial overlord of Cuba, imposing upon it the Platt Amendment, which granted the United States the legal right to interfere in Cuban political life and territorial control over the strategic naval base of Guantánamo in perpetuity (a base given renewed attention when the Bush administration sought to use it to evade the Geneva Conventions and American civil liberty protections for GWOT prisoners). Motivated, in part, by the seapower doctrine of Alfred Mahan, which stressed the significance of acquiring overseas naval bases, the United States also acquired the Hawaiian islands and Guam. Mahan and other prominent imperialists in the United States like Brook Adams, Henry Cabot Lodge and Theodore Roosevelt justified such imperialist expansionism in a variety of ways. Throughout his voluminous writings, Admiral Mahan argued in an institutionally self-serving way that the path to national greatness lay in commercial and naval expansionism. All truly great powers were naval powers. It was not necessary to acquire whole territories and formally occupy them (this, after all, was colonialism and the United States liked to think of itself as an anti-colonial nation); what the United States needed was an informal empire based on "open door" trade and a string of overseas naval bases that would give its navy the ability to project power in a troublesome region whenever it needed to do so.

Implicit within Mahan's naval expansionist creed, and even more explicit in the dovetailing visions of Lodge and Roosevelt, was a social Darwinian ideology that held that all states, peoples and so-called "races" were in a struggle for survival with each other and only the fittest and most aggressive survived. The supremacy of particular peoples and races, Theodore Roosevelt believed, was best expressed in war, an activity he romanticized intensely as manly, vigorous, exciting and fundamental to greatness. "There is no place in the world," he wrote, "for nations who have become enervated by soft and easy life, or who have lost their fiber of vigorous hardiness and manliness" (quoted in Beale, 1956: 52). Roosevelt's obsession with demonstrating "manliness" made him a crusading militarist, a forward-charging "rough rider" who gloried in his heroic exploits in Cuba during the brief Spanish–American war. Like many other imperialists of his time, Roosevelt was a white supremacist, one who believed in the imagined natural hierarchy of "races" with white Anglo-Saxons at the top and a whole series of "inferior races" like the Chinese, Latin Americans and Negroes well below them (Hunt, 1987). In the United States, the category of "race" was a rather flexible one which referred to nationality, language, culture and manners as much as it did to skin color and biological inheritance. Roosevelt's racism, in contrast to that later exalted by the Nazis, was more civilizational and ethnographic than it was biological and genetically determinist. Certain races could, with help and effort, be "raised up" to a higher level of civilization.

All of these different elements – seapower imperialism, bellicose masculinity, anti-colonial commercial expansionism and civilizational racism – came together when Theodore Roosevelt became president in 1901 after the assassination of William McKinley. Full of national pride and imperial hubris, Roosevelt argued that the United States was a "masterful race" which should "speak softly" but carry a "big stick" in the Pacific, Caribbean and Latin America. In Central America, Roosevelt practiced an aggressive form of geopolitical interventionism which gave birth to the state of Panama. Formerly a province of Columbia, Roosevelt's administration fermented an independence movement in the region in order to secure the territory necessary to construct a canal linking the Atlantic and Pacific oceans. Just after his resounding election to the presidency in 1905, Roosevelt sought to formalize his geopolitical thinking into a so-called "corollary" to the Monroe Doctrine (Reading 2), the grandiloquent declaration by President James Monroe in 1823 that European powers should not "extend their system to any portion of this hemisphere." Roosevelt's corollary sought to give notice that the American hemisphere was the special preserve of the United States. As, according to Roosevelt, the most civilized and superior state in the hemisphere, the United States had a right, indeed an obligation, to "exercise an international police power" in the region to keep troublesome and uncivilized states in line. Intervention in the affairs of unruly and immature states in order to enforce the rule of law and restore discipline was part of what Rudyard Kipling called "the white man's burden," the so-called "burden" that comes from being superior and more civilized than everyone else, an arrogant philosophy Roosevelt's corollary perfectly articulates.

KARL HAUSHOFER AND GERMAN *GEOPOLITIK*

The white supremacist sentiment that was common to the practice of British and American imperialist geopolitics found a distinct nationalist expression in Germany, where a school of German geopolitical thought was formalized after World War I by Karl Haushofer (1869–1946) (Murphy, 1997). Haushofer was a former military commander who became a political geographer at 50 after retiring from the German army with the rank of major general. Born in Munich, Haushofer's military career took him to Japan from 1908 to 1910 where he admired the national unity of a Japanese state that was strongly anti-democratic and increasingly militarist in orientation (Dorpalen, 1942). Haushofer, the military officer, in particular admired the discipline of Japanese life and the blind obedience and devotion with which the Japanese people followed their leaders. His stay in Japan provoked him to write a book and doctoral dissertation on the German influence on the development of the Japanese state (which was indeed significant).

During World War I, Haushofer served as a field commander for the German army on the Eastern front, with Rudolf Hess, later deputy leader of the Nazi Party, as his aide-de-camp. Devastated by Germany's defeat, Haushofer turned to academia and, with the help of friends, obtained a lecturing post in political geography at the University of Munich. Hess soon enrolled as one of his students. Munich, at this time, was a city of revolutionary and counterrevolutionary ferment. In 1919, a group of revolutionary socialists, rebelling against the wartime slaughter and material hardship brought down upon them by the Kaiser and his generals, established a socialist republic in Bavaria, a political experiment that was soon violently crushed by the military. In 1923, a new violently nationalist party called the National Socialist Workers Party (the Nazis for short), headquartered in Munich and made up largely of disaffected former soldiers, attempted to seize power in a Beer Hall Putsch. Hess, a senior member of the new party, fled in the wake of the failure and was hidden by Haushofer in his summer home in the Bavarian mountains. When Hess eventually gave himself up and was imprisoned, Haushofer visited him in Landsberg prison, where Hess introduced Haushofer to the leader of the Nazi Party, Adolf Hitler.

Like many soldiers, who define their masculinity in terms of martial glory, the German veterans of World War I had a difficult time coming to terms with the end of the war. Germany's defeat was experienced as personal defeat, its humiliation their own masculinity crisis. Haushofer, Hess and Hitler had a deep hatred of the peace treaty that took away Germany's colonies and part of its national territory after the war: the Treaty of Versailles. All felt that this treaty had emasculated Germany, reducing the territorial extent of this culturally advanced and populace nation, which they considered a natural world power, to a "narrow" territorial area in Central Europe. Unlike France and Britain, which were colony possessing powers, or the Soviet Union and the United States, which were space possessing powers, Germany was a strangulated state (Natter, 2004: 247–248). After what they described as the "geographical error" of Versailles, they believed that Germany's need for *Lebensraum* or living space was greater than ever. Consequently, they all worked, in their different ways, to overthrow the Treaty of Versailles and "make Germany a world power again" which they understood to mean making Germany a dominant *military* power capable of expanding territorially at the expense of its neighbors. Greatness, for these veterans whose masculinity was forged in the squalid trenches of World War I fighting over scraps of land, was all about territorial control and domination.

Karl Haushofer's crusade to overthrow the Treaty of Versailles led him to found the journal *Zeitschrift für Geopolitik* (*Journal of Geopolitics*) in 1924. This journal was to serve as the flagship for the new school of geography Haushofer and other likeminded geographers helped create: German *Geopolitik* (geopolitics). Like Mackinder in Great Britain, Haushofer believed that the leaders of the state should be educated in the geographical relationships he claimed governed international politics. Mixing the social Darwinist ideas of his intellectual hero, Friedrich Ratzel, and the ideas of Mackinder (Haushofer greatly admired Mackinder's writings, describing "The Geographical Pivot of History" as "a geopolitical masterwork"; Weigert, 1942: 116), Haushofer reduced the complexity of international relations to a few basic laws and principles which he tirelessly promoted in the *Zeitschrift* and numerous books. In a book on frontiers, Haushofer outlines the Ratzelian organic theory of the state and uses this to polemicize against the Treaty of Versailles. International politics was a struggle for survival between competing states. In order to survive, the German state must achieve *Lebensraum*. The best means of achieving this, following Mackinder (who did not anticipate his ideas being used by German strategists!), is for Germany to develop an alliance with the heartland power, the Soviet Union. Furthermore, Haushofer argued, Germany should align itself with Japan and strive to create a continental-maritime block stretching from Germany through Russia to Japan against the global maritime empires of France and Great Britain, empires Haushofer believed were weak and in decay.

In "Why Geopolitik?" (Reading 3), which was published in 1925, Haushofer claims that the reason Germany lost World War I was because its leaders did not study geopolitics. Geopolitics, for Haushofer, is the study of the "earth-boundedness" of political processes and institutions. Like Mackinder, he attributes special power to the god-like geopolitician, treating geopolitics as a faith that offers divine revelations. Geopolitics can make certain predictions. It can provide "realistic insight into the world picture as it presents itself from day to day." It will help "our statesmen . . . see political situations as they really are." Only the geopolitician can "see what is"! Haushofer's persistent emphasis on the need for geopolitical "training" is nothing more than a legitimation for the right-wing militarist foreign policy he and others promoted at the expense of the fragile democracy of the Weimar Republic. Haushofer justifies this "training" by declaring that our enemies study geopolitics so "we" had better start too!

ADOLF HITLER'S RACIST GEOPOLITICS

Haushofer discussed his ideas with Adolf Hitler at Landsburg prison, where Hitler enjoyed a rather comfortable imprisonment. He presided over a midday meal, had as many visitors as he wanted and spent much of his time outside in the garden. From July 1924 onwards, he began dictating *Mein Kampf* to Rudolf Hess and also to another secretary, Emil Maurice. Volume 1, dictated in prison, was published in 1925. Volume 2, which Hitler dictated in his villa on Obersalzberg after his early release from prison, was published at the end of 1926. The book did not sell widely until after the Nazis were handed power by conservative elements in the German state in 1933 fearful of communism and social revolution (Bullock, 1992: 140). Whereas Haushofer sought to advise leaders, Hitler sought to become the one that would restore Germany's greatness. Hitler had used the Beer Hall Putsch trial to project himself as the man of action and destiny who would lead "the revolution against the revolution" (i.e. a nationalist and militarist counterrevolution against communism and the "bourgeois" Weimar Republic).

Hitler's *Mein Kampf* is a racist tract in which Hitler outlines his crude social Darwinist vision of the world. He describes a racial struggle for survival between the pure and the impure (hybrid), the healthy and the parasitic, the national and the international, the noble and the treacherous. This basic set of binary categorizations is mapped onto the fundamental distinction Hitler makes between identity and difference, "Us" and "Them," the Self and the Other (see Table 2). Using two pseudoscientific racial categories with no basis in fact, "the Aryan" and "the Jew," Hitler defines a positive insider identity which he champions in opposition to a negative outsider identity. He invents "the German" in opposition to "the Jew." That people of Jewish faith and heritage could also be good German citizens was a

Identity	Difference
Self: "Us"	Other: "Them"
Insider, friend, citizen	Outsider, stranger, foreigner
Aryan German nation	Jewish Bolshevism
Rooted in traditional organic soil and society Idealized image of peasant society with "natural leaders"	Rootlessness; do not have any soil of their own and so are "parasitic" on the territory of others. Associated with the vices of modernity and urban life
A folkish community of the beautiful, healthy and racially pure	A collection of the impure, the hybrid and unhealthy, the dirty and the degenerate

Table 2   Hitler's racist map of identity and difference

contradiction in terms to the Nazis (even though many had fought and died during the war for Germany). As the historian Alan Bullock has noted, the identity category

> "the Jew" as one encounters it in the pages of *Mein Kampf* and Hitler's ravings bears no resemblance to flesh-and-blood human beings of Jewish descent: [It] is an invention of Hitler's obsessional fantasy, a Satanic creation, expressing his need to create an object on which he could concentrate his feelings of aggression and hatred.
>
> (Bullock, 1992: 145)

Like similar racist categories, "the Jew" is an eminently flexible archetype. "The Jew" could represent both an ultra-capitalist (bankers, financiers, industrial and department store owners) and an ultra-communist (a "Bolshevik," Marxist or German leftist), two totally opposite identities. Logical contradictions such as this are common in racist reasoning. The Other is whatever the racist decides it is. Both identities are present in the composite category, "Jewish Bolshevism" which represents "rootlessness," "internationalism" and "decay" in Hitler's worldview. The opposite of "Jewish Bolshevism" is "folkish nationalism," Hitler's racist version of German nationalism that imagined Germany as an idealized community of healthy and racially pure Aryan peasants rooted in the soil and ruled over by "natural leaders" like Hitler.

There are three fundamentally racist discourses of danger championed by Hitler in *Mein Kampf*. The first is the external threat posed to his idealized German nation by a supposed "international Jewish conspiracy" against all nation states, particularly Germany. The headquarters of this conspiracy are, again contradictorily, the radically different states of the United States (international finance capitalism) and the Soviet Union (communism). The second is the internal threat posed by German leftist organizations and political parties, and the general supposed "decay" of modern urban life (mis-cegenation, degenerate art, prostitution and mental illness). The third threat, which combines elements of all of Hitler's obsessions, is one that finds expression in Chapter 14 of Volume 2 of *Mein Kampf* (Reading 4). In this chapter, Hitler borrows from both Ratzel and Haushofer (without citation) to claim that Germany is currently an "impotent" nation without adequate territorial resources to feed its people. The Treaty of Versailles has "constricted" the German nation and left it without adequate space, especially in comparison to the other world powers. Since nations are competitive organic entities that gain nourishment from the soil, the German nation must begin pursuing *Lebensraum* or else face decay and further decline. In Hitler's worldview, size matters: great power is demonstrated by control of vast territorial spaces.

Hitler argues that it is up to the National Socialist movement (the Nazi Party) to "endeavor to eliminate the discrepancy between our population and our area," "to bring the land into consonance with the population." This "discrepancy," this need for harmony between land and people, is a "truth" generated by Hitler's appropriation of Ratzelian discourse, an example of the power of discourse to construct "the real" and, on the basis of this supposed "truth" or "reality" (in actuality, social constructions), to legitimize and justify certain political visions, in this case Hitler's militaristic ambitions. Because of his racism, Hitler has utter contempt for the Soviet Union, viewing it as a state in decay. Yet, paradoxically, the Jewish Bolshevism of the Soviet Union is a mortal threat. Again, we encounter a contradiction. The Other is weak and degenerate yet the Other is also an implacable and dangerous enemy. As he makes clear, Hitler does not want to return to the 1914 borders of Germany, to the territory of Germany before the Treaty of Versailles. Hitler's plans are much more radical. He scorns those, like Karl Haushofer, calling for an "Eastern orientation" or an alliance between Russia and Germany. Hitler's program is an "eastern policy in the sense of acquiring the necessary soil for our German people." From this one line, first published in 1926, we can see Hitler's megalomaniac desire to play god with the map of continental Europe and rearrange it completely. His imperialist vision was for the colonization of the East by a renewed German empire. The German "Aryan" master race would enslave the subhuman "Slavs" of the East. Hitler's "eastern policy" ultimately led to the genocidal war against the Soviet Union and "Jewish Bolshevism" that began in 1941 (Mayer, 1988). Ironically, one of the reasons for Hitler's downfall was his racist assumption that the Soviet Union would collapse in a few months after its invasion by the German army. It was not to be.

It is worth noting that there were important differences between the German geopolitics of Karl Haushofer and the Nazi geopolitics of Adolf Hitler. Haushofer nationalism was more conservative-aristocratic than counterrevolutionary fascist. Haushofer considered the British empire the ultimate enemy of Germany and urged an alliance with the Soviet Union, whereas Hitler admired the British empire and ultimately wanted to conduct a crusade against the Soviet Union and Jewish Bolshevism. In Haushofer's Ratzelian schema, space not race is the ultimate determinate of national destiny, whereas for Hitler race is more important than space. Racists believe that destiny is internal and biological not external and environmental (Bassin, 1987).

Nevertheless, these differences should not detract from the fundamental support Haushofer gave to Hitler and the Nazi regime both before and after it was handed power by the conservative establishment in 1933. Although he never became a Nazi Party member, Haushofer promoted Nazi ideology, writing a book called *National Socialist Thought in World Politics* to mark the Nazi ascent to power, even denouncing Jews despite the fact that his own wife was Jewish. Together with his son, Albrecht, Haushofer helped facilitate the German–Japanese cooperation that eventually resulted in the Anticomintern Pact of 1936. Both were advisors to Hitler during the Munich conference of 1938. The Nazi–Soviet pact of 1939 seemed to represent Haushofer's thinking. However, the strange flight of Rudolf Hess to England in May 1941 ended the influence of the Haushofer family (Heske, 1987). Karl was even imprisoned for eight weeks in Dachau after Hess's flight. Albrecht, too, was imprisoned. After his release, Albrecht maintained links with those aristocratic elements in the German establishment belatedly planning the overthrow of Hitler. Their attempt to assassinate Hitler on July 20, 1944 failed, however, and Albrecht Haushofer was imprisoned once again. Upon his release from Moabit prison in April 1945, he was murdered by a roaming SS squad.

Despite the limited role of the Haushofers in the last few years of the Nazi state, many sensational press stories in the allied countries, particularly the United States, presented him as the scientific brain behind the Nazi blueprint for "world conquest." Haushofer was said to run an enormous Institute of Geopolitics at the University of Munich which supposedly gathered information from all over the world.

This was then used by Haushofer and his colleagues to make predictions about the course of world politics and give advice to Nazi leaders about the most opportune times to invade countries and the like. All of this sensationalism, which Haushofer complained about after the war, was exaggerated and largely untrue (Ó Tuathail, 1996). Nevertheless, the admiration the German geopoliticians had for Mackinder was noticed by the allies and some of his works were reread and republished.

ISAIAH BOWMAN AND AMERICAN *LEBENSRAUM*

Fear of and fascination with the "new science of geopolitics" turned a spotlight on the most powerful American geographer at this time: Isaiah Bowman. Variously described in the press as "America's Haushofer" and President Franklin "Roosevelt's geographer," Isaiah Bowman had long been a power-broker in American Geography as well as a highly placed geographical "expert" serving the American state. Educated in physical geography at Harvard and Yale, Bowman's exploration and cartographic fieldwork in South America earned him academic notice and respectability. Like Mackinder, he was taken by the romance of exploration and conquest though he was more critical of the brutality it involved (Smith, 2003: 79). His appointment in 1914 as Director of the American Geographical Society provided him entree into the highest professional and social circles in the United States. Through deft policy entrepreneurship on his part, he became head of a project named the "Inquiry" that Woodrow Wilson and his advisors formed in 1917 to serve as a geographical intelligence agency for the American delegation heading off to the Paris Peace conference (the US state, at this time, had no central intelligence bureaucracy and only a small foreign policy staff). Bowman became Wilson's leading geographical advisor and was intimately involved in discussions over the redrawing of the borders of Poland and Yugoslavia.

The defeat of the League of Nations and death of President Wilson marked a temporary end to the Wilsonian project of having the United States promote a "new world order" organized around the principles of liberal democracy, national self-determination, open door trade and liberal capitalist investment rules. Bending recalcitrant European nationalisms to this American "Wilsonianism" or "liberal internationalism" proved extremely difficult in practice. Bowman discovered this in Paris in border negotiations over the messy geographies and sticky geopolitics of deeply contested places like Danzig/Gdansk (Germany/Poland) and Fiume/Rijeka (Italy/Croatia/Yugoslavia). To some Americans there was a fundamental clash between "European nationalism" and "American universalism" or between a territorial conception of power and a conception that places economic expansionism before territorial aggrandizement. Geopolitical storylines contrasting supposed primordial differences between the United States and Europe, however, tend to be overly simplifying (and self-serving for those pro-moting the storyline). For a start, the United States, as Haushofer and others pointed out, already had its own resource rich territorial base as its basis for world power; indeed it already had an informal empire in Central America, the Philippines and the Caribbean because of US military possessions and interventions there. Also, the notion that the Wilsonian creed represented universal not national interests is a particularly American conceit. Bowman's biographer, Neil Smith, argues that Wilsonianism served the interests of economic networks of power in the United States who stood to benefit from open trade and liberal capitalist investment conditions (Smith, 2003: 140). The idea that the United States values are universal ones, that America is an exceptional country because it is the "homeland of freedom," is a defining characteristic of American nationalism and geopolitical culture.

In the 1920s, isolationist discourses predominated in American foreign policy. Bowman and his friends wanted a more liberal internationalist foreign policy that would serve American self-interest while articulating Wilsonian ideals of "democracy" and "freedom" (of trade, markets and peoples). In 1921 Bowman helped transform a New York dinner club of establishment bankers, lawyers and diplomats into a more formal association called the Council on Foreign Relations. The association grew to become the most influential foreign policy club in the United States. Its journal *Foreign Affairs* (first published in 1922) became the voice of liberal internationalism and helped establish it as the alternative geopolitical tradition to isolationism in American geopolitical culture. Bowman also wrote a popular textbook *The New World* (first edition, 1921; fourth edition, 1928) which drew upon his work for the Inquiry to present what he considered an "objective" and "realistic" survey of the political geography of world affairs. But German geographers did not see it as such. To them, Bowman was one of the authors of the hated Treaty of Versailles. Around Haushofer, a small group began a three volume German riposte to *The New World* called *Macht und Erde* (*Power and Earth*).

Bowman's response to German *Geopolitik* in the 1930s was strangely mute (Smith, 2003: 283–289). It was only when the wartime press began to write that geopolitics actually began in the United States and started referring to him as "America's Haushofer" that he decided to publicly defend *The New World* and his geographical work. His 1942 essay "Geography versus Geopolitics" is worth critical examination for two reasons (Reading 5). First, it is a great example of the politics surrounding the very conceptualization of geopolitics, what is considered and named "geopolitics" and what is not. Bowman's very title tells us that his discursive strategy is one of contrasting that which he wants to call "geography" from that which he wants to call "geopolitics." How does he operationalize this classification? (Remember that the critical study of geopolitical discourse requires attention to how classification schemes work.) Very simply, in his storyline "geography" is scientific and rational whereas "geopolitics" is pseudo-scientific and dishonest. This distinction corresponds to a series of others that run throughout the text, that between democracy and state tyranny, the United States and Germany, a "good neighbor policy" (the actual name of FDR's Latin American policy) and a "bad neighbor policy" (Nazi foreign policy). For Bowman, geography, science, democracy and America are naturalized as equivalencies: all are on the same side. The opposite side is geopolitics which is condemned as "German perversion of truth." Yet, Bowman's storyline is undercut by the virulence of his own discourse (note carefully his use of adjectives like crooked, evil, dishonest, depraved, etc.). He does not blame Nazism for Germany's foreign policy but "German political and philosophical thinking and ambition for two hundred years"! Germans aim to "conquer the world"! Their theories of government have an "essential primitiveness." Is this discourse scientific or geopolitical? If we admit that it does not seem scientific then Bowman's distinctions collapse and he stands revealed as a geopolitician himself. His own descriptions are a crude form of geopolitical reasoning, a stealthy geopolitics that refuses to call itself "geopolitics." Indeed, he concedes as much in describing a changed geopolitical culture in wartime America and how the end of the United States' solitude means that "we are all strategists, statesmen, critics and devisers." The geographer is also a geopolitician.

This brings us to a second instructive feature of Bowman's essay: it is a good example of the ideology of having no ideology. Bowman begins with a quote from Alexander Hamilton that indicates that the world is divided into ideologists (those motivated by "some untoward bias") and non-ideologists, those who do not "entangle themselves in words" but, by implication, "tell it like it is." Bowman, of course, is part of the latter group. He represents himself as having no ideology just natural practicality. Consider, for example, how he defends *The New World*. He writes that it "interposed no ideological preconceived 'system' between a problem and its solution in a practical world . . . It sought to analyze real situations rather than justify any one of several conflicting nationalist policies." This recalls the "universal America" versus "nationalist Europe" distinction that Wilsonians used at the Paris Peace Conference. Bowman/the United States have no ideology but work in the "practical world" whereas German geopolitics/Europe view everything through nationalist bias and preconception. Their arguments "are only made up to suit the case for German aggression." The method of *The New World*, by contrast, was to "deal realistically with the political problems of the postwar world." Realists, as we noted in the introduction, are blind to the fact their so-called realism is dependent upon a naturalized and unquestioned geopolitical culture. In not acknowledging his own cultural embeddedness as an "ideological preconceived 'system'," Bowman was hiding his own embeddedness in the geopolitical tradition of American liberal internationalism.

Bowman's essay is revealing for what it does not discuss: his political work for the US state. At the time of its writing, Bowman was high level advisor to Franklin Roosevelt on a number of aspects of the American war effort. He chaired an updated version of the Inquiry called the M Project (the M stood for migration) that studied the question of wartime refugees though it led to no meaningful action to save Jews from their terrible plight in wartime Europe. Later he would be involved in wartime planning over the post-war future of Germany and the establishment of the United Nations. Neil Smith (2003) describes the particular geopolitical vision that motivated Bowman as "American lebensraum." Unlike German *lebensraum*, which was fixated on territorial power and control for the nation, American *lebensraum* was concerned with economic power and control for corporate interests in the post-war world. American policy makers like Bowman wanted to shape this order in ways that favored American economic interests, wedging open colonial markets previously denied to American business and structuring a political order that institutionalized free trade and liberal capitalist investment rules. This *lebensraum* did not deny the importance of a territorial order of power for this was a necessary prerequisite for economic order and stability. The United States would need to be involved in helping reconstruct states after the war. But the principal basis for American global power in the post-war period was to be commercial not colonial. Bowman's American *lebensraum* was distinctively different from that of the Nazis (Crampton and Ó Tuathail, 1996).

THE END OF GERMAN *GEOPOLITIK*

After the fall of Berlin and the end of the war, the American Jesuit priest Father Edmund Walsh, the founder of the School of Foreign Service at Georgetown University in Washington DC, was flown to Germany to interrogate Karl Haushofer on his teachings and possible influence on Nazi foreign policy. The question as to whether Haushofer should be tried for war crimes in Nuremberg had to be decided. Walsh and the American army found a frail and disillusioned old man. In a statement before Father Walsh and the American army on November 2, 1945, Haushofer tried to explain his teachings and writings within the context of post-war Germany. As might be expected, Haushofer's "Defense of German Geopolitics" (Reading 6) is a self-serving document in which he seeks to disassociate himself from the horrors of Nazism. He claims that he was interested in educating and training German youth about the world, that he occasionally overstepped the boundary separating pure and practical science but that a scholar "should have the right to stand at the side of his people with all his mental power." Pointing to his own family's suffering, he claims that he opposed "imperialistic plans of conquest," a highly questionable interpretation.

What is interesting and disturbing about Haushofer's "Defense," however, is his claim that much of what he did was "legitimate" geopolitics. Haushofer points out that his captives have acknowledged this, that many of his lectures correspond to what Walsh taught at Georgetown, that many British and American thinkers were "the basic inspirers of his teaching," and that the original goals of German geopolitics were quite similar to "legitimate American geopolitics"! Haushofer's basic defense is that the "legitimacy" of German geopolitics was corrupted by the Nazis. This line of argumentation is much too convenient for it seeks to save geopolitics and Haushofer's reputation by blaming everything on the Nazis. What Haushofer does not acknowledge and recognize is *the essentially contested and political nature of all geopolitical discourse*, whether it be German, American or British. In trying to take refuge in the concept of "legitimate" geopolitics, Haushofer is recycling the same ideological/objective distinction we have brought into question in this part of the book. Geopolitics is never objective up to a certain point, scientific to a certain borderline or legitimate up to a balance of a certain percentage as he suggests (and as his interrogators believed too). Geopolitical discourse emerges from particular geopolitical cultures and is conditioned by networks of power. At this time, it was invariably entwined with the dominant ideologies and culture of nationalist chauvinism in the states where it was produced. Haushofer is thus right to claim that he was merely doing what other geopoliticians (like Edmund Walsh and Isaiah Bowman in the United States) were doing in their particular countries. But this does not mean that Haushofer's geopolitics is morally equivalent to that of these other geopoliticians. Haushofer was guilty of propagandizing a militarist and imperialistic version of German nationalism. He was complicitous with many of the aims of Adolf Hitler and the Nazi Party. As such, he was guilty of lending support to one of the most murderous and brutal state regimes in the twentieth century.

Haushofer's "Defense" raises some fundamental questions about the practice of geopolitics as a whole. As a form of power/knowledge, geopolitics was clearly complicitous with many chauvinist, racist and imperialist ideologies in the first half of the twentieth century. It justified oppressive European colonial empires that were premised on white supremist assumptions, imperialist interventionism and, in Hitler's geopolitics, brought imperialist expansionism and racist brutality to the European continent. It encouraged statesmen to play god with the world political map and justified appalling state violence, the culmination of which was World War II.

Despite what Bowman's reasoning suggests, geopolitics did not go away after World War II and the fall of Nazi Germany. It was changing form. Late on Sunday night, March 10, 1946, Karl Haushofer and his wife Martha walked to a secluded hollow on their country estate in the Bavarian mountains. Both took an arsenic drink and then Karl helped his wife hang herself to make sure of death. Karl himself fell dead soon afterwards, his hands, as Edmund Walsh describes it, "clutching the Bavarian soil which he so passionately loved and so often described in his writings on *Lebensraum*" (Walsh, 1948: 34). A year later Halford Mackinder died in England. Six days after his death, President Harry Truman of the United States addressed a joint session of Congress and requested economic and military aid to help the governments of Greece and Turkey fight against the worldwide communist threat. The weakness of the British Empire was provoking the bold extension of American's imperial ambitions to Europe. Imperialist geopolitical discourses were soon being subsumed by newly emergent Cold War geopolitical discourses.

REFERENCES AND FURTHER READING

On Mackinder

Blouet, B. (1987) *Halford Mackinder: A Biography*. College Station, TX: Texas A&M.

Freeman, E.A. (1879) "Language and Race," *Internet Modern History Sourcebook*. Available at http://www.fordham.edu/halsall/mod/freeman-race.html.

Gilroy, P. (2000) *Against Race: Imagining Political Culture Beyond the Color Line*. Cambridge, MA: Harvard University Press.

Kearns, G. (1985) "Halford John Mackinder, 1861–1947." *Geographers Biobibliographical Studies*, 9: 71–86.

—— (1997) "The Imperial Subject: Geography and Travel in the Work of Mary Kingsley and Halford Mackinder," *Transactions, Institute of British Geographers*, NS, 22(4): 450–472.

—— (2003) "Imperial Geopolitics: Geopolitical Visions at the Dawn of the American Century," in *A Companion to Political Geography*, eds. J. Agnew, K. Mitchell and G. Toal. Oxford: Blackwell.

McClintock, A. (1995) *Imperial Leather: Race, Gender and Sexuality in the Colonial Conquest*. London: Routledge.

Mackinder, H.J. (1902) *Britain and the British Seas*. New York: D. Appleton.

—— (1904) "The Geographical Pivot of History," *Geographical Journal*, 23: 421–437.

—— (1907) "On Thinking Imperially," *Lectures on Empire*, ed. M.E. Sadler. London: Privately printed.

—— (1919) *Democratic Ideals and Reality*. New York: Henry Holt.

—— (1925) "The English Tradition and the Empire: Some Thoughts on Lord Milner's Credo and the Imperial Committees," *United Empire*, 16: 1–8.

Ó Tuathail, G. (1992) "Putting Mackinder in his Place: Material Transformations and Myth," *Political Geography Quarterly*, 11: 100–118.

Parker, W.H. (1982) *Mackinder: Geography as an Aid to Statecraft*. Oxford: Clarendon Press.

Ryan, J. (1998) *Picturing Empire: Photography and the Visualization of the British Empire*. Chicago, IL: University of Chicago Press.

On Theodore Roosevelt and American imperialism

Beale, H. (1956) *Theodore Roosevelt and the Rise of America to World Power*. New York: Collier.

Hunt, M. (1987) *Ideology and U.S. Foreign Policy*. New Haven, CT: Yale University Press.

LaFeber, W. (1963) *The New Empire: An Interpretation of American Expansion, 1860–1898*. Ithaca, NY: Cornell University Press.

Miller, N. (1992) *Theodore Roosevelt: A Life*. New York: Morrow.

Roosevelt, T. (1913) *Theodore Roosevelt: An Autobiography*. New York: Macmillan.

Williams, W.A. (1980) *Empire as a Way of Life*. New York: Oxford University Press.

—— (1962) *The Tragedy of American Diplomacy*. New York: Dell.

On Haushofer

Bassin, M. (1987) "Race Contra Space: The Conflict Between German Geopolitik and National Socialism," *Political Geography Quarterly*, 6: 115–134.

Dorpalen, A. (1942) *The World of General Haushofer: Geopolitics in Action*. New York: Farrar and Rinehart.

Heske, H. (1987) "Karl Haushofer: His Role in German Geopolitics and in Nazi Politics," *Political Geography Quarterly*, 6: 135–144.

Murphy, D. (1997) *The Heroic Earth: Geopolitical Thought in Weimar Germany, 1918–1933*. Kent, OH: Kent State University Press.

Natter, W. (2004) "Geopolitics in Germany, 1919–45: Karl Haushofer and the Zeitschrift für Geopolitik," in *A Companion to Political Geography*, eds. J. Agnew, K. Mitchell and G. Toal. Oxford: Blackwell.

Ó Tuathail, G. (1996) *Critical Geopolitics: The Politics of Writing Global Space*. London: Routledge.

Parker, G. (1985) *Western Geopolitical Thought in the Twentieth Century*. New York: St Martin's Press.

Walsh, E. (1948) *Total Power*. New York: Doubleday.

Weigert, H. (1942) *Generals and Geographers: The Twilight of Geopolitics*. London: Oxford University Press.

On Hitler and Nazi ideology

Bauman, Z. (1989) *Modernity and the Holocaust*. Cambridge: Polity.
Bullock, A. (1992) *Hitler and Stalin: Parallel Lives*. New York: Knopf.
Kershaw, I. (1987) *The "Hitler Myth": Image and Reality in the Third Reich*. New York: Oxford University Press.
—— (1999) *Hitler: 1889–1936 Hubris*. New York: Norton.
—— (2001) *Hitler: 1936–1945 Nemesis*. New York: Norton.
Mayer, A. (1988) *Why Did the Heavens Not Darken?* New York: Pantheon.

On Bowman and American *Lebensraum*

Crampton, A. and Ó Tuathail, G. (1996) "Intellectuals, Institutions and Ideology: The Case of Robert Strausz-Hupé and American Geopolitics," *Political Geography*, 15: 553–556.
Smith, N. (2003) *American Empire: Roosevelt's Geographer and the Prelude to Globalization*. Berkeley, CA: University of California Press.

"HANDS OFF!!"

Cartoon 1 Hands off!!
This British postcard dates from the beginning of the twentieth century and features the British lion rebuking the threatening advance of the German eagle towards the globe.

Source: Courtesy of Dr Peter Taylor's private collection

The Geographical Pivot of History

Halford J. Mackinder
from *Geographical Journal* (1904)

When historians in the remote future come to look back on the group of centuries through which we are now passing, and see them foreshortened, as we today see the Egyptian dynasties, it may well be that they will describe the last 400 years as the Columbian epoch, and will say that it ended soon after the year 1900. Of late it has been a commonplace to speak of geographical exploration as nearly over, and it is recognized that geography must be diverted to the purpose of intensive survey and philosophic synthesis. In 400 years the outline of the map of the world has been completed with approximate accuracy, and even in the polar regions the voyages of Nansen and Scott have very narrowly reduced the last possibility of dramatic discoveries. But the opening of the twentieth century is appropriate as the end of a great historic epoch, not merely on account of this achievement, great though it be. The missionary, the conqueror, the farmer, the miner, and, of late, the engineer, have followed so closely in the traveler's footsteps that the world, in its remoter borders, has hardly been revealed before we must chronicle its virtually complete political appropriation. In Europe, North America, South America, Africa, and Australasia there is scarcely a region left for the pegging out of a claim of ownership, unless as the result of a war between civilized or half-civilized powers. Even in Asia we are probably witnessing the last moves of the game first played by the horsemen of Yermak the Cossack and the shipmen of Vasco da Gama. Broadly speaking, we may contrast the Columbian epoch with the age which preceded it, by describing its essential characteristic as the expansion of Europe against almost negligible resistances, whereas mediaeval Christendom was pent into a narrow region and threatened by external barbarism. From the present time forth, in the post-Columbian age, we shall again have to deal with a closed political system, and nonetheless that it will be one of worldwide scope. Every explosion of social forces, instead of being dissipated in a surrounding circuit of unknown space and barbaric chaos, will be sharply re-echoed from the far side of the globe, and weak elements in the political and economic organism of the world will be shattered in consequence. There is a vast difference of effect in the fall of a shell into an earthwork and its fall amid the closed spaces and rigid structures of a great building or ship. Probably some half-consciousness of this fact is at last diverting much of the attention of statesmen in all parts of the world from territorial expansion to the struggle for relative efficiency.

It appears to me, therefore, that in the present decade we are for the first time in a position to attempt, with some degree of completeness, a correlation between the larger geographical and the larger historical generalizations. For the first time we can perceive something of the real proportion of features and events on the stage of the whole world, and may seek a formula which shall express certain aspects, at any rate, of geographical causation in universal history. If we are fortunate, that formula should have a practical value as setting into perspective some of the competing forces in current international politics. The familiar phrase about the westward march of empire is an empirical and fragmentary attempt of the kind. I propose this evening describing those physical features of the world which I believe to have been most co-ercive of human action, and presenting some of the chief phases of history as organically connected with them, even in the ages when they were unknown to geography. My aim will not be to discuss the influence of this or that kind of feature, or yet to make a study in regional geography, but rather to exhibit human history as part of the life of the world organism. I recognize that I can only arrive at one aspect of the truth, and I have no wish to stray into excessive materialism. Man and not nature initiates, but nature in large measure controls. My concern is with the general physical control, rather than the causes of universal history. It is obvious that only a first approximation to truth can be hoped for. I shall be humble to my critics.

The late Professor Freeman held that the only history which counts is that of the Mediterranean and European races. In a sense, of course, this is true, for it is among these races that have originated the ideas which have rendered the inheritors of Greece and Rome dominant throughout the world. In another and very important sense, however, such a limitation has a cramping effect upon thought. The ideas which go to form a nation, as opposed to a mere crowd of human animals, have usually been accepted under the pressure of a common tribulation, and under a common necessity of resistance to external force. The idea of England was beaten into the Heptarchy by Danish and Norman conquerors; the idea of France was forced upon competing Franks, Goths, and Romans by the Huns at Chalons, and in the Hundred Years' War with England; the idea of Christendom was born of the Roman persecutions, and matured by the Crusades; the idea of the United States was accepted, and local colonial patriotism sunk, only in the long War of Independence; the idea of the German Empire was reluctantly adopted in South Germany only after a struggle against France in comradeship with North Germany. What I may describe as the literary conception of history, by concentrating attention upon ideas and upon the civilization which is their outcome, is apt to lose sight of the more elemental movements whose pressure is commonly the exciting cause of the efforts in which great ideas are nourished. A repellent personality performs a valuable social function in uniting his enemies, and it was under the pressure of external barbarism that Europe achieved her civilization. I ask you, therefore, for a moment to look upon Europe and European history as subordinate to Asia and Asiatic history, for European civilization is, in a very real sense, the outcome of the secular struggle against Asiatic invasion.
[. . .]

For a thousand years a series of horse riding peoples emerged from Asia through the broad interval between the Ural mountains and the Caspian sea, rode through the open spaces of southern Russia, and struck home into Hungary in the very heart of the European peninsula, shaping by the necessity of opposing them the history of each of the great peoples around – the Russians, the Germans, the French, the Italians, and the Byzantine Greeks. That they stimulated healthy and powerful reaction, instead of crushing opposition under a widespread despotism, was due to the fact that the mobility of their power was conditioned by the steppes, and necessarily ceased in the surrounding forests and mountains.

A rival mobility of power was that of the Vikings in their boats. Descending from Scandinavia both upon the northern and the southern shores of Europe, they penetrated inland by the river ways. But the scope of their action was limited, for, broadly speaking, their power was effective only in the neighborhood of the water. Thus the settled peoples of Europe lay gripped between two pressures – that of the Asiatic nomads from the east, and on the other three sides that of the pirates from the sea. From its very nature neither pressure was overwhelming, and both therefore were stimulative. It is noteworthy that the formative influence of the Scandinavians was second only in significance to that of the nomads, for under their attack both England and France made long moves towards unity, while the unity of Italy was broken by them. In earlier times, Rome had mobilized the power of her settled peoples by means of her roads, but the Roman roads had fallen into decay, and were not replaced until the eighteenth century.
[. . .]
Mobility upon the ocean is the natural rival of horse and camel mobility in the heart of the continent. It was upon navigation of oceanic rivers that was based the Potamic stage of civilization, that of China on the Yangtze, that of India on the Ganges, that of Babylonia on the Euphrates, that of Egypt on the Nile. It was essentially upon the navigation of the Mediterranean that was based what has been described as the Thalassic stage of civilization, that of the Greeks and Romans. The Saracens and the Vikings held sway by navigation of the oceanic coasts.

The all important result of the discovery of the Cape road to the Indies was to connect the western and eastern coastal navigations of Euro-Asia, even though by a circuitous route, and thus in some measure to neutralize the strategical advantage of the central position of the steppe-nomads by pressing upon them in rear. The revolution commenced by the great mariners of the Columbian generation endowed Christendom with the widest possible mobility of power, short of a winged mobility. The one and continuous ocean enveloping the divided and insular lands is, of course, the geographical condition of ultimate unity in the command of the sea, and of the whole theory of modern naval strategy and policy as expounded by such writers as Captain Mahan and Mr. Spenser Wilkinson [1853–1937, an influential British military correspondent, reformer and first Professor of Military History at Oxford University in 1909]. The broad political effect was to reverse the relations of Europe and Asia, for whereas in the Middle Ages Europe was caged between an impassable desert to south, an unknown ocean to west, and icy or forested wastes to north and north-east, and in the east and south-east was constantly threatened by the superior mobility of the horsemen and camel-men, she now emerged upon the world, multiplying more than thirty fold the sea surface and coastal lands to which she had access, and wrapping her influence round the Euro-Asiatic land-power which had hitherto threatened her very existence. New Europes were created in the vacant lands discovered in the midst of the waters, and what Britain and Scandinavia were to Europe in the earlier time, that have America and Australia, and in some measure even Trans-Saharan Africa, now become to Euro-Asia, Britain, Canada, the United States, South Africa, Australia, and Japan are now a ring of outer and insular bases for sea-power and commerce, inaccessible to the land-power of Euro-Asia.

But the land-power still remains, and recent events have again increased its significance. While the maritime peoples of Western Europe have covered the ocean with their fleets, settled the outer continents, and in varying degree made tributary the oceanic margins of Asia, Russia has organized the Cossacks, and, emerging from her northern forests, has policed the steppe by setting her own nomads to meet the Tartar nomads. The Tudor century, which saw the expansion of Western Europe over the sea, also saw Russian power carried from Moscow through Siberia. The eastward swoop of the horsemen across Asia was an event almost as pregnant with political consequences as was the rounding of the Cape, although the two movements long remained apart.

It is probably one of the most striking coincidences of history that the seaward and the landward expansion of Europe should, in a sense, continue the ancient opposition between Roman and Greek. Few great failures have had more far-reaching consequences than the failure of Rome to Latinize the Greek. The Teuton was civilized and Christianized by the Roman, the Slav in the main by the Greek. It is the Romano-Teuton who in later times embarked upon the ocean; it was the Graeco-Slav who rode over the steppes, conquering the Turanian. Thus the modern land-power differs from the sea-power no less in the source of its ideals than in the material conditions of its mobility.

In the wake of the Cossack, Russia has safely emerged from her former seclusion in the northern forests. Perhaps the change of greatest intrinsic importance which took place in Europe in the last century was the southward migration of the Russian peasants, so that, whereas agricultural settlements formerly ended at the forest boundary, the centre of the population of all European Russia now lies south of that boundary, in the midst of the wheat-fields which have replaced the more western steppes. Odessa has here risen to importance with the rapidity of an American city.

A generation ago steam and the Suez Canal appeared to have increased the mobility of sea-power relatively to land-power. Railways acted chiefly as feeders to ocean-going commerce. But trans-continental railways are now transmuting the conditions of land-power, and nowhere can they have such effect as in the closed heartland of Euro-Asia, in vast areas of which neither timber nor accessible stone was available for road-making. Railways work the greater wonders in the steppe, because they directly replace horse and camel mobility, the road stage of development having here been omitted.

[. . .]

The Russian railways have a clear run of 6000 miles from Wirballen [today Virbalis on the Lithuania border with Kalingrad] in the west to Vladivostok in the east. The Russian army in Manchuria is as significant evidence of mobile land-power as the British army in South Africa was of sea-power. True, that the Trans-Siberian railway is still a single and precarious line of communication, but the century will not be old before all Asia is covered with railways. The spaces within the Russian Empire and Mongolia are so vast, and their potentialities in population, wheat, cotton, fuel, and metals so incalculably great, that it is inevitable that a vast economic world, more or less apart, will there develop inaccessible to oceanic commerce.

As we consider this rapid review of the broader currents of history, does not a certain persistence of geographical relationship become evident? Is not the pivot region of the world's politics that vast area of Euro-Asia which is inaccessible to ships, but in antiquity lay open to the horse-riding nomads, and is today about to be covered with a network of railways? There have been and are here the conditions of a mobility of military and economic power of a far-reaching and yet limited character. Russia replaces the Mongol Empire. Her pressure on Finland, on Scandinavia, on Poland, on Turkey, on Persia, on India, and on China replaces the centrifugal raids of the steppe men. In the world at large she occupies the central strategical position held by Germany in Europe. She can strike on all sides and be struck from all sides, save the north. The full development of her modern railway mobility is merely a matter of time. Nor is it likely that any possible social revolution will alter her essential relations to the great geographical limits of her existence. Wisely recognizing the fundamental limits of her power, her rulers have parted with Alaska; for it is as much a law of policy for Russia to own nothing over seas as for Britain to be supreme on the ocean.

Outside the pivot area, in a great inner crescent, are Germany, Austria, Turkey, India, and China, and in an outer crescent, Britain, South Africa, Australia, the United States, Canada, and Japan. In the present condition of the balance of power, the pivot state, Russia, is not equivalent to the peripheral states, and there is room for an equipoise in France. The United States has recently become an eastern power, affecting the European balance not directly, but through Russia, and she will construct the Panama Canal to make her Mississippi and Atlantic resources available in the Pacific. From this point of view the real divide between east and west is to be found in the Atlantic Ocean.

The oversetting of the balance of power in favour of the pivot state, resulting in its expansion over the marginal lands of Euro-Asia, would permit of the use of vast continental resources for fleet-building, and the empire of the world would then be in sight. This might happen if Germany were to ally herself with Russia. The threat of such an event should, therefore, throw France into alliance with the over-sea powers, and France, Italy, Egypt, India, and Korea would become so many bridge heads where the outside navies would support armies to compel the pivot allies to deploy land forces and prevent them from con-

centrating their whole strength on fleets. On a smaller scale that was what Wellington accomplished from his sea-base at Torres Vedras in the Peninsular War. May not this in the end prove to be the strategical function of India in the British Imperial system? Is not this the idea underlying Mr. Amery's conception that the British military front stretches from the Cape through India to Japan?

The development of the vast potentialities of South America might have a decisive influence upon the system. They might strengthen the United States, or, on the other hand, if Germany were to challenge the Monroe doctrine successfully, they might detach Berlin from what I may perhaps describe as a pivot policy. The particular combinations of power brought into balance are not material; my contention is that from a geographical point of view they are likely to rotate round the pivot state, which is always likely to be great, but with limited mobility as compared with the surrounding marginal and insular powers.

I have spoken as a geographer. The actual balance of political power at any given time is, of course, the product, on the one hand, of geographical conditions, both economic and strategic, and, on the other hand, of the relative number, virility, equipment, and organization of the competing peoples. In proportion as these quantities are accurately estimated are we likely to adjust differences without the crude resort to arms. And the geographical quantities in the calculation are more measurable and more nearly constant than the human. Hence we should expect to find our formula apply equally to past history and to present politics. The social movements of all times have played around essentially the same physical features, for I doubt whether the progressive desiccation of Asia and Africa, even if proved, has in historical times vitally altered the human environment. The westward march of empire appears to me to have been a short rotation of marginal power round the south-western and western edge of the pivotal area. The Nearer, Middle, and Far Eastern questions relate to the unstable equilibrium of inner and outer powers in those parts of the marginal crescent where local power is, at present, more or less negligible.

In conclusion, it may be well expressly to point out that the substitution of some new control of the inland area for that of Russia would not tend to reduce the geographical significance of the pivot position. Were the Chinese, for instance, organized by the Japanese, to overthrow the Russian Empire and conquer its territory, they might constitute the yellow peril to the world's freedom just because they would add an oceanic frontage to the resources of the great continent, an advantage as yet denied to the Russian tenant of the pivot region.

DISCUSSION [. . .]

Mr Amery [Leopold Amery (1873–1955) at the time a military correspondent for *The Times* of London, and later a Conservative MP (elected 1911), First Lord of the Admiralty (1922–24) and Colonial Secretary (1924–29)]: I think it is always enormously interesting if we can occasionally get away from the details of everyday politics and try to see things as a whole, and this is what Mr Mackinder's most stimulating lecture has done for us tonight. He has give us the whole of history and the whole of ordinary politics under one big comprehensive idea [. . .] I do not intend to make many more remarks, but there is just one point – a word of Mr Mackinder's suggested it to me. Horse and camel mobility has largely passed away; and it is now a question of railway mobility as against sea mobility.

I should like to say that sea mobility has gained enormously in military strength to what it was in ancient times, especially in the number of men that can be carried. In the old days the ships were mobile enough, but they carried few men, and the raids of the sea people were comparatively feeble. I am not suggesting anything political at the present time; I am merely stating a fact when I say that the sea is far better at conveying troops than anything, except fifteen or twenty parallel lines of railway. What I am coming to is this; that both the sea and the railway are going in the future – it may be near, or it may be somewhat remote – to be supplemented by the air as a means of locomotion, and when we come to that (as we are talking in broad Columbian epochs, I think I may be allowed to look forward a bit) – when we come to that, a great deal of this geographical distribution must lose its importance, and the successful powers will be those who have the greatest industrial basis. It will not matter whether they are in the centre of a continent or on an island; those people who have the industrial power and the power of invention and of science will be able to defeat all others. I will leave that as a parting suggestion.

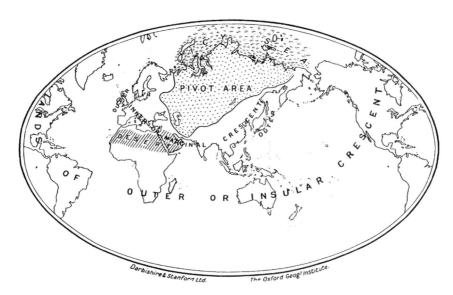

Map 1 The natural seats of power

Source: Mackinder 1904

The Roosevelt Corollary

Theodore Roosevelt

from *A Compilation of Messages and Papers of the Presidents* (1905)

[. . .] It is not true that the United States feels any land hunger or entertains any projects as regards the other nations of the Western Hemisphere save such as are for their welfare. All that this country desires is to see the neighboring countries stable, orderly, and prosperous. Any country whose people conduct themselves well can count upon our hearty friendship. If a nation shows that it knows how to act with reasonable efficiency and decency in social and political matters, if it keeps order and pays its obligations, it need fear no interference from the United States. Chronic wrongdoing or an impotence which results in a general loosening of the ties of civilized society, may in America, as elsewhere, ultimately require intervention by some civilized nation, and in the Western Hemisphere the adherence of the United States to the Monroe Doctrine may force the United States, however reluctantly, in flagrant cases of such wrongdoing or impotence, to the exercise of an international police power. If every country washed by the Caribbean Sea would show the progress in stable and just civilization which with the aid of the Platt amendment Cuba has shown since our troops left the island and which so many of the republics in both Americas are constantly and brilliantly showing, all question of interference by this Nation with their affairs would be at an end. Our interests and those of our southern neighbors are in reality identical. They have great natural riches, and if within their borders the reign of law and justice obtains, prosperity is sure to come to them. While they thus obey the primary laws of civilized society they may rest assured that they will be treated by us in a spirit of cordial and helpful sympathy. We would interfere with them only in the last resort, and then only if it became evident that their inability or unwillingness to do justice at home and abroad had violated the rights of the United States or had invited foreign aggression to the detriment of the entire body of American nations. It is a mere truism to say that every nation, whether in America or anywhere else, which desires to maintain its freedom, its independence, must ultimately realize that the right of such independence cannot be separated from the responsibility of making good use of it.

In asserting the Monroe Doctrine, in taking such steps as we have taken in regard to Cuba, Venezuela, and Panama, and in endeavoring to circumscribe the theater of war in the Far East, and to secure the open door in China, we have acted in our own interest as well as in the interest of humanity at large. There are, however, cases in which while our own interests are not greatly involved, strong appeal is made to our sympathies. [. . .] But in extreme cases action may be justifiable and proper. What form the action shall take must depend upon the circumstances of the case; that is, upon the degree of the atrocity and upon our power to remedy it. The cases in which we could interfere by force of arms as we interfered to put a stop to intolerable conditions in Cuba are necessarily very few.

Why Geopolitik?

Karl Haushofer
from *The World of General Haushofer* (1942)

While the theoretical foundations of Geopolitik were laid only in recent times, its practical application – the instinctive sense for geopolitical possibilities, the realization of its deep influence on political development – is as old as history itself. Geopolitical vision inspired daring leaders who guided their people along novel never-before-traveled roads. Powerful new states emerged because their creators, with the sensitivity of the true statesman, understood the geopolitical demands of the hour. Without such insight, violence and arbitrariness would have charted the course of history. Nothing with lasting value could have been created. All structures of state which might have been erected would sooner or later have crumbled into dust and oblivion before the eternal forces of soil and climate.

To be sure, the powerful will of a great and strong man may tear masses and nations away from soil-bound existence into roads other than nature had provided for them. But such actions are short-lived. In the end every people will sink back into its accustomed ways; its lasting earth-bound traits will eventually win out.

GEOPOLITIK AS EDUCATION IN STATECRAFT

Although our eyes can not penetrate the darkness of the future, scientific geopolitical analysis enables us to make certain predictions. Should we not therefore attempt to explore the field of Geopolitik more fully than we – and especially our diplomats – have thought necessary? To pose the question is to answer it. Our statesmen in particular ought to familiarize themselves with all those aspects of politics that can

be determined scientifically before piloting the destiny of state and nation into the mists of the unknown future. Jurisprudence and political science, which have been considered the sole prerequisites of education in statecraft, do not provide adequate training. A sound knowledge of geography and history is just as important. Above all, our future leaders must be schooled in geopolitical analysis.

Only this can give them the needed realistic insight into the world picture as it presents itself from day to day. Not by accident is the word "Politik" preceded by that little prefix "geo." This prefix means much and demands much. It relates politics to the soil. It rids politics of arid theories and senseless phrases which might trap our political leaders into hopeless Utopias. It puts them back on solid ground. Geopolitik demonstrates the dependence of all political developments on the permanent reality of the soil.

A whole body of literature has grown around this thesis. For the Alpine countries, Ratzel has traced the interdependence between politics and geographical environment in his *Alps as the Center of Historical Movements*. Krebs has given us an equally valuable work in his *Contributions to the Political Effects of Climate* in which he reveals the connection between lack of rain, aridity, and social and political unrest in East Asia. Kjellen, in his *Problem of the Three Rivers* (Rhine, Danube, Vistula), has shown us how the unhappy fate of Central Europe is inseparably tied up with the course of these rivers. And H.J. Mackinder, in his "Geographical Pivot of History," has attempted to review the entire world geopolitically and to forecast in 1904 what would happen between 1914 and 1924.

Why did our leading statesmen fail to see what this student of geopolitics realized as early as 1904?

Most likely because they lacked geopolitical training. In spite of excellent legal education and great administrative experience, they were unable to realize the effects of political-geographical trends. "Geographical ignorance may cost us dearly," warned Sir Thomas Holdich, one of England's most experienced students and drawers of boundaries.

GEOPOLITIK AND PRACTICAL POLITICS

Geopolitik has come to stay. We arrive at this conclusion from the fact that its application is gaining a growing following all over the world, while disregard of its teachings becomes increasingly dangerous. Some political successes can doubtless be attributed to geopolitical groundwork, among them the skillful selection of such English bases as Hong Kong, Singapore, and Penang. The reorganization of the Australian Commonwealth and the foundation of its new capital, Canberra, are likewise the result of geographical considerations. Geopolitically, even the choice of Tsingtao was a good one, provided one considers the establishment of a German base in China as geopolitically justifiable.

THE MISSION OF GEOPOLITIK

Geopolitik will serve our statesmen in setting and attaining their political objectives. It will present them with the scientific equipment of concrete facts and proven laws to help them see political situations as they really are. As an exact science, Geopolitik deserves serious consideration. Our leaders must learn to use all available tools to carry on the fight for Germany's existence – a struggle which is becoming increasingly difficult due to the incongruity between her food production and population density.

For our future foreign policy we therefore need Geopolitik. We need the same thorough training in this discipline as developed by England – though not under that name – with onesided purposefulness, as adopted by France [in the *Institut de France* and the *Ecole de Politique*], and as it is beginning to be used by Japan. Geopolitik is a child of geography; whoever takes up its study should therefore be trained geographically. To teach it requires first-hand knowledge; teachers of Geopolitik must know from practical experience not only the country they are teaching about but also the

one in which they are teaching. We must, moreover, study Geopolitik with a view to the present and future rather than to the past. As a nation governed by lawyers, we Germans have been too much under the influence of the *lex lata* [the law as it exists as opposed to the lex ferenda, what the law ought to be]. We considered politics more in terms of dead history than of living science: we looked back rather than ahead. In this manner we lost contact with the future. Making retrospective instead of precautionary future politics, we were left out of the realignment of the world when it occurred at the turn of the century.

This policy was doomed to failure. *Ducunt volentem, nolentem trahunt fata!* [Only those who are willing are guided by fate; the unwilling ones are dragged!] Nowhere does this maxim of Roman wisdom apply more truly than in the realm of politics. We learned our lesson.

[. . .]

Germany must emerge out of the narrowness of her present living space into the freedom of the world. We must approach this task well equipped in knowledge and training. We must familiarize ourselves with the important spaces of settlement and migration on earth. We must study the problem of boundaries as one of the most important problems of Geopolitik. We ought to devote particular attention to national self-determination, population pressure, living space, and changes in rural and urban settlement, and we must closely follow all shifts and transfers of power throughout the world.

The smaller the living space of a nation, the greater the need for a far-sighted policy to keep the little it can still call its own. A people must know what it possesses. At the same time, it should constantly study and compare the living spaces of other nations. Only thus will it be able to recognize and seize any possibility to recover lost ground.

"We must see foreign nations as they really are, not as we would like them to be." This occasional remark of Erich von Drygalski [Haushofer's academic mentor and thesis supervisor at the University of Munich] has served me as a beacon in my geopolitical work. Let us not stake our future foolishly on one card, let us not choose allies which others – better trained geopolitically – have considered doomed a half-century earlier. By prudent, courageous analysis of our world-political situation we shall always be able to preserve our sacred soil from shameful defeat. The admonitions "see what is," and "keep away from whatever our national honor cannot tolerate," are the pilot lights of our voyage. They are modest enough and even hardly sufficient to help our ship of state gain the open sea.

And yet – "I have neither men, arms, munitions, nor instructions [. . .]," the future commander of France's Army of the North wrote desperately to the Chief of National Defense on October 21, 1870. A victorious enemy was pressing him in front, and he was standing with his back against the wall – neutral Belgium that was already within gun range. Yet half a century later his grandsons stood east of the Rhine in a defenseless Germany, masters of the world's third largest colonial empire. During those fifty years France had taken up the study of geopolitics!

Eastern Orientation or Eastern Policy?

Adolf Hitler

from *Mein Kampf* (1942)

There are two reasons which induce me to submit to a special examination the relation of Germany to Russia: 1. Here perhaps we are dealing with the most decisive concern of all German foreign affairs; and 2. This question is also the touchstone for the political capacity of the young National Socialist movements to think clearly, and to act correctly.

[. . .]

If under foreign policy we must understand the regulation of a nation's relations with the rest of the world, the manner of this regulation will be determined by certain definite facts. As National Socialists we can, furthermore, establish the following principle concerning the nature of the foreign policy of a folkish state:

The foreign policy of the folkish state must safeguard the existence on this planet of the race embodied in the state, by creating a healthy, viable natural relation between the nation's population and growth on the one hand and the quantity and quality of its soil on the other hand.

As a healthy relation we may regard only that condition which assures the sustenance of a people on its own soil. Every other condition, even if it endures for hundreds, nay, thousands of years, is nevertheless unhealthy and will sooner or later lead to the injury if not annihilation of the people in question.

Only an adequately large space on this earth assures a nation of freedom of existence. [. . .]

Germany today is no world power. Even if our momentary military impotence were overcome, we should no longer have any claim to this title. What can a formation, as miserable in its relation of population to area as the German Reich today, mean on this planet?

In an era when the earth is gradually being divided up among states, some of which embrace almost entire continents, we cannot speak of a world power in connection with a formation whose political mother country is limited to the absurd area of five hundred thousand square kilometers.

From the purely territorial point of view, the area of the German Reich vanishes completely as compared with that of the so called world powers. Let no one cite England as a proof to the contrary, for England in reality is merely the great capital of the British world empire which calls nearly a quarter of the earth's surface its own. In addition, we must regard as giant states, first of all the American Union, then Russia and China. All are spatial formations having in part an area more than ten times greater than the present German Reich. And even France must be counted among these states. Not only that she complements her army to an ever-increasing degree from her enormous empire's reservoir of colored humanity, but racially as well, she is making such great progress in negrification that we can actually speak of an African state arising on European soil. [. . .]

Thus, in the world today we see a number of power states, some of which not only far surpass the strength of our German nation in population, but whose area above all is the chief support of their political power. Never has the relation of the German Reich to other existing world states been as unfavorable as at the beginning of our history two thousand years ago and again today. Then we were a young people, rushing headlong into a world of great crumbling state formations, whose last giant, Rome, we ourselves helped to fell. Today we find ourselves in a world of great power states in process of formation, with our own Reich sinking more and more into insignificance.

We must bear this bitter truth coolly and soberly in mind. We must follow and compare the German Reich through the centuries in its relation to other states with regard to population and area. I know that everyone will then come to the dismayed conclusion which I have stated at the beginning of this discussion: Germany is no longer a world power, regardless of whether she is strong or weak from the military point of view.

We have lost all proportion to the other great states of the earth, and this thanks only to the positively catastrophic leadership of our nation in the field of foreign affairs, thanks to our total failure to be guided by what I should almost call a testamentary aim in foreign policy, and thanks to the loss of any healthy instinct and impulse of self-preservation.

If the National Socialist movement really wants to be consecrated by history with a great mission for our nation, it must be permeated by knowledge and filled with pain at our true situation in this world; boldly and conscious of its goal, it must take up the struggle against the aimlessness and incompetence which have hitherto guided our German nation in the line of foreign affairs. Then, without consideration of "traditions" and prejudices, it must find the courage to gather our people and their strength for an advance along the road that will lead this people from its present restricted living space to new land and soil, and hence also free it from the danger of vanishing from the earth or of serving others as a slave nation.

The National Socialist movement must strive to eliminate the disproportion between our population and our area – viewing this latter as a source of food as well as a basis for power politics – between our historical past and the hopelessness of our present impotence. And in this it must remain aware that we, as guardians of the highest humanity on this earth, are bound by the highest obligation, and the more it strives to bring the German people to racial awareness so that, in addition to breeding dogs, horses, and cats, they will have mercy on their own blood, the more it will be able to meet this obligation.

[. . .]

We National Socialists must never under any circumstances join in the foul hurrah patriotism of our present bourgeois world. In particular it is mortally dangerous to regard the last pre-war developments as binding even in the slightest degree for our own course. From the whole historical development of the nineteenth century, not a single obligation can be derived which was grounded in this period itself. In contrast to the conduct of the representatives of this period, we must again profess the highest aim of all foreign policy, to wit: to bring the soil into harmony with the population. Yes, from the past we can only learn that, in setting an objective for our political activity, we must proceed in two directions: Land and soil as the goal of our foreign policy, and a new philosophically established, uniform foundation as the aim of political activity at home.

I still wish briefly to take a position on the question as to what extent the demand for soil and territory seems ethically and morally justified. This is necessary, since unfortunately, even in so called folkish circles, all sorts of unctuous big-mouths step forward, endeavoring to set the rectification of the injustice of 1918 as the aim of the German nation's endeavors in the field of foreign affairs, but at the same time find it necessary to assure the whole world of folkish brotherhood and sympathy.

I should like to make the following preliminary remarks: The demand for restoration of the frontiers of 1914 is a political absurdity of such proportions and consequences as to make it seem a crime. Quite aside from the fact that the Reich's frontiers in 1914 were anything but logical. For in reality they were neither complete in the sense of embracing the people of German nationality, nor sensible with regard to geo-military expediency. They were not the result of a considered political action, but momentary frontiers in a political struggle that was by no means concluded; partly, in fact, they were the results of chance.

As opposed to this, we National Socialists must hold unflinchingly to our aim in foreign policy, namely, to secure for the German people the land and soil to which they are entitled on this earth. And this action is the only one which, before God and our German posterity, would make any sacrifice of blood seem justified: before God, since we have been put on this earth with the mission of eternal struggle for our daily bread, beings who receive nothing as a gift, and who owe their position as lords of the earth only to the genius and the courage with which they can conquer and defend it; and before our German posterity in so far as we have shed no citizen's blood out of which a thousand others are not bequeathed to posterity. The soil on which someday German generations of peasants can beget powerful sons will sanction the investment of the sons of today, and will some day acquit the responsible statesmen of blood-guilt and sacrifice of the people, even if they are persecuted by their contemporaries. [. . .]

But we National Socialists must go further. The right to possess soil can become a duty if without extension of its soil a great nation seems doomed to destruction. And most especially when not some little nigger nation or other is involved, but the Germanic mother of life, which has given the present-day world its cultural picture. Germany will either be a world power or there will be no Germany. And for world power she needs that magnitude which will give her the position she needs in the present period, and life to her citizens. [. . .]

Never forget that the rulers of present-day Russia are common blood-stained criminals; that they are the scum of humanity which, favored by circumstances, overran a great state in a tragic hour, slaughtered and wiped out thousands of her leading intelligentsia in wild blood lust, and now for almost ten years have been carrying on the most cruel and tyrannical regime of all time. Furthermore, do not forget that these rulers belong to a race which combines, in a rare mixture, bestial cruelty and an inconceivable gift for lying, and which today more than ever is conscious of a mission to impose its bloody oppression on the whole world. Do not forget that the international Jew who completely dominates Russia today regards Germany, not as an ally, but as a state destined to the same fate. And you do not make pacts with anyone whose sole interest is the destruction of his partner. Above all, you do not make them with elements to whom no pact would be sacred, since they do not live in this world as representatives of honor and sincerity, but as champions of deceit, lies, theft, plunder, and rapine. [. . .]

In Russian Bolshevism we must see the attempt undertaken by the Jews in the twentieth century to achieve world domination. Just as in other epochs they strove to reach the same goal by other, though inwardly related processes. Their endeavor lies profoundly rooted in their essential nature. No more than another nation renounces of its own accord the pursuit of its impulse for the expansion of its power and way of life, but is compelled by outward circumstances or else succumbs to impotence due to the symptoms of old age, does the Jew break off his road to world dictatorship out of voluntary renunciation, or because he represses his eternal urge. He, too, will either be thrown back in his course by forces lying outside himself, or all

his striving for world domination will be ended by his own dying out. But the impotence of nations, their own death from old age, arises from the abandonment of their blood purity. And this is a thing that the Jew preserves better than any other people on earth. And so he advances on his fatal road until another force comes forth to oppose him, and in a mighty struggle hurls the heaven-stormer back to Lucifer.

The fight against Jewish world Bolshevization requires a clear attitude toward Soviet Russia. You cannot drive out the Devil with Beelzebub. If today even folkish circles rave about an alliance with Russia, they should just look around them in Germany and see whose support they find in their efforts. Or have folkish men lately begun to view an activity as beneficial to the German people which is recommended and promoted by the international Marxist press? Since when do folkish men fight with armor held out to them by a Jewish squire? [. . .]

If the National Socialist movement frees itself from all illusions with regard to this great and all-important task, and accepts reason as its sole guide, the catastrophe of 1918 can some day become an infinite blessing for the future of our nation. Out of this collapse our nation will arrive at a complete reorientation of its activity in foreign relations, and, furthermore, reinforced within by its new philosophy of life, will also achieve outwardly a final stabilization of its foreign policy. Then at last it will acquire what England possesses and even Russia possessed, and what again and again induced France to make the same decisions, essentially correct from the viewpoint of her own interests, to wit: A political testament.

The political testament of the German nation to govern its outward activity for all time should and must be:

Never suffer the rise of two continental powers in Europe. Regard any attempt to organize a second military power on the German frontiers, even if only in the form of creating a state capable of military strength, as an attack on Germany, and in it see not only the right, but also the duty, to employ all means up to armed force to prevent the rise of such a state, or, if one has already arisen, to smash it again. See to it that the strength of our nation is founded, not on colonies, but on the soil of our European homeland. Never regard the Reich as secure unless for centuries to come it can give every scion of our people his own parcel of soil. Never forget that the most sacred right on this earth is a man's right to have earth to till with his own hands, and the most sacred sacrifice the blood that a man sheds for this earth. [. . .]

Neither western nor eastern orientation must be the future goal of our foreign policy, but an eastern policy in the sense of acquiring the necessary soil for our German people. Since for this we require strength, and since France, the mortal enemy of our nation, inexorably strangles us and robs us of our strength, we must take upon ourselves every sacrifice whose consequences are calculated to contribute to the annihilation of French efforts toward hegemony in Europe. Today every power is our natural ally, which like us feels French domination on the continent to be intolerable.

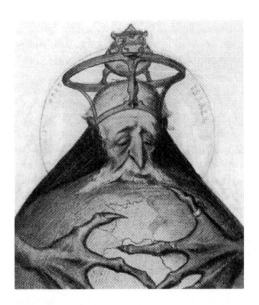

Cartoon 2 Rothschild
Anti-Semitism was never a specifically German phenomenon. This 1898 French cartoon illustrates the pervasive myth that the world was in the hands of Jewish bankers, personified by James (Jakob) Rothschild.

Source: C. Leandre 1898

Geography versus Geopolitics

Isaiah Bowman

from *Geographical Review* (1942)

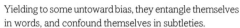

Yielding to some untoward bias, they entangle themselves in words, and confound themselves in subtleties.

(Alexander Hamilton, *The Federalist*)

The current discussion of German geopolitical writings involves the names, outlook, and reputation of certain American geographers, my own included. What was their position respecting geopolitics before general condemnation of Hitler and the Nazi program began? Did they foresee the evil consequences of German perversion of truth in the alleged new science of geopolitics which made use of the overlapping data of history, political science, and geography? It has recently been declared that American geopolitics was developed before it was taken up in Germany. The bad effect of this assertion touches more than personal or professional repute. It has given the question a national context.

Geopolitics presents a distorted view of the historical, political, and geographical relations of the world and its parts. It identifies no universal force or process like gravity. It is relative to the state to which it is applied. At least so say its advocates. Its arguments as developed in Germany are only made up to suit the case for German aggression. It contains, therefore, a poisonous self-destroying principle: when international interests conflict or overlap *might* alone shall decide the issue. Against "geopolitical needs" democracy opposes moral rights. Let us look first at the way in which this opposition arises.

THE MORAL BASIS OF DEMOCRACY

American democracy strives to achieve certain explicit purposes stated in a body of doctrine expressed in the first instance in our Declaration of Independence and subsequently in the Constitution with its amendments. It is at bottom the union of two principles (1) promotion of the general welfare through the consent of the governed and (2) respect for individual human rights. Doing evil things in the name of an alleged good cause is not the cornerstone of its philosophy. Democracy is an agreement upon purposes and a selection of means that a people's sense of justice approves. When the Reichstag in 1914 unanimously approved the German government's program, which had involved the violation of Belgian neutrality, it was expressing its sense of justice and giving its moral approval. By such approval it hoped to hasten the day of victory and peace. There you have also agreed purposes, a sense of justice, and moral approval! What was wrong with them?

Democracy starts with the individual. It believes in his general freedom to speak and act as he chooses provided he speaks the truth and acts so as not to harm the other members of his society: the natural and multiform "conflict of interests," greatly expanded in our complex modern societies, is resolved by laws passed by popular legislatures. There is no such thing in democracy as the worship of the state as an organism governed by "scientific" laws and rules applied by a dictator. On the contrary, it alleges that "the best society is that which increases spontaneity and life and variety" and that the state cannot itself produce an acceptable social life, it can only foster the forces which produce it. We hold, with Lord Acton, that liberty is possible only where there are other centers of organization than political. The state needs the voluntary support of its many-sided people, and "its business is to safeguard by harmonious regulation the rich various life of voluntary associations in the state."[1] Nonpolitical centers of organization in a democracy keep the public reminded of common purposes within a wide circle of diverse action and freedom. They promote lively and local discussion, independent thinking, and a richer context for majority decisions. [. . .]

The resulting democratic "culture" is not a thing imposed by government upon the individual or imposed upon other states by violence in the name of progress, or peace, or superiority, or the so-called "inevitabilities of geopolitics." It is one of many cultures possible in a world at peace, each suited to the genius of its people and the limitations of its total environment, geographical, political, economic, and social. Each distinctive culture has grown up within a given environmental framework that has left an indelible mark upon it. Both the richness and the peril of the modern world spring in part out of these circumstances.

THE BAD NEIGHBOR POLICY

It is the antithesis of voluntary cultural diversity, spontaneity, and respect for human rights and welfare that we find in the Nazi philosophy. The point of beginning with the German citizen today is the state; the line of his progress is violence to the individual; the goal of his policy is the enslavement of his neighbor. His is the bad-neighbor policy. We misrepresent German political thought if we suppose that the present war is merely a result of German reaction to the Treaty of Versailles. It is a result of German political and philosophical thinking and ambition for two hundred years. The Treaty of Versailles became a plausible pretext for reasserting an old philosophy. The record discloses that most republican leaders of postwar Germany gave only lip service to the principles of democracy and international cooperation. The solemn assurances given in October, 1918, when the defeat of Germany and her request for an armistice raised the question of democratic responsibility, lasted for fifteen years only. After that the government was no longer "free from arbitrary and irresponsible influence," and the promised responsibility of the Chancellor to the people came to an end.[2]

Versailles gave German leaders new examples of the frustration of German aims to conquer the world. The Nazi political program has its roots in something very deep in German life and history: a way of rationalizing greed and violence. Nothing has so clearly revealed the essential primitiveness of the German theories of government as the history of political thought in Germany for the past hundred years. Its "laws" of nation growth, its recent "science" of geopolitics which assumes that "political events depend upon the soil," its assertion that "determining forces which dictate the course of states" carry over into a mystical state of mind where "science ceases and belief begins" (Kjellen) – these are among the doctrines that are separated from democracy by an abyss so wide that today only war can bridge it.

Can any informed person now suppose that German leaders had a tenderer philosophy? The power makers, the architects of the German state, expressed themselves clearly and often on this theme. Hear the testimony of Bismarck on Alsace and Lorraine. He is not speaking in 1871 but in 1895:

> Their annexation was a geographical necessity. It is quite presumptuous to ask us to worry whether the Alsatians and Lorrainers want or do not want to be German. That is none of our business.[3]

If "the fatherland stands for war," as Treitschke concluded, the doctrines upon which Naziism are founded follow naturally. One cannot understand either the present-day Germany or the historical Germany who does not take the trouble to get at the root of the irreconcilable differences between them and us. No one can see the depravity of Nazi geopolitics who thinks that it is merely another way of reading political history and the political map.

We fight today a crooked and evil philosophy armed, in the case of Germany, with continental power. A whole nation has been deceived and reduced to intellectual servitude by hokum. In our future plans and dealings we must take this fact into account. It has taken war, the concentration camp, the hostage killings, plundered Dutch, Belgian, Greek, and Polish peoples (among many), fifth-column technique, and all the rest to convince America how implacable and far-reaching are the means which the exponents of that evil philosophy are willing to employ. [. . .]

DANGERS OF *MEIN KAMPF* REALIZED

In the very period in which this cooperative enterprise was being pushed forward vigorously, another set of social and political values was in process of formulation in Europe. By 1933 when Hitler took over control of Germany these values were fully deployed and exploited and became the basis of his program previously set out in detail in *Mein Kampf*. The background of the related discussion within Germany is of vital importance to us, both now and in our future dealings with the German people.

Slowly and almost against their will the American people became aroused to Nazi dangers. At first we thought of security in terms of the mollusk. The hemisphere was our shell. When danger became obvious, the public search began for the meaning of Nazi designs in terms of German political philosophy and historiography. When successive treacherous blows fell within the Western Hemisphere we could not fight back in Europe only or Japan only. We had to fight wherever there was fighting: our commitments suddenly became planetary. We and our sons began to sail great-circle courses of thought and action. The whole "wide improbable atlas" was opened daily as our military situation tied every neighborhood, large and small, to the rim of the world. Hitler's design was world dominion. We finally saw that our resistance must be as bold and far-flung as his design.

Thus all of us began to think geographically and to regard the map in terms of political ideas and systems. Port Moresby, Mayotte Island, and Dutch Harbor were regarded in terms that include all the lands and seas, the peoples and resources, the governments and ideologies that lie between. Suddenly we realized that even the remoter solitudes will not have their solitude restored after the war, and that victory this time means for America no resumption of something called "normal." We are obliged this time to think our way out as well as fight our way out of our international difficulties.

In the daily excitements that follow these realizations we are all strategists, statesmen, critics, and devisers. The boldness and imaginative quality that we urge upon our leaders find their counterpart in the rising flood of public comment on all international problems. This reflects commendable interest and enthusiasm in a free-speaking democracy. There is danger in it only when, under the guise of "science" or

institutional name or academic rank, wholly unsound and uncritical conclusions are set forth that purport to be based on "law," or reason, or trained judgment, or "the lessons of history." Geopolitics has migrated from Germany to America, not from America to Germany, and even the most ignorant and fantastic misconceptions and political immoralities have been widely disseminated in its name, and truth has been given spurious labels.

FOREWARNINGS RECALLED

[. . .]

In 1934 there was published the third volume of a trilogy of books entitled *Macht und Erde* (Power and Earth), prepared by the Work Group for the *Zeitschrift für Geopolitik*, founded by Karl Haushofer in 1924.[4] Haushofer was a contributing editor. The first edition of my book on problems in political geography, entitled *The New World*, had appeared in 1921, and Maull, one of the authors of *Macht und Erde* states that the trilogy was prepared as the German answer to *The New World*. [. . .]

The method of my book was to deal realistically with the political problems of the postwar world. Its philosophy was one of gradualness of change by rational means. It interposed no ideological preconceived "system" between a problem and its solution in a practical world in which historical accident, not design only, had played so large a part. It sought to analyze real situations rather than justify any one of several conflicting nationalistic policies. Its morality was a responsive and responsible world association based on justice as given fully in the first chapter of the fourth edition (1928). Looking at the competitive world, deeply shaken by the colossal losses of the war, it emphasized the need for "experimentation in the field of cooperative [planning]."

It was this point of view that was the object of attack by the advocates of geopolitics in Germany. The word "rational" means one thing to us and the opposite to the German geopolitical school. For gradualness they would substitute violence. By cooperation they mean that the cooperator eats the cooperee, on the theory of racial superiority. If there is to be a world association, a "new order," Germany must set its terms and impose its unique interpretation. If there is competition for resources and markets, the theory of *Lebensraum* gives Germany priority and justifies seizure. The only political experiment that has united Germany is war.

Deeply disturbed by the rapid growth in Germany of the pseudo science of geopolitics and alarmed by its territorial theories and implications as displayed in widest panorama in the *Zeitschrift für Geopolitik*, I attacked the school and its work in a group review in 1927. The review opens with the sentence "Political geography is still merely a term, not a science." Regional description and statistical and cartographic techniques are recognized as the special tools of the geographer in setting out the intimate life of communities. The review continues: "Some of the most important elements of culture seem not to get into the political geographies of continental Europe, namely, ethics, good manners, the elevation of fairness into a fine art, *cultured* living!"

Maull's *Politische Geographie* was specially selected for condemnation in my review because, as I then stated, "to put facts into a series, to invent mnemonic schemes is to achieve neither learning nor science."
[. . .]
In 1934 I said of doctrinaire writings in the international field in which geographical facts are marshaled to support political claims and philosophies:

> If the economics of Poland collides head-on with the economics of Germany we cannot merely turn to the map and rearrange its parts as if we were free to plant supine peoples upon vacant territory. The historical commitment is there and we cannot ignore it.[5]

Why can we not rearrange the map at will if we are strong enough to enforce our will? We can if we accept Treitschke's doctrine: "The triumph of the strong over the weak is the inexorable law of life." If we believe that there is an inescapable compulsion in strength to assert itself to the advantage of its owner, then we move ahead remorselessly to do what greed suggests and power makes possible.

In my view the "geographical-basis-of-power" idea of Ratzel, as set forth in the first edition of his political geography (1897), is completely unsound. In Germany it has become a ritual, something that one believes, something useful because it fits the national ambition to conquer and govern in the name of *Lebensraum*, a concept that has been expanded from its earlier purely descriptive economic meaning to one that gives territorial expansion a pseudoscientific justification.

Thus expanded it has become one more catchword in the jargon of Hitler's National Socialism. The relations of land and society are not capable of such isolated "scientific" expression. Society is a growing complex. "We deal with rapidly developing and diverse human societies in relation to an earth of which we have an ever-expanding knowledge."[6]

In the face of this perversion of fact to philosophy I advocated the study of real groups of *men* rather than easy book generalizations about *mankind*. Whether we are dealing with geographic relations, demographic data, or economic statistics, we are only in the fact-and-tool stage of investigation so far as national states and national policies are concerned. Scholarship alone supplies certain definite imperatives in policy making, notably in the fields of conservation, law, and public health, to mention but three examples; but the policies that are adopted represent the people's lethargy or will, foresight or the lack of it, justice or injustice, and the power or powerlessness of leaders in shaping public opinion. A national policy is the "diagonal of contending varieties" of the people's thought and action. It cannot be otherwise under the rule of "consent of the governed." The concept of justice did not come out of a library, however important libraries are in conserving the concept and disseminating and expanding knowledge about it. The several fields of scholarship furnish in and of themselves no end philosophy of politics, no guaranteed political design. They can, however, suggest possibilities and dangers in the realm of political relations, choices, moralities, purposes, and powers, beginning with the record of human experimentation.

AN APPRAISAL OF GEOPOLITICS

I shall not attempt to follow the details of the German theories of geopolitics or further document their evolution . . . because a recent book is available that every citizen should read. It is *Geopolitics: The Struggle for Space and Power* by Robert Strausz-Hupe. It appraises and reviews the philosophic background of German political thought. It is distinguished among recent monographs on the subject of geopolitics by the fact that its author demonstrates that he has read the German geographical and geopolitical writings which he analyzes! Moreover, he has an unfailing instinct for the weaknesses of the Haushofer school and its "science" of geopolitics.

Strausz-Hupe's most discriminating and useful remark is in the contrast he draws between the general ideas of Mackinder and the German political philosophy and its corresponding "system." Mackinder attempted to draw a lesson from history that might have implications bearing on state policies of the future.[7] He described trends of power that England could not ignore. In adopting Mackinder's view, continues Strausz-Hupe, Haushofer pushed his geopolitical dynamics to the point of absurdity by seeking a fixed end to world strife through control of an Eurasian heartland. Neither Mackinder nor Haushofer had theories that could stand up to the facts of air power and its relation to industrial strength. Such is the fate of all prophets in this unpredictable world. I might add that the *mind* of man is still a more important source of power than a heartland or a dated theory about it. It is always man that makes his history, however important the environment or the physical resources in setting bounds to the extension of power from any given center at a given time.

Important is Strausz-Hupe's observation that the permanence of boundaries depends "less upon the geographic virtues" than upon international understandings about them. This is a quite different thing from the geopolitical contention by Maull that a frontier is only a hiatus between power-political conditions, that is, a mere abstraction. This is not science, either geographical or political. This is the brute assertion of a man on his way to an object defined by greed. The objective reality, concludes Maull, is the growing state and its dynamic life. As such "it defies international law and treaties." In contrast is our rule of law that treaties stand until the parties in interest have negotiated a new instrument. [. . .]

GUARANTEE OF PERMANENT PEACE

[Postwar] dilemmas compel us to draw a wider circle around our national problems, now inextricably commingled with the problems of sixty other nations. English experience supplies a useful moral. Gladstone said in 1869 that England should have no joint interpreters:

England should keep entire in her own hands the means of estimating her own obligations upon the various states of facts as they arise; she should

not foreclose and narrow her own liberty of choice by declarations made to other Powers. . . .

England thought otherwise on the morning of September 3, 1939, when Neville Chamberlain reported that Germany had begun the invasion of Poland and announced: "We are at war." Intervention and withdrawal had marked the traditional policy of England in continental Europe. Thus we, too, occasionally emerge from our Western Hemisphere shell on the principle of limited liability only. This time we say that our emergence is permanent, that we must now make sure of our future, that we are only as imperishable as our resolution. These are polemical assertions, however. They are not inspired by divine revelations. Shall we be forever secure against "the resistless forces of rebirth," or escape the weakening effect of blind reliance upon "democracy" as a magical doctrine?

The tremulous balance of international forces will vex us at the end of the war. We shall be confused and fatigued by the complexities and responsibilities in which war has involved us. We shall want things certain and simple again: we once called it "normalcy." There is no sure "science" to bring us out of these new deeps of international difficulty. Geopolitics is simple and sure, but, as disclosed in German writings and policy, it is also illusion, mummery, an apology for theft. Scientific geography deepens the understanding. But, like history or chemistry, it has no ready-made formulas for national salvation through scientifically "demonstrated" laws. There are only two "laws" that will guarantee permanent peace in a world in which the choice lies between freedom and slavery: justice based on the doctrine of human rights, and the cooperative exercise of power to enforce justice.

NOTES

1 A.D. Lindsay, *The Essentials of Democracy*, Philadelphia, 1929.

2 Solf to Wilson, October 20, 1918. In *The German Delegation at the Paris Peace Conference* by Alma Luckau, New York, 1941, p. 144.

3 Address at Friedrichsruhe, April 24, 1895.

4 Otto Maull: Das Wesen der Geopolitik (*Macht und Erde*: Hefte zum Weltgeschehen, No. 1), Leipzig and Berlin, 1936, p. 23.

5 Isaiah Bowman, *Geography in Relation to the Social Sciences* (Report of the Commission on the Social Studies, American

History Association, Part 5), New York, Chicago, etc., 1934, p. 212.

6   Ibid., p. xi.

7   H.J. Mackinder, *Democratic Ideals and Reality: A Study in the Politics of Reconstruction*, New York, 1919; reissue, Henry Holt & Co., New York, 1942.

Defense of German Geopolitics

Karl Haushofer

from *Total Power: A Footnote to History* (1948)

Although not the originator of the technical term "geopolitics," nevertheless I have rightly been considered as the leading exponent of its manifestation in Germany. [. . .] The manner in which German geopolitics came into being is, by the same token, the justification for its appearance as a subject of higher teaching from 1919 onward; it was born of necessity. [. . .] It would be an inhuman and impossible demand to expect that a German scientist could disregard the inadequacy of the distribution of living space in central Europe, which had occurred in those times as a result of its overdeveloped industrialization and urbanization. To this must be added the dismemberment of central Europe by frontiers that could not last long and which, consequently, were geopolitically unjustifiable. For these reasons my book *Frontiers*, as well as other publications, was written.

What seemed most lacking in the resumption of the educational process for the training of German youth after the war was the capability to think in terms of wide space (in continents!) and the knowledge of the living conditions of others, namely of oceanic peoples. This broadness of thought, limited by a continental narrowness as well as by smallness in its world vision, became narrow-minded and lost in a welter of trivial controversies. It was cut off from the energizing breath of the sea and robbed of its overseas connections. [. . .] The knowledge, therefore, of the great ways of life that were essentially sea-minded – the British Empire, the United States of America, Japan, the Dutch East Indian Empire – was then even more inadequate than was the knowledge of the Near and Middle East, Eurasia, and the Soviet Union.

Therefore it seemed necessary for German geopolitics to provide knowledge about the empires that are spread over all the seas and about the Indo-Pacific space. By that means a counterweight was created against the pressure from within during the period 1919 to 1933. Later, this sense of pressure, under the tension of internal party conflicts, unfortunately served more and more to overshadow and obscure this necessary knowledge of other lands. In meeting this obligation the faculty of foreign sciences of the University of Berlin also served, together with the only Institute for Political Geography that existed in all Germany. This was directed by my son, Professor Albrecht Haushofer. There never was any institute for geopolitics in Munich. [. . .]

No normal understanding man of any other nation can deny that a German scholar also, after such a laborious career and with every aspiration for objectivity, should have the right to stand at the side of his people with all his mental power. This he does because of the findings in his domain of knowledge, because of conclusions arrived at honestly and legitimately in such a struggle for existence as prevailed during the years from 1919 to 1932.

Although I never claimed as my own the principle: "My country, right or wrong," in its complete consequences, nevertheless it has to be admitted that the borderline is easily crossed between pure science and practical science in such times of extreme tension. Therefore it happened (slipped in) that I occasionally overstepped those borders. This I also admitted and regretted openly to the interrogators; it was recognized on their part also that from 1933 onward I could work only under pressure, since my oral and written expressions were subject to four types of censorship.

Since the interrogators acknowledge that, in comparison with the United States conception of "legitimate" geopolitics, German geopolitics worked its way up to a balance of knowledge 60 to 70 per cent of which could be generally accepted as valid science, an exact differentiation will likewise have to be made between all that was printed about geopolitics before 1933 and after 1933.

If my whole scientific working material had not been broken up and in part carried off at the beginning of May by [the US army] I could point to numerous lectures, dating from the years 1919 to 1933, which correspond in their development, for example, with Scheme II "Methodology" of a course on geopolitics of the School of Foreign Service at Georgetown University in use there on 1 July 1944. Among my requisitioned papers was the collected and fully developed groundwork of my lectures.

All that was written and printed after 1933 was "under pressure" and must be judged accordingly. How the effects of this pressure (in which Rudolf Hess, who tried rather to protect, did not participate) eventually worked out can be proved by nearly three years either of imprisonment or of limitation on freedom imposed on my family, also by my own confinement in Dachau concentration camp, the murder of my eldest son by the Gestapo on 23 April 1945, the severe control over and later the suspension of the *Journal of Geopolitics*. In the Third Reich the party in power lacked any official organ receptive to or understanding the doctrines of geopolitics. Therefore they only used and wrongly understood catchwords which they did not even comprehend. Only Rudolf Hess, from the time when he was my pupil, before even the NSDAP [Nazi Party] ever existed, and the Minister for Foreign Affairs, Von Neurath, had a certain understanding for geopolitics without being able to apply it successfully. [. . .]

Those theories, originally deriving from Friedrich Ratzel (*The Earth and Life; Political Geography; Anthropogeography*) and from those who continued his theories in the United States (Semple) and in Sweden (Rudolf Kjellen), were formed to a larger extent from sources among English-speaking peoples than from continental peoples. They were presented to German circles in the form of the principle: "Let us educate our masters."

Mahan, Brook Adams, Joe Chamberlain [. . .]; Sir Thomas Holdrich (*The Creator of Frontiers*); Sir Halford Mackinder (*The Geographical Pivot of History*); Lord Kitchener (1909); later I[saiah] Bowman (*The New World*, and other writings) were the basic inspirers of my teachings and were quoted again and again. [. . .]

Imperialistic plans of conquest were never favored, neither by me in my writings nor in my lectures. As in my book on frontiers I also protested against the crippling of Germany through the border decisions of the Versailles Treaty, so in my public lecture activities I stood up for the Germans in South Tyrol. I welcomed the incorporation of Sudeten German territories, but I never approved of annexation of territories alien to our people and which had no German settlements.

I always regarded dreams of such annexations as dangerous dreams and therefore disapproved them.

The fact that thousands of German settlers were repatriated to Germany at much expense and suffering through VDA [Association of Germans Living Abroad] under my leadership, proves in the best manner that at that time, in any case, an occupation of those territories was not planned or, at least that the desirability of such an occupation was not known. If National Socialism had revealed, by the way it published its ideals in the early years of its development, that they included the conquest of alien-blooded peoples and their territories, it would have brought about its own retirement from power. This I stressed on every occasion, among others on 8 November 1938, and I opposed such plans of conquest. I believed in the promise of saturation made in 1938.

A truly equitable determination of frontiers which would satisfy everybody and which does not impose hardships on parts of any people is practically impossible because of the immense complicated overlapping of border languages and economic centers that have developed in the course of time, especially in eastern Europe. I, therefore, as well as my son Albrecht, and others of my pupils and co-workers tried in long discussions, without success, to work out completely just and lasting principles for such a delimitation of borders. In that, my efforts always were focused on the task of not creating irredentas in any form. Therefore it is self-evident that the charge of planning conquest, including carefully worked out maps to infiltrate into continents, such as South America, was manufactured from thin air. In such matters the sensation-loving press was raving without let or hindrance, even using detailed forgeries of maps. [. . .]

The book *Mein Kampf* I saw for the first time when the first edition was already in print. I refused to review this book because it had nothing to do with geopolitics. For me, at that time, it seemed to be one of the many ephemeral publications for purposes of agitation. It is self-evident that I had no part in its origin and I believe I am protected against the suspicion of participation, mentioned in the yellow press, if one makes a scientific comparison of my style of writing and the style of that book. I never saw Hitler alone. The last time I saw him was in the presence of witnesses on 8 November 1938, and I then had a sharp disagreement with him. From then on I was in disgrace. Since Rudolf Hess's flight in May 1941 I was exposed to the persecution of the Gestapo which ended only at the end of April 1945 with the murder of my eldest son because he shared the secret of 20 July 1944 [the plot against Hitler's life]. He also was in contact with English-speaking peoples. My friendship with Rudolf Hess had its origin in 1918 and is, in common with his attendance at my lectures at the University, four years older than the foundation of the National Socialist party. I saw Hitler for the first time in 1922, when he was one of the many popular platform orators who were then mushrooming from the overheated soil of the German people and from the multiplicity of societies and political movements. [. . .]

From autumn 1938 onward was the Way of Sorrow for German geopolitics. The individual fate of father and son is illustrated by my imprisonment and his death. This happened within the framework of the suffering of "political science" in all central Europe under the pressure of the autocracy of one party down to the misuse and misinterpretation [of geopolitics] by state officials. Despite all that, German geopolitics had originally – from 1919 to 1932 – goals quite similar to American geopolitics.

In the program of geopolitics, on its first appearance, one finds a statement saying that it aspired to be "the geographical conscience of the state." It should then, for instance, have demanded in 1938 that Germany be satisfied and grateful for the solution reached at Munich. When I actually tried to put this into effect – after my return from Italy and when I finally reached the head of the state on 8 November 1938 – I fell into his disfavor for it and never saw him again. Until that date, therefore, this representative of German geopolitics may well regard himself as a legitimate pre-defender, even in the sense of American geopolitics. The goal of German geopolitics originally had been, in common with legitimate American geopolitics – to achieve the possibility of excluding disorders in the future, like those of 1914 to 1918, through mutual understanding of peoples and their potentialities to develop on the basis of their cultural foundations and living space; also to obtain for minorities the highest measure of justice and politico-cultural autonomy – as was the case in Estonia, for instance, and for a time seemed to be accomplished in Transylvania.

This presupposed a geographically correct picture of the world; it required mutuality, moreover, and respect of one nationality and race by others as well as recognition of the human right to "personality." It demanded the highest degree of indulgence and tolerance, of which my lectures and activities were replete, for instance, from 1919 to 1932. [. . .]

In the memorandum which was written as answers to the questioning of General Eisenhower's staff and which lay before the interrogators, I specified in detail that an international geopolitics could become one of the best means to prevent future world catastrophes. It would have to be built on a lively exchange of ideas and persons, of professors, teachers, assistants, and students.

In the spirit of its name and by the political art of its leadership it could restore to due honor the "sacrament of the earth," the holiness of the soil which supports humanity. German geopolitics, between the earthquakes of 1914 to 1919 and from 1938 to 1945, endeavored to build a road toward this exalted goal.

Granting that errors and mistakes accompanied the course of geopolitics, they can be turned to profit by the wisdom of that saying in the English language: "All human progress resolves itself into the building of new roads."

PART TWO

Cold War Geopolitics

INTRODUCTION TO PART TWO

Gearóid Ó Tuathail

Questions of geography were always deeply implicated in the Cold War that developed between the Soviet Union and the United States after World War II. By the end of the war, the states of Eastern Europe had become part of a Soviet sphere of influence. Stalin's regime had brutally ruled the Soviet Union since the late 1920s, imprisoning up to 8 million of its own citizens by 1940 in Gulag prison camps (Applebaum, 2003: 580). It was determined to create a security zone for itself to prevent yet another invasion of its territory by Western powers. This had happened immediately after the Bolshevik revolution and again when Hitler invaded in June of 1941. An estimated 27 million people died defending their homes against Hitler's racist crusade, the Red Army driving back the German war machine until it was destroyed and Berlin captured. Peace with security was thus foremost on Stalin's mind, which he decided was peace through Soviet domination in Eastern and Central Europe.

The United States' experience in World War II was considerably different. Its national territory and civilian population escaped the horrific destruction and indiscriminate mass murder of the total war waged on the European continent, in North Africa and in Asia. With all its leading competitors in ruins, the United States was the single most powerful state in the world, a state with supreme confidence in its nationalist myths and ideals (see Reading 11). During the war, the American stance had combined a maximalist statement of its political and economic ideals with a minimalist program of war aims. Once the war was over, the American state – led by an inexperienced president in Harry Truman who suddenly had an awesome weapon, the atomic bomb, at his command – found it difficult to resist envisioning the world according to its ideals of political democracy and capitalist economics. The United States' leaders claimed that its ideals were universal ideals, its aspirations not self-serving but those of humanity. A clash between the Soviet Union and the United States over the future of Eastern Europe was probably inevitable; a Cold War between both powers, however, was not.

TRUMAN, KENNAN AND THE ORIGINS OF THE COLD WAR

Why the antagonism between these states developed in such a way as to eventually divide the European continent in two was a consequence of the geopolitical discourse that became dominant in the United States in 1946 and 1947 and the reaction it provoked from the Stalinist regime. While certain groups within the Truman administration favored diplomacy and a *realpolitik* approach to Stalin, others championed an implacable view of the Soviet Union as an inherently expansionist power. It was this latter view that became the dominant geopolitical orthodoxy. An early defining statement of this essentialist conception of the Soviet Union was provided by the United States' chargé d'affaires in Moscow, George Kennan (1904–2005). In February 1946, Kennan, ill and bedridden, dictated an 8000 word communiqué to Washington that became known as the Long Telegram. In it Kennan expounded his conception of the Soviet Union as an historically and geographically determined power with an unfolding necessity to constantly expand. This, Kennan argued, was the essence of the Soviet Union and nothing really could be done about it. Most significantly, no deals can or should be struck with the Soviet Union.

Kennan's views were pounced upon by more hard-line anti-communist elements in the Truman administration and widely circulated. Kennan himself was recalled to Washington to head up a new Policy Planning agency within the "national security state" being created by the Truman administration at this time (Yergin, 1978). The Truman administration's attitude to the Soviet Union became more belligerent as the Soviet Union sought to manipulate internal politics in various Eastern European states to its own advantage. In March 1946 former British Prime Minister Winston Churchill strengthened the hard-line forces in the Truman administration by charging that an "iron curtain" has descended across the continent of Europe. Churchill had his own agenda of preempting an anti-imperialist alliance of the United States and the Soviet Union against the British empire in favor of an Anglo-American anti-communist alliance that would commit the United States to aid the British empire in a joint struggle against the Soviet Union (Taylor, 1990). To the Soviets, this alliance appeared to be already in existence when over the course of 1946 the Soviet army was eased out of Iran, a country on its borders, while Anglo-American oil companies gained control over that country's profitable and strategic oil reserves.

Mutual suspicion and antagonism deepened in 1947 when the British government informed the Americans that it could no longer afford to aid the reactionary Greek monarchy trying to re-establish itself in power after the war. The need for the Truman administration to convince a reluctant US Congress to provide aid to the corrupt Greek monarchy in its fight against leftist guerrillas and to Turkey in a lingering squabble with the Soviets over control of the Dardanelles provided the occasion for a speech in which President Truman outlined what became known as the Truman Doctrine (Reading 7).

The Truman Doctrine is the first significant public statement of American Cold War geopolitics. In it Truman uses the local situation of the civil war in Greece and the long-standing dispute over the Dardanelles to enunciate a more universal struggle between freedom and totalitarianism across the globe. Dwelling not on the geographical specificity of the conflicts in question, Truman's speech strives to articulate abstract and absolutist truths. In a dramatic crescendo, Truman declares: "At the present moment in world history nearly every nation must choose between alternative ways of life." This "choice," however, is not a free choice but a worldwide struggle between two ways of life which are simplistically represented by Truman as freedom versus totalitarianism.

Truman's rhetorical leap from the local to the universal, from the particular to the absolute, became an abiding characteristic of American Cold War geopolitical discourse. Within such a discourse, the geographical complexities of particular places and specific conflicts were displaced by Manichean categories and formulaic oppositions between universal concepts. Through the use of earth-labeling categories like "the free world" and "the enslaved world," the geographical kaleidoscope of the map becomes the geopolitical monochrome of good versus evil, capitalism versus communism, the West versus the East, America versus the Soviet Union. All places and conflicts were interpreted within the binary terms of this Manichean map.

Geopolitical discourse build around a simplistic storyline of good versus evil is a legacy of the influence of religion on American geopolitical culture. Wilsonianism, the previous attempt by the United States to craft a discourse for itself as a world power, reflected the influence of evangelical Protestantism (Wilson's parents were missionaries in China). Truman's discourse draws upon a similar tradition in American geopolitical culture and was instantly popular among Americans who knew little about the world but believed, out of religious conviction, that their country was a territorial expression of virtue ("God's country") whereas the Soviet bloc was a territorial expression of godless evil ("Satan's domain"). Not all members of Truman's cabinet thought in such simplistic terms but they knew they needed to talk that way to persuade a majority of the US Congress to underwrite their plans for a permanent global role for the United States in the post-war world. For example, Truman's sophisticated Secretary of State, Dean Acheson, explained the geopolitical significance of Greece before Congress thus:

> Like apples in a barrel infected by one rotten one, the corruption of Greece would infect Iran and all to the east. It would also carry infection to Africa through Asia Minor and Egypt, and to Europe through Italy and France, already threatened by the strongest domestic Communist parties in Western Europe.
>
> (Acheson, 1969: 219)

That geographically embedded states could be represented as mere "apples in a barrel" was indicative of the triumph of an anti-geographical form of reasoning in Cold War geopolitical discourse. The geographical specificity and complexity of particular conflicts, such as that in Greece or Turkey, were subordinate to the "higher truth" of the struggle between freedom and totalitarianism across the world map.

George Kennan was privately critical of the crude and alarmist tone of the Truman Doctrine which successfully scared Congress, as Truman hoped it would, into providing aid to the embattled Greek and Turkish governments. Many professional foreign policy experts considered Truman's declaration dangerous because it contained no rational calculation of means and ends for US foreign policy. His statement that "it must be the policy of the United States to support free peoples who are resisting attempted subjugation by armed minorities or by outside pressures" placed no geographic limits on US foreign policy. Implicitly, the Truman Doctrine envisioned a worldwide anti-communist crusade: an unlimited totalitarian threat required an unlimited global commitment by the United States.

While George Kennan was cognizant of this danger (and became even more so as he got older), an essay written by him called "The Sources of Soviet Conduct" and published a few months after the Truman Doctrine in the main journal of the Council on Foreign Relations, *Foreign Affairs*, reinforced rather than questioned the crude geopolitical vision articulated in Truman's speech (Reading 8). Published initially under the pseudonym "Mr X" before Kennan's identity was disclosed, this essay is the intellectual foundation of the post-war American foreign policy of "containment" of the Soviet Union. Expanding ideas he had developed in the Long Telegram and elsewhere, Kennan argued that Soviet communism was the ideology of a maladjusted group of fanatics who had seized power in 1917 and were driven by a perpetual insecurity to destroy "all competing power" both inside and outside the country. Communist ideology is ultimately a "fig leaf" for Kennan, the primordial sources of Soviet conduct being internal to and determined by Russian history and geography: "From the Russian-Asiatic world out of which they had emerged they [Soviet communists] carried with them a skepticism as to the possibilities of permanent and peaceful coexistence of rival forces" (Keenan 1947: 570). Soviet communist caution and flexibility are precepts

> fortified by the lessons of Russian history: of centuries of obscure battles between nomadic forces over the stretches of a vast unfortified plain. Here caution, circumspection, flexibility and deception are the valuable qualities; and their value finds natural appreciation in the Russian or the oriental mind.
>
> (Keenan 1947: 576)

This form of reasoning is remarkably crude and deterministic. Soviet communists are insecure fanatics. Their ideology, in tandem with the primordial patterns of Russian history and geography, has produced a Soviet state that is inherently expansionist. These 'essential truths' dominate historical contingencies and geographical particularities. Yet, the historian Anders Stephanson (1989) has argued that Soviet foreign policy in fact

> varied substantially over time in both magnitude and target (as Kennan should have known), depending precisely on which powers seemed to pose the greatest danger . . . Who is actually out there doing, doing what to whom, were important questions to Moscow.
>
> (Stephanson 1989: 76)

The position Kennan articulates, however, absolves Western leaders and intellectuals of statecraft from actually engaging with the practical specifics of Soviet foreign policy at particular times and places. It promotes retreat to absolute truths and geopolitical slogans about communists being fanatics and the Soviet Union as relentlessly expansionist. Kennan's argument, in other words, objectifies the Soviet Union as a predetermined expansionist entity that needs containment "by the adroit and vigilant application of counter-force at a series of constantly shifting geographical and political points." Kennan's argument ironically precludes his own profession: diplomacy. Since the Soviets are supposedly fanatics, there is no real possibility of dialogue and negotiation with them. They are Other.

Kennan's call for "a policy of firm containment, designed to confront the Russians with unalterable counter-force at every point where they show signs of encroaching upon the interests of a peaceful and stable world" echoes the unlimited rhetoric found in the Truman Doctrine. Kennan's conclusion about the Soviet challenge being "a test of the over-all worth of the United States as a nation among nations" evokes long-standing American myths of manifest destiny and national exceptionalism and reverberates with the global anti-communist crusade envisioned by Truman (see Reading 12). The unlimited and universalist nature of this crusade, however, was profoundly unsettling to some. In a series of newspaper articles that subsequently became the book *The Cold War* that named the era, the political journalist Walter Lippmann described the "X" article's recommendations as a "strategic monstrosity" (Lippmann, 1947: 18). It makes no distinctions between places and commits the United States to confront the Russians with counterforce "at every point" across the globe, "instead of at those points which we have selected because, there at those points, our kind of sea and air power can best be exerted" (Lippmann 1947: 19). Furthermore, it gives a "blank check" from the American people to its military institutions and to those regimes the US government decides are allies in its global crusade against communism. Lippmann concludes by emphasizing diplomacy, noting that for "a diplomat to think that rival and unfriendly powers cannot be brought to a settlement is to forget what diplomacy is all about" (1947: 60).

Lippmann's worst fears, however, were largely realized as diplomacy became sidelined and containment militarism became the guiding principle of US foreign policy. The Soviet response to the hardening Western attitude and the Marshall Plan aid program for select states in Europe was to fall back on its own Manichean vision of the world. As articulated by the Soviet intellectual of statecraft Andrei Zhdanov in September 1947, the world was divided into "two camps," an "imperialist and antidemocratic camp" led by the United States with the British Empire as its leading ally versus an "anti-imperialist and democratic camp" led by the Soviet Union and the "new democracies" in Eastern Europe (Reading 9). These "new democracies," however, were in reality Soviet-inspired regimes prohibited from participating in the Marshall Plan by Moscow. As the Cold War between the United States and the Soviet Union deepened, their domestic political structure became increasingly repressive and Stalinist.

THE COLD WAR AS A GEOPOLITICAL SYSTEM

Frozen on the map over the next four decades, the Cold War came to describe a geopolitical system with two constituent geopolitical orders, each of which was characterized by a particular organization of domestic, allied and "Third World" space. The very earth label "Third World" is a product of the Cold War's division of global space into a First World of capitalist states, a Second World of communist states, and a Third World of developing states where the United States and the Soviet Union competed for influence. The geopolitical order established by the Americans after World War II was geographically more extensive than the Soviet order. First, the US domestic political order was organized around a "vital center" that held that the United States needed to pursue global power to check the "Soviet threat" (Wolfe, 1984). Exaggerated visions of this threat mobilized an American nationalism in the cause of globalism and containment militarism. Those on the left critical of "vital center" politics were constantly red-baited by politicians on the right like Joseph McCarthy and Richard Nixon. Patriotism became defined as God-fearing anti-communism, and politicians from Truman to Kennedy and Nixon to Reagan rode anti-communist crusades all the way to the White House.

American Cold War geopolitical discourse also had an important economic dimension. The Cold War, according to Wolfe (1982), was central to the creation of a consensual "politics of growth" in post-war America. Through exaggeration of the Soviet threat, American intellectuals of statecraft were able to transform the US state from a reluctant isolationist power into a crusading interventionist power dedicated to promoting an open world economy and safeguarding the free enterprise system. "Containment" became an unquestioned imperative within American foreign policy. Cold War visions of "containment" were also extended into American domestic life and popular culture. Figures like Ronald Reagan, president of the Screen Actors Guild from 1947 to 1952, for example, sought to enforce the cultural authority of a conservative white establishment by "blacklisting" those whose ideas challenged this hegemonic cultural order (Campbell, 1990; May, 1989).

Second, the establishment and modernization of a global structure of extended deterrence, by means of the North Atlantic Treaty Organization (NATO) in Western Europe and the Mutual Security Treaty with Japan, helped incorporate and subordinate the United States' major capitalist allies into an American-led military order. The economic reconstruction and recovery of Western Europe and Japan were facilitated by generous aid from the US state and its promotion of an open capitalist world economy, the type of "American lebensraum" envisioned by Bowman (who died in 1950). A convergence of interests among the ruling classes in all three regions facilitated the establishment of an American "empire by invitation" (Lundestad, 1990).

Third, the US national security state acquired the means and power in the space demarcated as the "Third World" to intervene and attack people, and states its leaders are considered a threat to their version of "American" values, institutions and economic interests (Chomsky, 1991; Kolko, 1988). The general proclivity of the US state for unilateralist interventionism to oppose radical social revolution was already manifest in Central America and the Caribbean since the late nineteenth century. This proclivity became a global one after World War II and led the US national security state, through the work of the Central Intelligence Agency (CIA) and other groups, to intervene in the domestic politics of many states, in some instances like Iran in 1953, Guatemala in 1956 and Chile in 1973, aiding the overthrow of democratically elected governments. The United States also massively intervened in civil wars in Korea and Vietnam against what it perceived as a worldwide communist threat.

The geopolitical order established by the Soviet communist elite in the wake of World War II was largely confined to Eastern Europe and the Soviet Union. Its order was defined by, first, the domination of domestic politics by the Communist Party and political life by state institutions and the secret police.

Patriotism was so entwined with communism, which was akin to an official state religion, that anyone who questioned it was branded an agent of the "imperialist West." Dissident intellectuals and many others were persecuted and sent to forced labor camps, internal exile, and mental homes. Recent calculations suggest that the total number of prisoners in the Soviet Gulag reached 28.7 million (Applebaum, 2003: 581). Just as the United States built a huge military-industrial complex to support its national security state, so also did the Soviet Union; indeed, its state structure and institutions became even more militarized than those of the United States. Second, the Soviet geopolitical order was characterized by the maintenance of a system of extended deterrence in Eastern Europe by means of pro-Moscow ruling communist elites and the military structures of the Warsaw Pact Organization. Because it did not have nearly the resources and wealth of the capitalist West, the Soviet state intervened erratically in the "Third World," selectively sponsoring a few radical states like Egypt (for a period), North Korea, Vietnam and Cuba.

Europe was the principal theater where both competing geopolitical orders faced each other and the site of its greatest militarization. Ironically both superpowers came to share a mutual interest in the Cold War as a system because it guaranteed their mutual positions on the European continent. Cox (1990) notes:

> Historically . . . the Cold War served the interests of both the USSR and the United States. For this reason neither sought to alter the nature of the relationship once it had been established. Their goal, therefore, was not so much victory over the other as the maintenance of balance. In this sense the Cold War was more of a carefully controlled game with commonly agreed rules than a contest where there could be clear winners and losers.
>
> (Cox 1990: 31)

Yet there were real winners and losers *but within not between* the respective geopolitical orders. Evaluated in terms of war, death and destruction, the Cold War saw the Soviet state wage war in its geopolitical zone against popular uprisings in Poland, Hungary, Czechoslovakia and Afghanistan. Warsaw Pact states had their independent identity crushed and subordinated to the security interests of the Kremlin. The United States, with the help of certain allies, sought to police radical movements in its zone and waged war against radical social change in the "Third World." From Vietnam to Afghanistan, the Cold War was far from being an "imaginary war" (Kaldor, 1990) or a "long peace" (Gaddis, 1987).

That the United States became involved in civil wars in Korea and Vietnam, locations thousands of miles from the United States and of questionable strategic value in themselves, was a consequence of a geographically unspecified commitment to containment in US Cold War geopolitical culture. Truman's universal crusade against communism and Kennan's call for firm containment "at every point" made the task of demarcating US interests difficult to sustain. Secretary of State Dean Acheson tried to define a "defensible perimeter" of the United States which excluded the Asiatic mainland, including Korea, yet within a year the United States was at war in Korea against what was represented as the latest front in a worldwide communist challenge. The same happened with Vietnam where Presidents Eisenhower, Kennedy and Johnson declared on various occasions that the United States should not become militarily involved.

Yet, the universalist nature of America's anti-communist crusade and the simplemindedness of its domino theory reasoning confounded these expressions of intent not to get involved. The hysterical vision of Southeast Asia falling like a row of dominoes to communism was absurd geographically yet nevertheless dominant geopolitically. The "best and brightest" US intellectuals of statecraft pushed the

United States to become militarily involved in the Vietnamese civil war. The Secretary of Defense in the Kennedy and Johnson administration, Robert McNamara, was so involved in running the Vietnam war that it became known as "McNamara's war." Despite being one of the smartest men of his generation, McNamara never questioned the simple-minded geopolitical discourse that propelled the United States into the middle of the Vietnamese civil war. Decades later, McNamara would eventually reflect on his role in the conflict. Reading 10 examines McNamera's account of US geopolitical discourse at this time and finds it wanting because he ignores how power relations shaped Cold War geopolitical culture.

Reasoning equivalent to the domino theory can also be found in Soviet geopolitical discourse, though here the dominoes or satellite states were geographically much closer to the Soviet Union. The attempt by the communist leaders of Czechoslovakia to institute a series of reforms designed to address their deteriorating economic situation became a matter of concern for Soviet and other Eastern European leaders in 1968. In order to stimulate the economy, Czechoslovak reformers led by Alexander Dubcek instituted a series of measures that loosened the firm dictatorship of the Communist Party over the economy and state. The result was increasing cultural liberalization, what reformers celebrated as the Prague Spring, but what nervous communist bureaucratic dictators in East Germany, Poland and the Soviet Union described as the "Czechoslovakian disease." Fearing that this "disease" of political reform and cultural liberalization would spread, the Red Army invaded Czechoslovakia on August 20 with the support of smaller units from Poland, East Germany, Hungary and Bulgaria ("five allied socialist countries"). More than 80 Czechoslovak citizens were killed and several hundred wounded during a month of clashes following the invasion.

The justification for the invasion became known as the "Brezhnev Doctrine," a geopolitical statement originally published as an article in the official Soviet Communist newspaper *Pravda* by Politburo leader Leonid Brezhnev under the pseudonym "Kovalev" (Reading 11). In this article, Brezhnev articulates the limits within which the communist satellite states of Eastern Europe must operate, effectively spelling out the subordination of the geographically diverse Eastern European communist dictatorships to the Soviet geopolitical order. Any decision these states make "must damage neither socialism in their country nor the fundamental interests of the other socialist countries nor the worldwide communist movement." State communist leaders who exercise their nominal state sovereignty and national independence in a way that deviates from these principles are guilty of "one-sidedness" and "revisionism," code words for unacceptable independent thinking. Throughout the article, the Soviet invasion is justified by resort to polarized geopolitical discourse. It's "Us" against "Them" and any group that tries to promote greater democracy and loosen the dictatorship of the Communist Party in Eastern Europe is ultimately aiding the enemy, which in Soviet geopolitical discourse is "world imperialism" and, echoing the line that West Germany is still innately fascist and expansionist, "West German revanchists." In the binary logic of Soviet geopolitical discourse, any questioning of Cold War categories or promotion of "neutrality" is objectively "antisocialist" and "counterrevolutionary." Remarkably, Brezhnev claims that the Red Army and its support units in Czechoslovakia are "not interfering in the country's internal affairs" but helping the Czechoslovak people exercise their "inalienable right to decide their destiny themselves." However, there is an all important qualification: "after profound and careful consideration, without intimidation by counterrevolutionaries, without revisionist and nationalist demagoguery." In other words, the Red Army is merely helping the Czechoslovak people exercise their self-determination in a way that the Soviet Union's leadership judges to be ideologically and geopolitically correct. Such is the thin apologism for military interventionism and geopolitical domination.

The argument against the "loss of geography" evident in such geopolitical reasoning is developed further by Ó Tuathail and Agnew in their study of geopolitics and discourse (Reading 12). A foundational essay in the establishment of critical geopolitics, this reading outlines four theses on geopolitical reasoning and international politics. It then turns to analyze the practical geopolitical reasoning found in American foreign policy historically, concluding with a deconstruction of ways in which Kennan represents the Soviet Union in his Long Telegram and "X" article.

THE SECOND COLD WAR

Because of its Vietnam experience, the domestic political consensus around the policy of containment militarism that the US state had pursued since the late 1940s was subject to increasing challenge and critique. The old "absolute truths" of the Soviet Union as an implacably expansionist power with which one could not negotiate, and of a worldwide communist conspiracy directed from Moscow, had propelled the United States into Vietnam and sent thousands of its soldiers to their death. In a bid to adjust to the changed conditions of world power in the 1970s and the breakdown of foreign policy consensus, the Nixon administration, with Henry Kissinger as the president's leading intellectual and also practitioner of statecraft, pursued a policy of détente or peaceful coexistence with the Soviet Union, and accommodation with communist China. Rather than continue the militarist policy of driving for military superiority over the Soviets, the Nixon–Kissinger administration recognized that both states now had the weapon systems to destroy each other. Acknowledging the reality of "mutually assured destruction" (MAD), the Nixon administration promoted the doctrine of nuclear deterrence and sought to negotiate limited arms control agreements with the Soviets. Nevertheless, Nixon and Kissinger continued the United States' long-standing crusade against perceived leftist ("pro-Soviet") governments in the "Third World," involving the United States in assassinations, coups d'état and illegal wars in places like Chile, Angola and Cambodia, many of which were devastated by the resultant upheaval.

Not everyone within the US national security community agreed with the Nixon administration's policy of détente towards the Soviet Union. A distinct group of Cold Warriors in Congress, lead by Henry ("Scoop") Jackson whose ties to the Boeing Corporation were so close he was known as "the senator from Boeing" (he actually represented Washington state where Boeing had its corporate headquarters), claimed that the Soviets were using détente to trick the West into reducing its defense expenditures. They had supporters within the Nixon administration and, after Nixon was forced to resign in disgrace, within the new Ford administration. Former Nixon confidante and Secretary of Defense in the Ford administration, Donald Rumsfeld worked to undermine Kissinger's arms control measures with the Soviets. Former Rumsfeld aide, Richard Cheney, became Ford's chief of staff and quietly supported the campaign against détente. Critics charged that the CIA had underestimated Soviet strength and that the Soviets were engaging in a massive military buildup in a new bid for world domination. To ameliorate the critics Ford's CIA Director, George H.W. Bush, appointed a team of outside experts to review classified data and draw up its own conclusions about Soviet intentions. The group, called Team B, was led by hard-line anti-Soviet historian Richard Pipes and included Paul Wolfowitz. They duly reported what they already believed: it was possible to interpret available intelligence data as demonstrating that the Soviet Union was seeking global military superiority and was using détente as a means of achieving that end. Pre-conviction trumped professional intelligence analysis. The bottom line was that US defense spending should increase dramatically.

The hard right critics of détente in the 1970s were part of a 'neoconservative' tendency in American geopolitical culture that was distinct from traditional conservatism in three ways. First, neoconservatives were moral absolutists who rejected political realism on the grounds that it compromised with what they considered "evil" in the world (principally the Soviet Union but also China and North Korea).

Détente, for them, was "moral relativism," the position that all viewpoints are equally valid. This concept was an invention of the German émigré philosopher Leo Strauss, a leading neoconservative thinker, who used it as a simplifying foil to justify his moral absolutism. Second, neoconservatives embraced big government spending for worthy global crusades. An expansive state apparatus was required to fight evil and "tyranny" across the globe. Third, neoconservatives consistently favored military means over diplomacy in dealing with geopolitical crises. Neoconservatives were suspicious of professional diplomats and the intelligence bureaucracies, believing them to be complacent about the machinations of communist evil (thus their push for Team B).

When the liberal Democrat Jimmy Carter was elected to the White House in 1976, neoconservatives organized themselves in opposition and recruited some high profile Cold Warriors including Paul Nitze as well as Daniel Pipes and Richard Pearle (Senator Jackson's most influential Congressional aide). Calling themselves "The Committee on the Present Danger" (CPD), after a similar hard right militarist group in the 1950s, the group went public at a press conference on November 11, days after Carter's election, with a manifesto called "Common Sense and the Common Danger" (Reading 13).

The CPD's manifesto is significant for its attempt to reassert the "absolute truths" of containment militarism at precisely the moment when it seemed that the US political system was moving beyond this creed. Recycling the traditional discourses of danger from the early Cold War – "Our country is in a period of danger, and the danger is increasing" – the CPD asserted that the "threats we face" are "more subtle and indirect than was once the case" but, somewhat contradictorily, they are nevertheless massive, worldwide and unparalleled. Reasserting the "good versus evil" rhetoric that is part of American geopolitical culture, the manifesto declares that the "principal threat to our nation, to world peace, and to the cause of human freedom is the Soviet drive for dominance based upon an unparalleled military buildup." As one might expect from intellectuals of statecraft with ties to a domestic military-industrial complex eager to continue expanding whatever the world geopolitical situation, the CPD manifesto calls for a massive military buildup on the part of the United States to check the global communist threat and build a "strong foundation" (code for military superiority) from which to supposedly negotiate "hardheaded and verifiable agreements" with the Soviets.

Throughout the Carter years, the intellectuals of statecraft associated with the CPD worked hard to criticize and undermine the foreign policy of the Carter administration. The perceived failures of this administration's policy in Iran and elsewhere, together with the Soviet invasion of Afghanistan in 1979, were represented by the CPD as a consequence of its straying from the timeless truths of Cold War militarism. While these arguments were spurious, they nevertheless had the effect the CPD wanted. The "iron triangle" of politicians, generals and defense contractors were very successful in massively boosting US defense spending by the late Carter administration.

The CPD's return to the binary rhetoric of the Cold War shaped the geopolitical thinking of the Reagan administration. Ronald Reagan was a member of the founding board of directors of the committee and his championing of "good versus evil" rhetoric appealed to evangelical voters who organized strongly to defeat Carter. Many CPD members, including Paul Nitze and Richard Pearle, took up important policy positions within the Reagan administration. Reagan's strong anti-Communism helped define the period as the "Second Cold War" (Halliday, 1983). Military spending soared. In dollar terms, the defense budget almost doubled between 1979 and 1983, from 5.1 per cent of GNP to 6.6 per cent (Sherry, 1995: 401). His administration initiated a series of aggressive military policies across the world. In Central America, for example, the Reagan administration used the CIA to sponsor a proxy army against the Sandinistas in Nicaragua that had in 1979 overthrown the US supported dictator of that country. The administration also provided increased military aid to El Salvador's ruling class, enabling it to fight a guerrilla insurgency and murder those organizing for social justice within the country (Ó Tuathail, 1986). The Reagan Doctrine of actively supporting counterrevolutionary guerrillas fighting so-called "pro-Soviet" governments around the world was zealously pursued by CIA Director William Casey and ideological helpers like Lt Colonel Oliver North. Operating as a shadow government unto themselves, they provided training, weapons and money to militias from Afghanistan to Angola. American popular geopolitics at this time celebrated this culture of heroic militarism and self-righteous nationalism, with post-Vietnam remasculization themes particularly evident in the *Rambo* and *Missing in Action* film franchises (Jeffords, 1989).

BEYOND THE COLD WAR IN EUROPE

In Western Europe, the Reagan administration began a new round in the militarization of the continent by pursuing the deployment of so-called "limited" nuclear weapons systems like the cruise and Pershing II missiles onto European soil. This effort by NATO to introduce medium range nuclear weapons targeting East Germany and other Central European states provoked resistance across European civil society as peace movements emerged in a number of countries in mobilization against the deployment. Working together across national frontiers and the East–West divide, many thousands of leading dissident intellectuals throughout Europe signed an Appeal for European Nuclear Disarmament (END) first launched on April 28, 1980 by a group of sponsors and the Russell Peace Foundation (Reading 14). Like the CPD manifesto, this document is also preoccupied with danger but not the danger posed by a territorial Other as irreducible enemy. Rather, the END manifesto is concerned with the danger posed by escalating Cold War militarism and its advocates and apologists. Seeking a way beyond the mutually reinforcing militarism of the East–West confrontation, it argued for a European-wide dis-identification and de-alignment from the military blocs dividing Europe. De-alignment was a conscious refusal of the "us" versus "them" polarities offered by Cold War discourse. "We must commence to act as if a united, neutral and pacific Europe already exists. We must learn to be loyal, not to 'East' and 'West,' but to each other, and we must disregard the prohibitions and limitations imposed by any national state." The END manifesto became the charter of the European peace movement that tried, in organizing annual conventions in different European cities, to foster "detente from below" among citizen groups and diverse social movements including progressive Christians, feminists, greens, trade unionists and democratic leftist groups. In the United States, a less challenging movement to freeze nuclear weapons garnered enough popular support to provoke some politicians to question the wisdom of the Reagan administration's nuclear buildup (Sherry, 1995).

While the de-alignment ideas associated with the END appeal had little immediate effect on the military policies of NATO and the Reagan administration, for the "limited strike" American missiles that were deployed did nevertheless percolate, via dissident Eastern European intellectuals, through to a new generation of Soviet bureaucrats seeking to save the communist system from stagnation, corruption and imperial overstretch (most evident in the Soviet Union's disastrous military campaign in Afghanistan). The new breed of communist politician who came to champion these new ideas was Mikhail Gorbachev. Gorbachev's foreign policy was radical for it deliberately set out to deprive the Reagan administration of its convenient "enemy image" of the Soviet Union as an "evil empire." Launching a policy of *glasnost* or "openness" in Soviet society in 1986, Gorbachev envisioned a radical restructuring and renewal (*perestroika*) of the Soviet Union based on modernized and humane communist principles. Declaring that "no country enjoys a monopoly of the truth," he signaled the end of the Brezhnev Doctrine as the geopolitical principles governing the Soviet Union's relationship with the Eastern European communist regimes (Walker, 1993: 290). One Soviet commentator humorously dubbed the new geopolitical philosophy the "Sinatra Doctrine" (evoking Frank Sinatra's famous song "My Way"), the principle being that each Eastern European state can and should find its own way to reform and change without Soviet interventionism. While Gorbachev's self-interested attempt to save the communist system and the Soviet Empire from the top ultimately failed, his "new political thinking" in Soviet foreign policy helped undermine Cold War geopolitical discourse.

THE END OF THE COLD WAR AS THE END OF HISTORY?

From the perspective of neoconservatives, it was the Reagan administration's military buildup that "won" the Cold War. Yet such a view is difficult to sustain for without Gorbachev's push to end the Cold War peacefully, it might have been quite different, indeed horrifyingly different given the Reagan administration's deep militarism. Certainly neoconservative intellectuals of statecraft within the Reagan administration, like Richard Pearle, and within the subsequent Bush administration, like Assistant Secretary of Defense Paul Wolfowitz, were deeply suspicious of Gorbachev and demonized his anti-militarist thinking about security in Europe. Yet, Gorbachev's new thinking was not easily demonized, especially when he backed up his words with concrete anti-militarist policies. Gorbachev's concerted push for arms reductions (not just arms control) and his refusal to intervene to save the communist dictatorships in Eastern Europe in the historic autumn of 1989 resulted in the fall of the Berlin Wall and the beginning of the end of the Cold War in Europe. The profound geopolitical consequences of his radical new policies eventually provoked a counter-reaction by hardliners within the Soviet military-industrial complex in August 1991, an attempted coup whose failure spiraled into the consequent dissolution of the Soviet Union.

Even before these remarkable events, the idea that the West had triumphed in the Cold War was being discussed in American geopolitical culture. A striking example is Francis Fukuyama's 1989 essay "The End of History?" (Reading 15). Fukuyama worked at the time as deputy director of policy planning in the Bush administration, having previously worked in the Reagan administration. While Fukuyama's position recalled the memory of George Kennan, his essay received attention because of its timeliness, triumphalist thesis, and the neoconservative network around the journal *The National Interest*, its original place of publication. Funded by wealthy endowments sympathetic to neoconservatism, the journal *The National Interest* was first established by Irving Kristol, the dean of the American neo-conservative movement, in 1985 (Atlas, 1989). (A decade later Kristol's son, William Kristol, established a neoconservative weekly, *The Weekly Standard*, with financing from Rupert Murdoch.) Fukuyama's essay was presented in the summer of 1989 issue as a major philosophical statement on the direction of history.

Fukuyama's provocative thesis has been described as "self-congratulation raised to the status of philosophy" (Atlas, 1989: 42). Drawing upon the writings of the early nineteenth-century German philosopher Friedrich Hegel as interpreted in the 1930s by the Paris-based Russian émigré philosopher Alexandre Kojève, Fukuyama claims that we are now witnessing the end of History (capital H) as a struggle over ideas and principles. Arguing that Hegel had proclaimed the end of History in 1806 with the triumph of Napoleon (over the Prussian monarchy at the Battle of Jena) and the universal principles of the French revolution (claims contested by Hegel scholars), Fukuyama uses Kojève's argument that Western Europe and the United States represent the "universal homogeneous state" that Hegel first identified in the Napoleonic state to assert that the West is the culmination of historical progress. In Fukuyama's version of Hegel's arguments, the "end of History" is that point at which humanity has actualized the universal truths first expressed in the French revolution, the "principles of liberty and equality." History does not literally end because most states are still struggling to reach and actualize these universal truths. However, a few vanguard states have reached and actualized these universal truths. Unlike most states, the "universal homogeneous state" has reached the pinnacle of historical evolution. It is "homogeneous" because "all prior contradictions" (like race and class) "are resolved and all human needs are satisfied. There is no struggle or conflict over 'large' issues, and consequently no need for generals and statesmen; what remains is primarily economic activity."

In making this sweeping argument, Fukuyama ignores some inconvenient details and simplifies others. For example, he conflates the American and French revolution but there were differences. The American Constitution's acceptance of slavery, for example, is a salutary reminder of the gap between the rhetoric of liberty and its actual practice. The Napoleonic state was an imperial dictatorship, with secret police and considerable press censorship. It too hardly represented the "principles of liberty and equality" in practice. Fukuyama accepts the ideology of liberalism but does not worry too much about the contradictions in practice (his phrase "man's universal rights to freedom" reveals a blindness to feminist critiques of the universal subject as "man"). The conceit driving the article is that Western capitalist states are at the "vanguard of humanity." This is an enlarged version of the durable conceit that the American state represents universal principles, that it embodies "the hopes of humanity" (see Bush quote below).

Fukuyama later elaborated his thesis in greater detail in a subsequent book (Fukuyama, 1992). Many critics challenged his use of both Hegel and Kojève, noting that Fukuyama has them making arguments that they do not actually make (Ryan, 1992). Fukuyama himself can claim that his arguments are more complex than his admirers and critics allow. His article is important as an early neoconservative attempt to recast Cold War discourse in the light of the imminent collapse of communism in Europe. Fukuyama's method is similar to what we find in other geopoliticians. He reads world politics as if it were a self-evident spectacle before a detached observer making no allowance for his own embeddedness in the conceits of American geopolitical culture (i.e. the "god trick"). Fukuyama considers his own state and community as the consummation of history, the fulfillment of human historical destiny. All other states are struggling to attain the "end of History" state the West has reached. The West is "post-historical," the rest of the world still struggling in the "historical." Fukuyama's central conceit leads him to make a series of observations that capture the emergent geopolitical culture of the 1990s. For example, he claims that "it matters very little what strange thoughts occur to people in Albania or Burkino Faso, for we are interested in what one could in some sense call the common ideological heritage of mankind." This disdain for the Balkans and Africa was unfortunately mirrored in the practice of Western powers in the face of genocide in Bosnia and Rwanda in the 1990s. Despite being a nominally Marxist state, Fukuyama claims that China can no longer "act as a beacon of illiberal forces around the world." China has embraced the market; in the 1990s it would become a major trading partner of the United States. The Soviet Union is also no longer an alternative to liberalism for Gorbachev's "democratization and decentralization principles are highly subversive of the fundamental precepts of both Marxism and Leninism." The ideological transformations in both states lead Fukuyama to proclaim the passing of Marxism-Leninism as a "living ideology of world historical significance."

As a map of meaning designed to make sense of world politics in the early 1990s, Fukuyama's scheme is flawed in two significant ways. First, it is an ethnocentric schema which fails to acknowledge the serious problems – what Fukuyama after Hegel would call "contradictions" – that beset Western states. This triumphalist complacency is a function both of the imprecise concepts Fukuyama uses – principally the notions of the "universal homogeneous state" and "liberalism" which are self-idealized and imaginary concepts rather than historical ones – and those he does not use – principally capitalism but also militarism, racism, and patriarchy. The "universal homogeneous state" in Fukuyama stretches historically from Napoleon to NATO, including along the way such radically different states like the United States, South Korea, Japan and even, after agricultural reform, China! To categorize certain states as "liberal" does not really tell us very much about the specific geographical structure of states and the contradictions of their particular historical versions of liberalism, its compromise with nationalism, militarism, patriarchy and racial supremacy in different states. Western states are far from being universally liberal or homogeneous. As events over the following decade and a half revealed, there were deep divides within the so-called West that were made manifest by American unilateralism and militarism.

In ignoring capitalism and suggesting a receding of the class issue, Fukuyama ignores the turbulent "creative destruction" wrought by the globalization of capitalism and the marked increase in income inequality across the West, particularly in the United Kingdom and the United States. To assume that the Western modernity represented by the European Union and the United States represents a stable culmination of civilization bereft of serious "contradictions" (like class, race, identity, environment and globalization issues) is more ideological statement than analysis. The claim that "the egalitarianism of modern America represents the essential achievement of the classless society envisioned by Marx" (part of his dismissal of the questions of class and race) reveals a commitment to idealized self-images over empirical evidence and concrete historical realities.

Second, Fukuyama's assumption that the declining appeal of Marxist-Leninist ideology and the supposed spread of the liberalism of the "universal homogeneous state" (what he calls "post-historical consciousness") will lead to the receding of international conflict between states and the growing "Common Marketization" of world politics was unduly optimistic (consider, for example, his remarks on Burma). Again, the problem is that Fukuyama's abstract philosophical narrative leads him to sweeping conclusions that elide the messy territoriality of world politics. While Marxist-Leninist ideology was declining, anti-Western, anti-capitalist and anti-modern ideologies were far from dead, as the new millennium revealed. One need not endorse political realist or neorealist assumptions about the interstate system to argue that states are complex entities motivated by forces and ideologies which sometimes propel them into conflict with each other (e.g. China and Taiwan). Fukuyama's assumption that international life for those who have reached the end of history (Western Europe and North America) is far more preoccupied with economics than with politics and strategy is consonant with the conceit that liberal democratic states are peaceful not bellicose. This observation overlooks Cold War militarism and the fact that Western Europe, the site of the European version of the "universal homogeneous state," was the most militarized place on the planet by the late 1980s. The contrast Fukuyama offers is a false one, for economics is not so easily separated from politics, ideology, and military power. All these forms of social power are mutually entwined with each other in practice.

While Fukuyama was declaring victory in the Cold War, the Bush administration was responding cautiously. The fall of the Berlin Wall, the collapse of the Soviet bloc dictatorships, the Soviet coup of August 1991, and the dramatic dissolution of the Soviet Union that followed took Cold War politicians and Western national security institutions by surprise. Few had predicted such a dramatic collapse of communism in Europe, the disintegration of the Warsaw Pact and the implosion of the Soviet Union. This intelligence failure was partly a function of the fact that Cold War discourse served more of a binding ideological function for the West than it served as a description of the Soviet Union. It was not possible to think of the Soviet Union as weak and vulnerable when one was already ideologically and institutionally committed to it as an overarching threat and implacable enemy.

Yet, rather than the end of the Cold War leading to a critical reassessment of the national security agencies and intellectuals that had built the Soviet Union into such an awesome threat, bureaucratic and ideological relegitimation were the order of the day. Initially, the Bush administration pronounced "uncertainty," "unpredictability," "instability" and "chaos" as the new threats (Ó Tuathail, 1992). The Iraqi invasion of the tiny petroleum kingdom of Kuwait in August 1990 provided an occasion for renewing the imperative for US leadership in world affairs. A "new world order" was possible. The subsequent war to evict the Iraqi military from Kuwait in early 1991 by an international coalition led by the United States provided Bush with a platform to re-proclaim America's geopolitical creed as the familiar (con)fusion of American nationalism and liberal universalism:

> For two centuries, America has served the world as an inspiring example of freedom and democracy. For generations, America has led the struggle to preserve and extend the blessing of liberty. And, today, in a rapidly changing world, American leadership is indispensable. Americans know that leadership brings burdens and sacrifices. But we also know why the hopes of humanity turn towards us. We are Americans. We have a unique responsibility to do the hard work of freedom.
>
> (Bush, 1991)

The new geopolitical storyline was of the United States as the "sole remaining superpower," the "indispensable" leader in world affairs. It was a storyline that would define the geopolitics of the late 1990s, as the Clinton administration struggled with crises in the Middle East and Southeast Europe, and the geopolitics of the early twenty-first century as Bush's son, George W. Bush, became president and sought to finish the war his father had begun against Saddam Hussein's Iraq in 1991.

REFERENCES AND FURTHER READING

On Truman, Kennan and the origins of the Cold War

Acheson, D. (1969) *Present at the Creation*, New York: Norton.
Kennan, G. (1947) "The Sources of Soviet Conduct," *Foreign Affairs*, 25: 568–582.
—— (1951) *American Diplomacy*. Chicago, IL: University of Chicago Press.
Lippmann, W. (1947) *The Cold War*. New York: Harper & Row.
Stephanson, A. (1989) *Kennan and the Art of Foreign Policy*. Cambridge, MA: Harvard University Press.
Taylor, P. (1990) *Britain and the Cold War*. London: Pinter.
Yergin, D. (1978) *Shattered Peace: The Origins of the Cold War and the National Security State*. Boston, MA: Houghton Mifflin.

On the Cold War as a geopolitical system

Applebaum, A. (2003) *Gulag: A History*. New York: Random House.
Campbell, D. (1990) *Writing Security*. Minneapolis, MN: University of Minnesota.
Chomsky, N. (1991) *Deterring Democracy*. London: Verso.
Cox, M. (1990) "From the Truman Doctrine to the Second Superpower Detente: The Rise and Fall of the Cold War," *Journal of Peace Research*, 27: 25–41.
Gaddis, J. (1987) *The Long Peace*. New York: Oxford University Press.
Halberstam, D. (1972) *The Best and the Brightest*. London: Penguin.
Halliday, F. (1983) *The Making of the Second Cold War*. London: Verso.
Kaldor, M. (1990) *The Imaginary War*. Oxford: Basil Blackwell.

Kolko, G. (1988) *Confronting the Third World*. New York: Pantheon.

Lundestad, G. (1990) *The American "Empire."* London: Oxford University Press.

May, L. (ed.) (1989) *Recasting America: Culture and Politics in the Age of the Cold War*. Chicago, IL: University of Chicago Press.

Ó Tuathail, G. (1991) "The Bush Administration and the 'End' of the Cold War: A Critical Geopolitics of U.S. Foreign Policy in 1989," *Geoforum*, 23: 437–452.

Simons, T. (1991) *Eastern Europe in the Postwar World*. New York: St Martin's Press.

Walker, M. (1993) *The Cold War: A History*. New York: Holt.

Wolfe, A. (1982) *America's Impasse: The Rise and the Fall of the Politics of Growth*. Boston, MA: South End Press.

—— (1984) *The Rise and Fall of the "Soviet Threat."* Boston, MA: South End Press.

On the Second Cold War

Dalby, S. (1990) *Creating the Second Cold War*. New York: Guilford.

Jeffords, S. (1989) *The Remasculinization of America*. Bloomington, IN: Indiana University Press.

Litwak, R. (1984) *Detente and the Nixon Doctrine*. London: Cambridge University Press.

Ó Tuathail, G. (1986) "The Language and Nature of the 'New' Geopolitics: The Case of US–El Salvador Relations," *Political Geography Quarterly*, 5: 73–85.

Saunders, J. (1983) *Peddlers of the Crisis: The Committee on the Present Danger and the Politics of Containment*. Boston, MA: South End Press.

Sherry, M. (1995) *In the Shadow of War*. New Haven, CT: Yale University Press.

Beyond the Cold War in Europe

Gorbachev, M. (1988) *Perestroika: New Thinking for our Country and the World*. New York: Harper & Row.

Kaldor, M. (ed.) (1991) *Europe from Below*. London: Verso.

Kaldor, M. and Falk, R. (eds) (1987) *Dealignment: A New Foreign Policy Perspective*. New York: Basil Blackwell.

Kaldor, M., Holden, G. and Falk, R. (eds) (1989) *The New Detente: Re-Thinking East–West Relations*. London: Verso.

McGwire, M. (1991) *Perestroika and Soviet National Security*. Washington, DC: Brookings Institution.

Thompson, E.P. (1985) *The Heavy Dancers*. New York: Pantheon.

Thompson, E.P. and Smith, D. (eds) (1980) *Protest and Survive*. New York: Penguin.

Thompson, E.P. et al. (1982) *Exterminism and Cold War*. London: Verso.

The end of the Cold War and the "New World Order"

Atlas, J. (1989) "What is Fukuyama Saying?" *New York Times Magazine*, October 22.

Bush, G. (1991) State of the Union Address, January 29, 1991. *Public Papers of the Presidents of the United States: George H. W. Bush*.

Fukuyama, F. (1992) *The End of History and the Last Man*. New York: Free Press.

Ó Tuathail, G. (1992) "The Bush Administration and the 'End' of the Cold War," *Geoforum*, 23: 437–452.

Ryan, A. (1992) "Professor Hegel Goes to Washington," *New York Times Book Review*, March 26.

Cartoon 3 Cold Warriors anonymous

The "addiction" of both the superpowers to geopolitical interventionism during the Cold War is satirized by Matt Wuerker. Here he represents Gorbachev's "new political thinking" as an admission of guilt to a councilor. President Bush, however, is represented as still "in denial" about the US's addiction to geopolitical interventionism.

Source: M. Wuerker

7

The Truman Doctrine

President Harry Truman

from *Public Papers of the Presidents of the United States* (1947)

The gravity of the situation which confronts the world today necessitates my appearance before a joint session of the Congress. The foreign policy and the national security of this country are involved. One aspect of the present situation, which I wish to present to you at this time for your consideration and decision, concerns Greece and Turkey.

The United States has received from the Greek Government an urgent appeal for financial and economic assistance. Preliminary reports from the American Economic Mission now in Greece and reports from the American Ambassador in Greece corroborate the statement of the Greek Government that assistance is imperative if Greece is to survive as a free nation. I do not believe that the American people and the Congress wish to turn a deaf ear to the appeal of the Greek Government.
[. . .]
The very existence of the Greek state is today threatened by the terrorist activities of several thousand armed men, led by Communists, who defy the Government's authority at a number of points, particularly along the northern boundaries. A commission appointed by the United Nations Security Council is at present investigating disturbed conditions in Northern Greece and alleged border violations along the frontier between Greece on the one hand and Albania, Bulgaria, and Yugoslavia on the other.

Meanwhile, the Greek Government is unable to cope with the situation. The Greek Army is small and poorly equipped. It needs supplies and equipment if it is to restore the authority to the Government throughout Greek territory. Greece must have assistance if it is to become a self-supporting and self-respecting democracy. The United States must supply this assistance. We have already extended to Greece certain types of relief and economic aid but these are inadequate. There is no other country to which democratic Greece can turn. No other nation is willing and able to provide the necessary support for a democratic Greek Government.

The British Government, which has been helping Greece, can give no further financial or economic aid after March 31. Great Britain finds itself under the necessity of reducing or liquidating its commitments in several parts of the world, including Greece. We have considered how the United Nations might assist in this crisis. But the situation is an urgent one requiring immediate action, and the United Nations and its related organizations are not in a position to extend help of the kind that is required. [. . .]

Greece's neighbor, Turkey, also deserves our attention. The future of Turkey as an independent and economically sound state is clearly no less important to the freedom-loving peoples of the world than the future of Greece. The circumstances in which Turkey finds itself today are considerably different from those of Greece. Turkey has been spared the disasters that have beset Greece. And during the war, the United States and Great Britain furnished Turkey with material aid.

Nevertheless, Turkey now needs our support. Since the war Turkey has sought financial assistance from Great Britain and the United States for the purpose of effecting that modernization necessary for the maintenance of its national integrity. That integrity is essential to the preservation of order in the Middle East.

The British Government has informed us that, owing to its own difficulties, it can no longer extend financial or economic aid to Turkey. As in the case of Greece, if Turkey is to have the assistance it needs, the United States must supply it. We are the only country able to provide that help. I am fully aware of the broad implications involved if the United States extends assistance to Greece and Turkey, and I shall discuss these implications with you at this time.

One of the primary objectives of the foreign policy of the United States is the creation of conditions in which we and other nations will be able to work out a way of life free from coercion. This was a fundamental issue in the war with Germany and Japan. Our victory was won over countries which sought to impose their will, and their way of life, upon other nations. [. . .]

The peoples of a number of countries of the world have recently had totalitarian regimes forced upon them against their will. The Government of the United States has made frequent protests against coercion and intimidation, in violation of the Yalta Agreement, in Poland, Rumania, and Bulgaria. I must also state that in a number of other countries there have been similar developments.

At the present moment in world history nearly every nation must choose between alternative ways of life. The choice is too often not a free one. One way of life is based upon the will of the majority, and is distinguished by free institutions, representative government, free elections, guarantees of individual liberty, freedom of speech and religion, and freedom from political oppression. The second way of life is based upon the will of a minority forcibly imposed upon the majority. It relies upon terror and oppression, a controlled press and radio, fixed elections, and the suppression of personal freedoms.

I believe that it must be the policy of the United States to support free peoples who are resisting attempted subjugation by armed minorities or by outside pressures. I believe that we must assist free peoples to work out their own destinies in their own way. I believe that our help should be primarily through economic and financial aid, which is essential to economic stability and orderly political processes.

The world is not static and the status quo is not sacred. But we cannot allow changes in the status quo in violation of the Charter of the United Nations by such methods as coercion, or by such subterfuges as political infiltration. In helping free and independent nations to maintain their freedom, the United States will be giving effect to the principles of the Charter of the United Nations. It is necessary only to glance at a map to realize that the survival and integrity of the Greek nation are of grave importance in a much wider situation. If Greece should fall under the control of an armed minority, the effect upon its neighbor, Turkey, would be immediate and serious. Confusion and disorder might well spread throughout the entire Middle East.

Moreover, the disappearance of Greece as an independent state would have a profound effect upon those countries in Europe whose peoples are struggling against great difficulties to maintain their freedoms and their independence while they repair the damages of war. It would be an unspeakable tragedy if these countries, which have struggled so long against overwhelming odds, should lose that victory for which they sacrificed so much. Collapse of free institutions and loss of independence would be disastrous not only for them but for the world. Discouragement and possibly failure would quickly be the lot of neighboring peoples striving to maintain their freedom and independence.

Should we fail to aid Greece and Turkey in this fateful hour, the effect will be far reaching to the West as well as to the East. We must take immediate and resolute action. I therefore ask the Congress to provide authority for assistance to Greece and Turkey in the amount of $400,000,000 for the period ending June 30, 1948. [. . .]

In addition to funds, I ask the Congress to authorize the detail of American civilian and military personnel to Greece and Turkey, at the request of those countries, to assist in the tasks of reconstruction, and for the purpose of supervising the use of such financial and material assistance as may be furnished. I recommend that authority also be provided for the instruction and training of selected Greek and Turkish personnel.

Finally, I ask that the Congress provide authority which will permit the speediest and most effective use, in terms of needed commodities, supplies, and equipment, of such funds as may be authorized. [. . .]

This is a serious course upon which we embark. I would not recommend it except that the alternative is much more serious. The United States contributed $341,000,000,000 toward winning World War II. This is an investment in world freedom and world peace. The assistance that I am recommending for Greece and Turkey amounts to little more than one-tenth of 1 percent of this investment. It is only common sense that we should safeguard this investment and make sure that it was not in vain.

The seeds of totalitarian regimes are nurtured by misery and want. They spread and grow in the evil soil of poverty and strife. They reach their full growth when the hope of a people for a better life has died. We must keep that hope alive. The free peoples of the world look to us for support in maintaining their freedoms. If we falter in our leadership, we may endanger the peace of the world – and we shall surely endanger the welfare of our own Nation. Great responsibilities have been placed upon us by the swift movement of events. I am confident that the Congress will face these responsibilities squarely.

8

The Sources of Soviet Conduct

George F. Kennan
from *Foreign Affairs* (1947)

The political personality of Soviet power as we know it today is the product of ideology and circumstances: ideology inherited by the present Soviet leaders from the movement in which they had their political origin, and circumstances of the power which they now have exercised for nearly three decades in Russia. There can be few tasks of psychological analysis more difficult than to try to trace the interaction of these two forces and the relative role of each in the determination of official Soviet conduct. Yet the attempt must be made if that conduct is to be understood and effectively countered.

[. . .]

The circumstances of the immediate post-Revolution period – the existence in Russia of civil war and foreign intervention, together with the obvious fact that the Communists represented only a tiny minority of the Russian people – made the establishment of dictatorial power a necessity. The experiment with "war Communism" and the abrupt attempt to eliminate private production and trade had unfortunate economic consequences and caused further bitterness against the new revolutionary regime. While the temporary relaxation of the effort to communize Russia, represented by the New Economic Policy, alleviated some of this economic distress and thereby served its purpose, it also made it evident that the "capitalistic sector of society" was still prepared to profit at once from any relaxation of governmental pressure, and would, if permitted to continue to exist, always constitute a powerful opposing element to the Soviet regime and a serious rival for influence in the country. Somewhat the same situation prevailed with respect to the individual peasant who, in his own small way, was also a private producer.

Lenin, had he lived, might have proved a great enough man to reconcile these conflicting forces to the ultimate benefit of Russian society, though this is questionable. But be that as it may, Stalin, and those whom he led in the struggle for succession to Lenin's position of leadership, were not the men to tolerate rival political forces in the sphere of power which they coveted. Their sense of insecurity was too great. Their particular brand of fanaticism, unmodified by any of the Anglo-Saxon traditions of compromise, was too fierce and too jealous to envisage any permanent sharing of power. From the Russian-Asiatic world out of which they had emerged they carried with them a skepticism as to the possibilities of permanent and peaceful coexistence of rival forces. Easily persuaded of their own doctrinaire "rightness," they insisted on the submission or destruction of all competing power. Outside of the Communist Party, Russian society was to have no rigidity. There were to be no forms of collective human activity or association which would not be dominated by the Party. No other force in Russian society was to be permitted to achieve vitality or integrity. Only the Party was to have structure. All else was to be an amorphous mass. [. . .]

II

[. . .] The Soviet concept of power, which permits no focal points of organization outside the Party itself, requires that the Party leadership remain in theory the sole repository of truth. For if truth were to be found elsewhere, there would be justification for its expression in organized activity. But it is precisely that which the Kremlin cannot and will not permit.

The leadership of the Communist Party is therefore always right, and has been always right ever since in 1929 Stalin formalized his personal power by announcing that decisions of the Politburo were being taken unanimously.

On the principle of infallibility there rests the iron discipline of the Communist Party. In fact, the two concepts are mutually self-supporting. Perfect discipline requires recognition of infallibility. Infallibility requires the observance of discipline. And the two together go far to determine the behaviorism of the entire Soviet apparatus of power. But their effect cannot be understood unless a third factor be taken into account: namely, the fact that the leadership is at liberty to put forward for tactical purposes any particular thesis which it finds useful to the cause at any particular moment and to require the faithful and unquestioning acceptance of that thesis by the members of the movement as a whole. This means that truth is not a constant but is actually created, for all intents and purposes, by the Soviet leaders themselves. It may vary from week to week, from month to month. It is nothing absolute and immutable – nothing which flows from objective reality. It is only the most recent manifestation of the wisdom of those in whom the ultimate wisdom is supposed to reside, because they represent the logic of history. The accumulative effect of these factors is to give to the whole subordinate apparatus of Soviet power an unshakeable stubbornness and steadfastness in its orientation. This orientation can be changed at will by the Kremlin but by no other power. Once a given party line has been laid down on a given issue of current policy, the whole Soviet governmental machine, including the mechanism of diplomacy, moves inexorably along the prescribed path, like a persistent toy automobile wound up and headed in a given direction, stopping only when it meets with some unanswerable force. The individuals who are the components of this machine are unamenable to argument or reason which comes to them from outside sources. Their whole training has taught them to mistrust and discount the glib persuasiveness of the outside world. Like the white dog before the phonograph, they hear only the "master's voice." And if they are to be called off from the purposes last dictated to them, it is the master who must call them off. Thus the foreign representative cannot hope that his words will make any impression on them. The most that he can hope is that they will be transmitted to those at the top, who are capable of changing the party line. But even those are not likely to be swayed by any normal logic in the words of the bourgeois representative. Since there can be no appeal to common purposes, there can be no appeal to common mental approaches. For this reason, facts speak louder than words to the ears of the Kremlin; and words carry the greatest weight when they have the ring of reflecting, or being backed up by, facts of unchallengeable validity.

But we have seen that the Kremlin is under no ideological compulsion to accomplish its purposes in a hurry. Like the Church, it is dealing in ideological concepts which are of long-term validity, and it can afford to be patient. It has no right to risk the existing achievements of the revolution for the sake of vain baubles of the future. The very teachings of Lenin himself require great caution and flexibility in the pursuit of Communist purposes. Again, these precepts are fortified by the lessons of Russian history: of centuries of obscure battles between nomadic forces over the stretches of a vast unfortified plain. Here caution, circumspection, flexibility and deception are the valuable qualities; and their value finds natural appreciation in the Russian or the oriental mind. Thus the Kremlin has no compunction about retreating in the face of superior force. And being under the compulsion of no timetable, it does not get panicky under the necessity for such retreat. Its political action is a fluid stream which moves constantly, wherever it is permitted to move, toward a given goal. Its main concern is to make sure that it has filled every nook and cranny available to it in the basin of world power. But if it finds unassailable barriers in its path, it accepts these philosophically and accommodates itself to them. The main thing is that there should always be pressure, increasing constant pressure, toward the desired goal. There is no trace of any feeling in Soviet psychology that that goal must be reached at any given time.

These considerations make Soviet diplomacy at once easier and more difficult to deal with than the diplomacy of individual aggressive leaders like Napoleon and Hitler. On the one hand it is more sensitive to contrary force, more ready to yield on individual sectors of the diplomatic front when that force is felt to be too strong, and thus more rational in the logic and rhetoric of power. On the other hand it cannot be easily defeated or discouraged by a single victory on the part of its opponents. And the patient persistence by which it is animated means that it can be effectively countered not by sporadic acts which represent the momentary whims of democratic opinion but only by intelligent long-range policies on the part of Russia's adversaries – policies no less steady in their purpose, and no less variegated and resourceful in their application, than those of the Soviet Union itself.

In these circumstances it is clear that the main element of any United States policy toward the Soviet Union must be that of a long-term, patient but firm and vigilant containment of Russian expansive tendencies. It is important to note, however, that such a policy has nothing to do with outward histrionics: with threats or blustering or superfluous gestures of outward "toughness." While the Kremlin is basically flexible in its reaction to political realities, it is by no means unamenable to considerations of prestige. Like almost any other government, it can be placed by tactless and threatening gestures in a position where it cannot afford to yield even though this might be dictated by its sense of realism. The Russian leaders are keen judges of human psychology, and as such they are highly conscious that loss of temper and of self-control is never a source of strength in political affairs. They are quick to exploit such evidences of weakness. For these reasons, it is a sine qua non of successful dealing with Russia that the foreign government in question should remain at all times cool and collected and that its demands on Russian policy should be put forward in such a manner as to leave the way open for a compliance not too detrimental to Russian prestige.

III

In the light of the above, it will be clearly seen that the Soviet pressure against the free institutions of the Western world is something that can be contained by the adroit and vigilant application of counter-force at a series of constantly shifting geographical and political points, corresponding to the shifts and maneuvers of Soviet policy, but which cannot be charmed or talked out of existence. The Russians look forward to a duel of infinite duration, and they see that already they have scored great successes. It must be borne in mind that there was a time when the Communist Party represented far more of a minority in the sphere of Russian national life than Soviet power today represents in the world community.

But if ideology convinces the rulers of Russia that truth is on their side and that they can therefore afford to wait, those of us on whom that ideology has no claim are free to examine objectively the validity of that premise. The Soviet thesis not only implies complete lack of control by the West over its own economic destiny, it likewise assumes Russian unity, discipline and patience over an infinite period. Let us bring this apocalyptic vision down to earth, and suppose that the Western world finds the strength and resourcefulness to contain Soviet power over a period of ten to fifteen years. What does that spell for Russia itself?

The Soviet leaders, taking advantage of the contributions of modern technique to the arts of despotism, have solved the question of obedience within the confines of their power. Few challenge their authority; and even those who do are unable to make that challenge valid as against the organs of suppression of the state.

The Kremlin has also proved able to accomplish its purpose of building up in Russia, regardless of the interests of the inhabitants, an industrial foundation of heavy metallurgy, which is, to be sure, not yet complete but which is nevertheless continuing to grow and is approaching those of the other major industrial countries. All of this, however, both the maintenance of internal political security and the building of heavy industry, has been carried out at a terrible cost in human life and in human hopes and energies. [. . .]

Here is a nation striving to become in a short period one of the great industrial nations of the world while it still has no highway network worthy of the name and only a relatively primitive network of railways. Much has been done to increase efficiency of labor and to teach primitive peasants something about the operation of machines. But maintenance is still a crying deficiency of all Soviet economy. Construction is hasty and poor and in vast sectors of economic life it has not yet been possible to instill into labor anything like that general culture of production and technical self-respect which characterizes the skilled worker of the West.

It is difficult to see how these deficiencies can be corrected at an early date by a tired and dispirited population working largely under the shadow of fear and compulsion. And as long as they are not overcome, Russia will remain economically a vulnerable, and in a certain sense an impotent, nation, capable of exporting its enthusiasms and of radiating the strange charm of its primitive political vitality but unable to back up those articles of export by the real evidences of material power and prosperity.

IV

It is clear that the United States cannot expect in the foreseeable future to enjoy political intimacy with the Soviet regime. It must continue to regard the Soviet Union as a rival, not a partner, in the political arena. It must continue to expect that Soviet policies will reflect no abstract love of peace and stability, no real faith in the possibility of a permanent happy coexistence of the Socialist and capitalist worlds, but rather a cautious, persistent pressure toward the disruption and weakening of all rival influence and rival power.

Balanced against this are the facts that Russia, as opposed to the Western world in general, is still by far the weaker party, that Soviet policy is highly flexible, and that Soviet society may well contain deficiencies which will eventually weaken its own total potential. This would of itself warrant the United States entering with reasonable confidence upon a policy of firm containment, designed to confront the Russians with unalterable counter-force at every point where they show signs of encroaching upon the interests of a peaceful and stable world.

But in actuality the possibilities for American policy are by no means limited to holding the line and hoping for the best. It is entirely possible for the United States to influence by its actions the internal developments, both within Russia and throughout the international Communist movement, by which Russian policy is largely determined. This is not only a question of the modest measure of informational activity which this government can conduct in the Soviet Union and elsewhere, although that, too, is important. It is rather a question of the degree to which the United States can create among the peoples of the world generally the impression of a country which knows what it wants, which is coping successfully with the problems of its internal life and with the responsibilities of a World Power, and which has a spiritual vitality capable of holding its own among the major ideological currents of the time. To the extent that such an impression can be created and maintained, the aims of Russian Communism must appear sterile and quixotic, the hopes and enthusiasm of Moscow's supporters must wane, and added strain must be imposed on the Kremlin's foreign policies. For the palsied decrepitude of the capitalist world is the keystone of Communist philosophy. Even the failure of the United States to experience the early economic depression which the ravens of the Red Square have been predicting with such complacent confidence since hostilities ceased would have deep and important repercussions throughout the Communist world.

By the same token, exhibitions of indecision, disunity and internal disintegration within this country have an exhilarating effect on the whole Communist movement. At each evidence of these tendencies, a thrill of hope and excitement goes through the Communist world; a new jauntiness can be noted in the Moscow tread; new groups of foreign supporters climb on to what they can only view as the bandwagon of international politics; and Russian pressure increases all along the line in international affairs.

It would be an exaggeration to say that American behavior unassisted and alone could exercise a power of life and death over the Communist movement and bring about the early fall of Soviet power in Russia. But the United States has it in its power to increase enormously the strains under which Soviet policy must operate, to force upon the Kremlin a far greater degree of moderation and circumspection than it has had to observe in recent years, and in this way to promote tendencies which must eventually find their outlet in either the break-up or the gradual mellowing of Soviet power. For no mystical Messianic movement – and particularly not that of the Kremlin – can face frustration indefinitely without eventually adjusting itself in one way or another to the logic of that state of affairs.

Thus the decision will really fall in large measure in this country itself. The issue of Soviet–American relations is in essence a test of the over-all worth of the United States as a nation among nations. To avoid destruction the United States need only measure up to its own best traditions and prove itself worthy of preservation as a great nation.

Surely, there was never a fairer test of national quality than this. In the light of these circumstances, the thoughtful observer of Russian–American relations will find no cause for complaint in the Kremlin's challenge to American society. He will rather experience a certain gratitude to a Providence which, by providing the American people with this implacable challenge, has made their entire security as a nation dependent on their pulling themselves together and accepting the responsibilities of moral and political leadership that history plainly intended them to bear.

9

Soviet Policy and World Politics

Andrei Zhdanov
from *The International Situation* (1947)

The end of World War II brought with it big changes in the world situation. The military defeat of the bloc of fascist states, the character of the war of liberation from fascism, and the decisive role played by the Soviet Union in the vanquishing of the fascist aggressors sharply altered the alignment of forces between the two systems – the socialist and the capitalist – in favor of socialism.

What is the essential nature of these changes? The principal outcome of World War II was the military defeat of Germany and Japan – the two most militaristic and aggressive of the capitalist countries. [. . .]

[Second], the war immensely enhanced the international significance and prestige of the USSR. [. . .]

[Third], the capitalist world has also undergone a substantial change. Of the six so-called great imperialist powers (Germany, Japan, Great Britain, the USA, France, and Italy), three have been eliminated by military defeat. France has also been weakened and has lost its significance as a great power. As a result, only two great imperialist world powers remain – the United States and Great Britain. But the position of one of them, Great Britain, has been undermined. The war revealed that militarily and politically British imperialism was not so strong as it had been. [. . .]

[Fourth], World War II aggravated the crisis of the colonial system, as expressed in the rise of a powerful movement for national liberation in the colonies and dependencies. This has placed the rear of the capitalist system in jeopardy. The peoples of the colonies no longer wish to live in the old way. The ruling classes of the metropolitan countries can no longer govern the colonies on the old lines. [. . .]

Of all the capitalist powers, only one – the United States – emerged from the war not only unweakened, but even considerably stronger economically and militarily. The war greatly enriched the American capitalists. [. . .] But the end of the war confronted the United States with a number of new problems. The capitalist monopolies were anxious to maintain their profits at the former high level, and accordingly pressed hard to prevent a reduction of the wartime volume of deliveries. But this meant that the USA must retain the foreign markets which had absorbed American products during the war, and moreover, acquire new markets, inasmuch as the war had substantially lowered the purchasing power of most of the countries [to do this] [. . .] the United States proclaimed a new frankly predatory and expansionist course. The purpose of this new, frankly expansionist course is to establish the world supremacy of American imperialism. [. . .]

The fundamental changes caused by the war on the international scene and in the position of individual countries have entirely changed the political landscape of the world. A new alignment of political forces has arisen. The more the war recedes into the past, the more distinct become two major trends in postwar international policy, corresponding to the division of the political forces operating on the international arena into two major camps; the imperialist and anti-democratic camp, on the one hand, and the anti-imperialist and democratic camp, on the other. The principal driving force of the imperialist camp is the USA. Allied with it are Great Britain and France. [. . .] The cardinal purpose of the imperialist camp is to strengthen imperialism, to hatch a new imperialist war, to combat socialism and democracy, and to support reactionary and anti-democratic pro-fascist regimes and movements everywhere.

The anti-fascist forces comprise the second camp. This camp is based on the USSR and the new democracies. It also includes countries that have broken with imperialism and have firmly set foot on the path of democratic development, such as Rumania, Hungary, and Finland. [. . .]

Soviet foreign policy proceeds from the fact of the coexistence for a long period of the two systems – capitalism and socialism. From this it follows that cooperation between the USSR and countries with other systems is possible, provided that the principle of reciprocity is observed and that obligations once assumed are honored. Everyone knows that the USSR has always honored the obligations it has assumed. Britain and America are pursing the very opposite policy in the United Nations. They are doing everything they can to renounce their commitments and to secure a free hand for the prosecution of a new policy, a policy which envisages not cooperation among the nations, but the hounding of one against the other, violation of the rights and interests of democratic nations, and the isolation of the USSR. [. . .] The strategical plans of the United States envisage the creation in peacetime of numerous bases and vantage grounds situated at great distances from the American continent and designed to be used for aggressive purposes against the USSR and the countries of the new democracy. [. . .]

Economic expansion is an important supplement to the realization of America's strategical plan. American imperialism is endeavoring like a usurer to take advantage of the postwar difficulties of the European countries, in particular of the shortage of raw materials, fuel, and food in the Allied countries that suffered most from the war, to dictate to them extortionate terms for any assistance rendered. With an eye to the impending economic crisis, the United States is in a hurry to find new monopoly spheres of capital investment and markets for its goods. American economic "assistance" pursues the broad aim of bringing Europe into bondage to American capital. The more drastic the economic situation of a country is, the harsher are the terms which the American monopolies endeavor to dictate to it. [. . .]

Lastly, the aspiration to world supremacy and the anti-democratic policy of the United States involve an ideological struggle. The principal purpose of the ideological part of the American strategical plan is to deceive public opinion by slanderously accusing the Soviet Union and the new democracies of aggressive intentions, and thus representing the Anglo-Saxon bloc in a defensive role, and absolving it of responsibility for preparing a new war. [. . .]

The unfavorable reception which the Truman doctrine was met with accounts for the necessity of the appearance of the Marshall Plan which is a more carefully veiled attempt to carry through the same expansionist policy. The vague and deliberately guarded formulations of the Marshall Plan amount in essence to a scheme to create a bloc of states bound by obligations to the United States, and to grant American credits to European countries as recompense for their renunciation of economic, and then of political, independence.

The dissolution of the Comintern, which conformed to the demands of the development of the labor movement in the new historical situation, played a positive role. The dissolution of the Comintern once and for all disposed of the slanderous allegation of the enemies of Communism and the labor movement that Moscow was interfering in the internal affairs of other states, and that the Communist Parties in the various countries were acting not in the interests of their nations, but on orders from outside. [. . .]

In the course of the four years that have elapsed since the dissolution of the Comintern (1943), the Communist Parties have grown considerably in strength and influence in nearly all the countries of Europe and Asia. [. . .] But the present position of the Communist Parties has its shortcomings. Some comrades understood the dissolution of the Comintern to imply the elimination of all ties, of all contact, between the fraternal Communist Parties. But experience has shown that such mutual isolation of the Communist Parties is wrong, harmful and, in point of fact, unnatural. The Communist movement develops within national frameworks, but there are tasks and interests common to the parties of various countries. We get a rather curious state of affairs [. . .] the Communists even refrained from meeting one another, let alone consulting with one another on questions of mutual interest to them, from fear of the slanderous talk of their enemies regarding the "hand of Moscow." [. . .] There can be no doubt that if the situation were to continue it would be fraught with most serious consequences to the development of the work of the fraternal parties.

The need for mutual consultation and voluntary coordination of action between individual parties has become particularly urgent at the present junction when continued isolation may lead to a slackening of mutual understanding, and at times, even to serious blunders.

10

A Geopolitical Discourse with Robert McNamara

Gearóid Ó Tuathail

from *Geopolitics* (2000)

Robert McNamara is a former Secretary of Defense who served in the administrations of Presidents Kennedy and Johnson. In 1968 he left the Johnson administration to become President of the World Bank, a position he held until his retirement in 1981. Since then he has served on the board of various corporations, lent his voice to calls for a rethinking of America's nuclear strategy and, after considerable reluctance with respect to Vietnam, pursued international dialogues on the events and conflicts that occupied much of his time when in office.[1] Between 1987 and 1992 McNamara participated in five major oral history conferences on the Cuban Missile crisis that added significantly to the historical record. For McNamara, the conferences revealed the crisis as even more dangerous that he believed, underscoring the need to rethink the role of nuclear weapons in international politics.[2] Prompted by the questions of biographers, McNamara finally broke his silence on the Vietnam War in his well-publicized memoir *In Retrospect* in which he famously wrote, 'we were wrong, terribly wrong. We owe it to future generations to explain why.'[3] Remarkably, McNamara subsequently threw himself into an even more ambitious set of conferences with his former Vietnamese enemies between November 1995 and February 1998 that resulted in the work *Argument Without End: In Search of Answers to the Vietnam Tragedy.*[4] The volume summarizes the results of the U.S.–Vietnamese meetings and reproduces selected transcripts of the fascinating dialogue between the parties. Although composed by multiple authors, McNamara is the dominant author of the volume, framing all chapters with his own introduction and conclusion.

McNamara's career has inspired considerable journalistic analyses and scholarly historical evaluation. In the Kennedy administration, McNamara's work at the Pentagon received considerable acclaim in the press and in political circles, leading some to even suggest his name for Vice-President in Kennedy's anticipated re-election campaign. However, as the delusions and deceptions that sustained the Vietnam War in America deepened, the popular portrait of McNamara turned fiercely critical. The press dubbed the war 'McNamara's war' (a description he initially embraced) and he came to personify the arrogance and illusions that drove it. Pulitzer prize winning journalist David Halberstam's 1973 book *The Best and the Brightest* codified this critique in a devastating picture of McNamara as an arrogant technocrat with an obsession with numbers and control.[5] Deborah Shapley's 1993 *Promise and Power: The Life and Times of Robert McNamara* was not the first biography of McNamara but it is the definitive work so far, a work of solid scholarship.[6] Written with the help of a series of McNamara interviews, Shapley's work does not undermine the critical portrait of McNamara but does provide a fuller account of his public career, and the events and personalities that shaped it. The work, however, lacks a theory of power and is somewhat underdeveloped in its consideration of the evolution of McNamara's intellectual thought. Paul Hendrickson's 1996 award winning work *The Living and the Dead: Robert McNamara and the Five Lives of a Lost War* weaves a detailed Jungian influenced psychological portrait of McNamara's time in office with the tragic and scarred lives of five ordinary Americans damaged by the Vietnam War.[7] A beautifully written book, his themes expand and deepen the journalistic indictment of Halberstam with even greater attention to the nuances and deceptions of McNamara's tenure (McNamara did not co-operate with Hendrickson's study). Yet, powerful and honest as this work is, it perpetuates the personification of the war in McNamara's character at the expense of more structural explanations of the conflict and forces that drove it. Our meeting on the 11th of October 2000 was by design a brief one in his office at the Corning Corporation suite in downtown Washington D.C. [. . .]

GEOPOLITICAL POWER/KNOWLEDGE AND VIETNAM

For political geographers, one of the most interesting passages in McNamara's *In Retrospect* is where he admits that the top policy-makers in the Kennedy and later Johnson administration really knew very little about Vietnam as a place and region.[8] Rather, Vietnam was constituted and known as a location within the terms of the Cold War geopolitics. The place itself was overwhelmed by the role it came to occupy within a Cold War geopolitical script, a role that was neither inevitable nor, in hindsight, justified. McNamara's use of the term 'geopolitics' is somewhat confusing for he first appears to use it in opposition to 'Cold War' thinking at the global scale. He sets up his knowledge about Vietnam by beginning at the global scale and with the origins of the Cold War:

> My thinking about Southeast Asia in 1961 differed little from that of many Americans of my generation who had served in World War II and followed foreign affairs by reading the newspapers but lacked expertise in geopolitics and Asian affairs. Having spent three years helping turn back German and Japanese aggression only to witness the Soviet takeover of Eastern Europe following the war, I accepted the idea advanced by George F. Kennan, in his famous July 1947 'X' article in *Foreign* Affairs, that the West, led by the United States, must guard against Communist expansion through a policy of containment. I considered this a sensible basis for decisions about national security and the application of Western military force (p. 30).

From this premise McNamara proceeds to elaborate on the deeply anti-geographic assumption of Cold War ideology (though not described as such by McNamara), namely that Communism was monolithic. Again he de-personalizes his knowledge by contextualizing it within his interpretation of the dominant geopolitical consciousness of the time, a consciousness focused on power struggles at the global scale:

> Like most Americans, I saw Communism as monolithic. I believed the Soviets and Chinese were cooperating in trying to extend their hegemony. In hindsight, of course, it is clear that they had no unified strategy after the late 1950s. But their split grew slowly and only gradually became apparent. At the time, Communism still seemed on the march. Mao Zedong and his followers had controlled China since 1949 and had fought with North Korea against the West; Nikita Khrushchev had predicted Communist victory through 'wars of national liberation' in the Third World and had told the West, 'We will bury you.' His threat gained credibility when the USSR launched *Sputnik* in 1957, demonstrating its lead in space technology. The next year Khrushchev started turning up the heat on West Berlin. And now Castro had transformed Cuba into a Communist beachhead in our hemisphere. We felt beset and at risk. This fear underlay our involvement in Vietnam (p. 30).

Vietnam, in other words, was a location within a global game, part of a script whose previously significant locations were China, Korea, Cuba and West Berlin. It is only at this point that McNamara personalizes his account:

> I did not see the Communist danger as overwhelming, as did many people on the right. It was a threat I was certain could be dealt with, and I shared President Kennedy's sentiment when he called on America and the West to bear the burden of a long twilight struggle. 'Let every nation know,' he said in his inaugural address, 'whether it wishes us well or ill, that we shall pay any price, bear any burden, meet any hardship, support any friend, oppose any foe to assure the survival and the success of liberty' (p. 31).

McNamara then outlines what he did know about Vietnam at the time, that 'Ho Chi Minh, a Communist, had begun efforts to free the country from French rule after World War I.' McNamara recounts Japanese occupation, Ho Chi Minh's declaration of Vietnam's independence after Japan's surrender, and 'that the United States had acquiesced to France's return to Indochina for fear that a Franco-American split would make it harder to contain Soviet expansion in Europe' (p. 31). The exceedingly dubious premises of this scalar geopolitics – a campaign to block Vietnamese independence in order to shore up France and thus Western Europe – led the U.S. to subsidize the French colonial apparatus in Vietnam. Here McNamara switches scales from global to regional geopolitics describing a U.S. subsidized 'French military action' against 'Ho's forces, which were in turn supported by the Chinese.' But that regional geopolitics was almost inseparable from global geopolitics is evident as McNamara adds: 'And I knew that the United States viewed Indochina as a necessary part of our containment policy – an important bulwark in the Cold War' (p. 31).

The crucial (mis)identification of the political movements and struggles characterizing Asian politics in the 1950s by American policy-makers is noted by McNamara as he discusses the geopolitical orthodoxy of the time, that which seemed 'common sense' and 'obvious':

> It seemed obvious that the Communist movement in Vietnam was closely related to guerrilla insurgencies in Burma, Indonesia, Malaya, and the Philippines during the 1950s. We viewed these conflicts not as nationalistic movements – as they largely appear in hindsight – but as signs of a unified Communist drive for hegemony in Asia. This way of thinking had led Dean Acheson, President Truman's secretary of state, to call Ho Chi Minh 'the mortal enemy of native independence in Indochina' (p. 31).

McNamara then discusses how this Cold War geopolitical orthodoxy was given expression by Eisenhower:

> I also knew that the Eisenhower administration had accepted the Truman administration's view that Indochina's fall to Communism would threaten U.S. security. Although it had appeared unwilling to commit U.S. combat forces in the region, it had sounded the warning of the Communist threat there clearly and often. In April 1954, President Eisenhower made his famous prediction that if

Indochina fell, the rest of Southeast Asia would 'go over very quickly' like a 'row of dominoes.' He had added, 'The possible consequences of the loss are just incalculable to the free world.' That year our country assumed responsibility from France for protecting Vietnam south of the 1954 partition line (p. 31).

McNamara records Kennedy's commitment to this geopolitical orthodoxy, transitioning from what he 'knew' and was 'aware' of to what he himself believed (i.e. that very same orthodoxy):

> I was aware, finally, that during his years in the Senate, John F. Kennedy had echoed Eisenhower's assessment of Southeast Asia. 'Vietnam represents the cornerstone of the Free World in Southeast Asia,' he had said in a widely publicized speech in 1956. 'It is our offspring. We cannot abandon it, we cannot ignore its needs.' Two developments after I became secretary of defense reinforced my way of thinking about Vietnam: the intensification of relations between Cuba and the Soviets, and a new wave of Soviet provocations in Berlin. Both seemed to underscore the aggressive intent of Communist policy. In that context, the danger of Vietnam's loss and, through falling dominoes, the loss of all Southeast Asia made it seemed reasonable to consider expanding the U.S. effort in Vietnam (p. 32).

After this highlight summary of the dominant geopolitical narrative that rendered world politics meaningful to U.S. decision-makers, McNamara then conceded the disjuncture between this 'global' geopolitical knowledge about Vietnam as a Cold War 'bulwark' or 'domino,' and 'local' geographical knowledge of Vietnam as a place with a particular culture, history, people and sets of values. Again his argument is not one about personal ignorance alone but of relative elite decision-maker ignorance about Vietnam:

> None of this made me anything close to an East Asian expert, however. I had never visited Indochina, nor did I understand or appreciate its history, language, culture, or values. The same must be said, to varying degrees, about President Kennedy, Secretary of State Dean Rusk, National Security Adviser McGeorge Bundy, military adviser

Maxwell Taylor, and many others. When it came to Vietnam, we found ourselves setting policy for a region that was terra incognita (p. 32).

This lack of knowledge is accounted for in the following manner by McNamara:

> Worse, our government lacked experts for us to consult to compensate for our ignorance. When the Berlin crisis occurred in 1961 and during the Cuban Missile Crisis in 1962, President Kennedy was able to turn to senior people like Llewellyn Thompson, Charles Bohlen, and George Kennan, who knew the Soviets intimately. There were no senior officials in the Pentagon or State Department with comparable knowledge of Southeast Asia. I knew of only one Pentagon officer with counter-insurgency experience in the region – Col. Edward Lansdale, who had served as an adviser to Ramon Magsaysay in the Philippines and Diem in South Vietnam. But Lansdale was relatively junior and lacked broad geopolitical expertise (p. 32).

McNamara then goes on to make an argument that is, ironically, also made by David Halberstam: the U.S.'s lack of expertise was a legacy of McCarthyism. 'Without men of sophisticated, nuanced insights,' 'we badly misread China's objectives and mistook its bellicose rhetoric to imply a drive for regional hegemony. We also totally underestimated the nationalist appeal of Ho Chi Minh's movement. We saw him first as a Communist and only second as a Vietnamese nationalist' (p. 33).

McNamara's account of the geopolitical discourse that was used to constitute 'Vietnam' as a particular type of location, drama and 'stake' within a larger regional and global power struggle between Communism and the 'free world' raises a series of interesting questions. For critical geopoliticians, his account is a useful illustration of how Cold War geopolitics worked in an anti-geographical way. The scale of the determination of meaning was the global visions and paranoid fantasies of the geopolitical culture of the United States in its worldwide competition with the Soviet Union. Vietnam was never simply about Vietnam; the local and national scales of the country were always inevitably nested within a larger geo-political game at the regional and global scale. Here images like the 'domino theory' and the worldwide struggle against a 'monolithic Communism' 'on the

march' overwhelmed the local geography and history of a place like Vietnam.

The argument can, indeed, be taken further. It could be argued that the paranoid simplicities of Cold War discourse could work only by actively ignoring or suppressing local geographical and historical knowledge. It is this *question of the relationship of power to knowledge and ignorance* that provoked some controversy after McNamara's book was published. A number of critics disputed McNamara's claim that the U.S. government 'lacked expertise for us to consult to compensate for our ignorance.' One of these critics was Louis Sarris, a Vietnamese affairs analyst for the State Department in the 1960s, who wrote in the *New York Times* that McNamara's above claim is untrue. 'In fact, there was, from the earliest days of our involvement in Vietnam, a number of reliable analyses in the State Department, the C.I.A. and even the Defense Intelligence Agency, let alone information from American officials and journalists in the field and academic and military experts.'[9] Sarris places the question of power and authority not knowledge at the center of his argument: 'The basic problem was the unwillingness of McNamara and other top policy makers to accept the relevance of information with which they personally disagreed.' Sarris details what he considered the capitulation of Dean Rusk and the State Department to McNamara's Pentagon over the production of knowledge about the military situation in Vietnam. Under an agreement between the two, according to Sarris, the State Department would stop issuing independent assessments of the military picture in Vietnam. State Department appraisals of the war 'were off limits, and in most cases our reports were kept within the department' (p. 396).

Sarris's point about institutional power and knowledge dependency flows inside the U.S. state provoked a letter of response by McNamara in which he clarified his argument by narrowing it considerably. Sarris is wrong, he contended, because top U.S. decision-makers like himself did not have 'experienced senior advisors' who 'had associated both socially and professionally' with the top Vietnamese leadership like Thompson, Bohland and Kennan supposedly did with the Soviets. Elaborating the point later in a foot-note in *Argument Without End*, McNamara wrote: 'Who knew Ho Chi Minh? Who knew many – or any – of his colleagues? Where was there a single individual who both had a nuanced understanding of the North Vietnamese and their Chinese allies and the friendship and respect of President Kennedy and Johnson? The answer is obvious: There was no one.'[10] In refining his argument to such a degree, McNamara actually strengthens the connection between knowledge and power that is implicit in Sarris's contentions. The knowledge that would have counted, for McNamara, was the knowledge of those with social entrée to the world of Vietnamese Communist Party elites and U.S. presidents, a potential social power circle that was impossibly small.

This dispute between, on the one hand, McNamara's argument that 'we lacked experienced senior advisors' with knowledge of the Vietnamese and, on the other hand, the argument that what the president's advisors knew reflected the power structure they represented and the power they wielded is particularly consequential in considering *Argument Without End*. McNamara's commitment to a naturalistic scientific epistemology is evident in his framing of the 'basic question' of the Vietnamese–U.S. dialogues: 'In the light of what now can be learned from the historical record, what U.S. and Vietnamese decisions might have been different and what difference would they have made on the course of the war *if each side had judged the other side's intentions and capabilities more accurately?*' (p. 17, emphasis in original). The question is a matter of the fullness and accuracy of truth, and not a question of the relationship of knowledge to power for McNamara. The central thesis McNamara seeks to advance in the volume is that the Vietnam War was a tragic consequence of 'missed opportunities' and 'misperceptions' on *both sides*. The 'joint responsibility' is a contentious theme but one that McNamara insists upon. The initial Vietnamese position, as articulated by Tran Quang Co, first deputy Foreign Minister of Vietnamese government, is that 'McNamara's argument about the outbreak of the war being a result of "misjudgments, miscalculations and misinformation" about the other side is only applicable to the U.S. side. Vietnam had no choice but to fight' (p. 49).

Throughout the volume McNamara returns again and again to the 'misjudgments, miscalculations and misinformation' formula, suggesting that an 'accurate' perception would have avoided the 'tragedy.' In reviewing the volume in the *New York Review of Books*, the Asian historian Jonathan Mirsky writes:

> It is irritating to read Mr McNamara's repeated breast-beating about 'American ignorance of the history, language and culture of Vietnam.' A quick reading of the Pentagon Papers shows how well informed some of the intelligence agencies were from the late 1940s on. Apart from the volumous Chinese and French literature on Vietnam's history of fighting foreigners, which were available in translation, there were specialists in American universities, some of them writing in this journal during Mr McNamara's time in office, who disputed Washington's justification for the war. While not accurate in every detail, their analyses, if taken seriously in the White House, would have arrested, if not stopped, the war.[11]

Mirsky's point poses the same issue as Sarris: was it 'lack of knowledge' that accounted for the Vietnam tragedy as McNamara apparently contends, or the impermeability of a Cold War 'mindset' to 'counter-evidence' and 'heterodox knowledge' that is to blame? I would argue that much more consideration needs to be given to the latter. The whole political culture and power structure that sustained and perpetuated Cold War geopolitical discourse needs to be addressed. Historian Ernest May suggests the power of the contextual political climate when he argues that 'given the assumptions generally shared by Americans in the 1960s, it seems probable that any collection of men or women would have decided as did members of the Kennedy and Johnson administrations.'[12] Put as a variant of Heidegger's famous aphorism about language, geopolitical discourses have people rather than people having geopolitical discourses.

Since the question of geopolitical power/ knowledge is complex and contentious, I was unable to do little more than raise it briefly with McNamara. In response to critics like Sarris McNamara stated that 'they did not understand that I did not have contact with people in the bowels of the State Department. I had contact with [Llewellyn] "Tommy" Thompson, with Dean Rusk, with Chester Bowles, with George Ball.' He reiterated his claim that Thompson was the unsung hero of the Cuban Missile Crisis – because he told Kennedy he was wrong to assume Khrushchev would not remove the missiles from Cuba – and that 'there was no "Tommy" Thompson in relation to Vietnam and nobody has suggested the name of anybody' (p. 429). He vigorously rejected my suggestion that 'accurate knowledge' about Vietnam was in circulation at the time but was 'not let in' by top decision-makers. While there was much knowledge about Vietnam 'out there,' he noted, 'the people

Kennedy depended upon, the Tommy Thompson equivalents, were wrong' (McNamara did note, at this point, that one of the few who was right was John Kenneth Galbraith).

I tried a different tack to get at the question of power and knowledge suggesting that his claim that the U.S. could have withdrawn from Vietnam in late 1963 or late 1964 'went against the Cold War.' Kennedy's foreign-policy decision makers could have come to a decision to withdraw from Vietnam in their internal deliberations but these conclusions faced a geo-political power structure and culture that would not readily have accepted this intention to withdraw. Put differently, in order to get elected Kennedy had demonstrated himself a 'true believer' in Cold War discourse. To withdraw from Vietnam would have seen him having to fight the hegemony of the Cold War discourse he himself practiced and was defined by. McNamara conceded that this was 'half right' but responded, as he does in *Argument Without End*, that 'it is the responsibility of leaders to lead, not follow. It is their responsibility to resist the pressures of the majority if the majority is misinformed or fails to understand and properly evaluate the full range of options open to our country' (p. 397). He stressed the importance of strong leadership, adding that it would not have been easy. But 'if I'd know then what I know now I am absolutely positive I could have helped Kennedy or Johnson turn the public around. The fact is the media, the Congress, the academicians, the political leaders were all in favor of the US program in Vietnam, all through 1966 and most through 1967. They could have been turned around by a strong president without any question, in my mind.'

Robert McNamara's public career is a complex geopolitical life. In his memoir he wrote of President Johnson as 'towering, powerful, paradoxical figure' quoting Walt Whitman's poetry to the effect that he 'contained multitudes' (p. 98). He could have written similar words about himself (though he was very different from Johnson). McNamara is also a towering, powerful and paradoxical figure with more than a few contradictions and multitudes of his own. That he has much to answer for is without doubt. That he has also tried to answer, even if it is in ways critics find objectionable, is also beyond doubt. His is an examined life, and will undoubtedly continue to be so.

NOTES

1   McNamara's first publication was *The Essence of Security: Reflections in Office* (New York: Harper and Row, 1968) which collected his Cold War liberal thinking on security as development and not simply weapons systems. He wrote that 'for too long we have come to identify security with exclusively military phenomena and most particularly with military hardware. It just isn't so, and we need to accommodate ourselves to the facts of the matter if we want to see security survive and grow in the southern half of the globe' (p. 150). The problem, however, is that this was not the foreign policy practice of the administrations McNamara served. McNamara also collected his World Bank philosophy into two volumes, and wrote *Blundering Into Disaster* (New York: Pantheon, 1986) as an intervention into the debate about the nuclear weapons in the mid-1980s.

2   See James G. Blight and David Welsh, *On The Brink: Americans and Soviets Reexamine the Cuban Missile Crisis*. Second edition (New York: Hill and Wang, 1990).

3   Robert McNamara, *In Retrospect: The Tragedy and Lessons of Vietnam* (New York: Vintage, 1995), p. xx.

4   Robert McNamara, James Blight, Robert Brigham, Thomas Biersteker and Col. Herbert Schandler, *Argument Without End: In Search of Answers to the Vietnam Tragedy* (New York: Public Affairs, 1999).

5   David Halberstam, *The Best and the Brightest* (New York: Penguin, 1973).

6   Deborah Shapley, *Promise and Power: The Life and Times of Robert McNamara* (Boston: Little Brown, 1993).

7   Paul Hendrickson, *The Living and the Dead: Robert McNamara and the Five Lives of A Lost War* (New York: Vintage, 1997).

8   '[I]t is very hard, today, to recapture the innocence and confidence with which we approached Vietnam in the early days of the Kennedy administration. We knew very little about the region.' *In Retrospect*, p. 39.

9   Louis G. Sarris, 'McNamara's War and Mine,' *New York Times*, Op-Ed, 5 September 1995. Reprinted in *In Retrospect*, pp. 391–396.

10   *Argument Without End*, p. 429.

11   Jonathan Mirsky, 'The Never-Ending War,' *New York Review of Books*, XLVII, 9 (1999), pp. 54–63.

12   Ernest May, *'Lessons' of the Past: The Use and Misuse of History in American Foreign Policy* (New York: Oxford University Press, 1973), pp. 120–121. Quoted in *Argument Without End*, p. 7.

The Brezhnev Doctrine

Leonid Brezhnev
from *Pravda* (1968)

SOVEREIGNTY AND THE INTERNATIONALIST OBLIGATION OF SOCIALIST COUNTRIES

In connection with the events in Czechoslovakia the question of the relationship and interconnection between the socialist countries' national interests and their internationalist obligations has assumed particular urgency and sharpness. The measures taken jointly by the Soviet Union and other socialist countries to defend the social gains of the Czechoslovak people are of enormous significance for strengthening the socialist commonwealth, which is the main achievement of the international working class.

At the same time it is impossible to ignore the allegations being heard in some places that the actions of the five socialist countries contradict the Marxist-Leninist principle of sovereignty and the right of nations to self-determination. Such arguments are untenable primarily because they are based on an abstract, nonclass approach to the question of sovereignty and the right of nations to self-determination.

There is no doubt that the peoples of the socialist countries and the Communist Parties have and must have freedom to determine their country's path of development. However, any decision of theirs must damage neither socialism in their country nor the fundamental interests of the other socialist countries nor the worldwide workers' movement, which is waging a struggle for socialism. This means that every Communist Party is responsible not only to its own people but also to all the socialist countries and to the entire Communist movement. Whoever forgets this in placing sole emphasis on the autonomy and independence of Communist Parties lapses into one-sidedness, shirking his internationalist obligation.

The Marxist dialectic opposes one-sidedness; it requires that every phenomenon be examined in terms of both its specific nature and its overall connection with other phenomena and processes. Just as, in V. I. Lenin's words, someone living in a society cannot be free of that society, so a socialist state that is in a system of other states constituting a socialist commonwealth cannot be free of the common interests of that commonwealth.

The sovereignty of individual socialist countries cannot be counterposed to the interests of world socialism and the world revolutionary movement. V. I. Lenin demanded that all Communists "struggle *against* petty national narrowness, exclusivity and isolation, and for taking into account the whole, the overall situation, for subordinating the interests of the particular to the interests of the general" (*Complete Collected Works* [in Russian], Vol. XXX, p. 45).

Socialist states have respect for the democratic norms of international law. More than once they have proved this in practice by resolutely opposing imperialism's attempts to trample the sovereignty and independence of peoples. From this same standpoint they reject left-wing, adventurist notions of "exporting revolution" and "bringing bliss" to other peoples. However, in the Marxist conception the norms of law, including the norms governing relations among socialist countries, cannot be interpreted in a narrowly formal way, outside the general context of the class struggle in the present-day world.

Socialist countries resolutely oppose the export and import of counterrevolution. Each Communist Party is free in applying the principles of Marxism-Leninism and socialism in its own country, but it cannot deviate from these principles (if, of course, it remains a Communist Party). In concrete terms this means primarily that every Communist Party cannot fail to take into account in its activities such a decisive fact of our time as the struggle between the two antithetical social systems – capitalism and socialism. This struggle is an objective fact that does not depend on the will of people and is conditioned by the division of the world into two antithetical social systems. "Every person," V. I. Lenin said, "must take either this, our, side or the other side. All attempts to avoid taking sides end in failure and disgrace" (Vol. XLI, p. 401).

It should be stressed that even if a socialist country seeks to take an "extrabloc" position, it in fact retains its national independence thanks precisely to the power of the socialist commonwealth – and primarily to its chief force, the Soviet Union – and the might of its armed forces. The weakening of any link in the world socialist system has a direct effect on all socialist countries, which cannot be indifferent to this. Thus, the antisocialist forces in Czechoslovakia were in essence using talk about the right to self-determination to cover demands for so-called neutrality and the CSR's [Czechoslovak Socialist Republic] withdrawal from the socialist commonwealth. But implementation of such "self-determination," i.e., Czechoslovakia's separation from the socialist commonwealth, would run counter to Czechoslovakia's fundamental interests and would harm the other socialist countries. Such "self-determination," as a result of which NATO troops might approach Soviet borders and the commonwealth of European socialist countries would be dismembered, in fact infringes on the vital interests of these countries' peoples, and fundamentally contradicts the right of these peoples to socialist self-determination. The Soviet Union and other socialist states, in fulfilling their international duty to the fraternal peoples of Czechoslovakia and defending their own socialist gains, had to act and did act in resolute opposition to the antisocialist forces in Czechoslovakia.

Comrade W. Gomulka, First Secretary of the Central Committee of the Polish United Workers' Party, used a metaphor to illustrate this point:

> To those friends and comrades of ours from other countries who believe they are defending the just cause of socialism and the sovereignty of peoples by denouncing and protesting the introduction of our troops in Czechoslovakia, we reply: If the enemy plants dynamite under our house, under the commonwealth of socialist states, our patriotic, national and internationalist duty is to prevent this using any means that are necessary.

People who "disapprove" of the actions taken by the allied socialist countries ignore the decisive fact that these countries are defending the interests of worldwide socialism and the worldwide revolutionary movement. The socialist system exists in concrete form in individual countries that have their own well-defined state boundaries and develops with regard for the specific attributes of each such country. And no one interferes with concrete measures to perfect the socialist system in various socialist countries. But matters change radically when a danger to socialism itself arises in a country. World socialism as a social system is the common achievement of the working people of all countries, it is indivisible, and its defense is the common cause of all Communists and all progressive people on earth, first and foremost the working people of the socialist countries.

The Bratislava statement of the Communist and Workers' Parties on socialist gains says that "it is the common internationalist duty of all socialist countries to support, strengthen and defend these gains, which were achieved at the cost of every people's heroic efforts and selfless labor."

What the right-wing, antisocialist forces were seeking to achieve in Czechoslovakia in recent months was not a matter of developing socialism in an original way or of applying the principles of Marxism-Leninism to specific conditions in that country, but was an encroachment on the foundations of socialism and the fundamental principles of Marxism-Leninism. This is the "nuance" that is still incomprehensible to people who trusted in the hypocritical cant of the antisocialist and revisionist elements.Under the guise of "democratization" these elements were shattering the socialist state step by step; they sought to demoralize the Communist Party and dull the minds of the masses; they were gradually preparing for a counterrevolutionary coup and at the same time were not being properly rebuffed inside the country.

The Communists of the fraternal countries naturally could not allow the socialist states to remain idle in the name of abstract sovereignty while the country was endangered by anti-socialist degeneration.

The five allied socialist countries' actions in Czechoslovakia are consonant with the fundamental interests of the Czechoslovak people themselves. Obviously it is precisely socialism that, by liberating a nation from the fetters of an exploitative system, ensures the solution of fundamental problems of national development in any country that takes the socialist path. And by encroaching on the foundations of socialism, the counterrevolutionary elements in Czechoslovakia were thereby undermining the basis of the country's independence and sovereignty.

The formal observance of freedom of self-determination in the specific situation that had taken shape in Czechoslovakia would signify freedom of "self-determination" not for the people's masses and the working people, but for their enemies. The antisocialist path, the "neutrality" to which the Czechoslovak people were being prodded, would lead the CSR straight into the jaws of the West German revanchists and would lead to the loss of its national independence. World imperialism, for its part, was trying to export counterrevolution to Czechoslovakia by supporting the antisocialist forces there.

The assistance given to the working people of the CSR by the other socialist countries, which prevented the export of counterrevolution from the outside, is in fact a struggle for the Czechoslovak Socialist Republic's sovereignty against those who would like to deprive it of this sovereignty by delivering the country to the imperialists.

Over a long period of time and with utmost restraint and patience, the fraternal Communist Parties of the socialist countries took political measures to help the Czechoslovak people to halt the antisocialist forces' offensive in Czechoslovakia. And only after exhausting all such measures did they undertake to bring in armed forces.

The allied socialist countries' soldiers who are in Czechoslovakia are proving in deeds that they have no task other than to defend the socialist gains in that country. They are not interfering in the country's internal affairs, and they are waging a struggle nor in words but in deeds for the principles of self-determination of Czechoslovakia's peoples, for their inalienable right to decide their destiny themselves after profound and careful consideration, without intimidation by counterrevolutionaries, without revisionist and nationalist demagoguery.

Those who speak of the "illegality" of the allied socialist countries' actions in Czechoslovakia forget that in a class society there is and can be no such thing as nonclass law. Laws and the norms of law are subordinated to the laws of the class struggle and the laws of social development. These laws are clearly formulated in the documents jointly adopted by the Communist and Workers' Parties.

The class approach to the matter cannot be discarded in the name of legalistic considerations. Whoever does so and forfeits the only correct, class-oriented criterion for evaluating legal norms begins to measure events with the yardsticks of bourgeois law. Such an approach to the question of sovereignty means, for example, that the world's progressive forces could not oppose the revival of neo-Nazism in the FRG [Federal Republic of Germany], the butcheries of Franco and Salazar or the reactionary outrages of the "black colonels" in Greece, since these are the "internal affairs" of "sovereign states." It is typical that both the Saigon puppets and their American protectors concur completely in the notion that sovereignty forbids supporting the struggle of the progressive forces. After all, they shout from the housetops that the socialist states that are giving aid to the Vietnamese people in their struggle for independence and freedom are violating Vietnam's sovereignty. Genuine revolutionaries, as internationalists, cannot fail to support progressive forces in all countries in their just struggle for national and social liberation. The interests of the socialist commonwealth and the entire revolutionary movement and the interests of socialism in Czechoslovakia demand full exposure and political isolation of the reactionary forces in that country, consolidation of the working people and consistent fulfillment of the Moscow agreement between the Soviet and Czechoslovak leaders.

There is no doubt that the actions taken in Czechoslovakia by the five allied socialist countries in Czechoslovakia, actions aimed at defending the fundamental interests of the socialist commonwealth and primarily at defending Czechoslovakia's independence and sovereignty as a socialist state, will be increasingly supported by all who really value the interests of the present-day revolutionary movement, the peace and security of peoples, democracy and socialism.

12

Geopolitics and Discourse

Practical Geopolitical Reasoning in American Foreign Policy

Gearóid Ó Tuathail and John Agnew
from *Political Geography Quarterly* (1992)

The Cold War, Mary Kaldor recently noted, has always been a discourse, a conflict of words, "capitalism" versus "socialism" (Kaldor, 1990). Noting how Eastern Europeans always emphasize the power of words, Kaldor adds that the way we describe the world, the words we use, shape how we see the world and how we decide to act. Descriptions of the world involve geographical knowledge and Cold War discourse has had a regularized set of geographical descriptions by which it represented international politics in the post-war period. The simple story of a great struggle between a democratic "West" against a formidable and expansionist East has been the most influential and durable geopolitical script of this period. This story, which today appears outdated, was a story which played itself out not in Central Europe but in exotic "Third-World" locations, from the sands of the Ogaden in the Horn of Africa, to the mountains of El Salvador, the jungles of Vietnam and the valleys of Afghanistan. Of course, the plot was not always a simple one. It has been complex and nuanced, making the post-war world a dynamic, dramatic and some-times ironic one. [. . .] Yet the story was a compelling one which brought huge military-industrial complexes into existence on both sides of the "East-West" divide and rigidly disciplined the possibilities for alternative political practices throughout the world. All regional conflicts, up until very recently, were reduced to its terms and its logic. Now with this story's unraveling and its geography blurring, it is time to ask how did the Cold War in its geopolitical guise come into existence and work?

This paper [. . .] attempts to establish a conceptual basis for answering [this question]. It seeks to outline a re-conceptualization of geopolitics in terms of discourse and apply this to the general case of American foreign policy. Geopolitics, some will argue, is, first and foremost, about practice and not discourse; it is about actions taken against other powers, about invasions, battles and the deployment of military force. Such practice is certainly geopolitical but it is only through discourse that the building up of a navy or the decision to invade a foreign country is made meaningful and justified. It is through discourse that leaders act, through the mobilization of certain simple geographical understandings that foreign-policy actions are explained and through ready-made geo-graphically infused reasoning that wars are rendered meaningful. How we understand and constitute our social world is through the socially structured use of language. Political speeches and the like afford us a means of recovering the self-understandings of influential actors in world politics. They help us understand the social construction of worlds and the role of geographical knowledge in that social construction.

The paper is organized into two parts. The first part attempts to sketch a theory of geopolitics by employing the concept of discourse. Four suggestive theses on the implications of conceptualizing geopolitics in discursive terms are briefly outlined. The second part addresses the question of American geopolitics and provides an account of some consistent features of the practical geopolitical reasoning by which American foreign policy has sought to write a geography of international politics. This latter part involves a detailed analysis of two of the most famous texts on the origins of the Cold War: George Kennan's "Long Telegram" of 1946 and his "Mr X" article in 1947. The irony of these influential geopolitical representations of the USSR is that they were not concrete geographical representations but over determined and ahistorical abstractions. It is the anti-geographical quality of geopolitical reasoning that this paper seeks to illustrate.

GEOPOLITICS AND DISCOURSE

Geopolitics, as many have noted, is a term which is notoriously difficult to define. In conventional academic understanding geopolitics concerns the geography of international politics, particularly the relationship between the physical environment (location, resources, territory, etc.) and the conduct of foreign policy. Within the geopolitical tradition the term has a more precise history and meaning. A consistent historical feature of geopolitical writing, from its origins in the late nineteenth century to its modern use by Colin Gray and others, is the claim that geopolitics is a foil to idealism, ideology and human will. This claim is a long-standing one in the geopolitical tradition which from the beginning was opposed to the proposition that great leaders and humans will alone determine the course of history, politics and society. Rather, it was the natural environment and the geographical setting of a state which exercised the greatest influence on its destiny.

By its own understandings and terms, geopolitics is taken to be a domain of hard truths, material realities and irrepressible natural facts. Geopoliticians have traded on the supposed objective materialism of geopolitical analysis. According to Gray (1988: 93), "geopolitical analysis is impartial as between one or another political system or philosophy". It addresses the base of international politics, the permanent geopolitical realities around which the play of events in international politics unfolds. These geopolitical realities are held to be durable, physical determinants of foreign policy. Geography, in such a scheme, is held to be a non-discursive phenomenon: it is separate from the social, political and ideological dimensions of international politics.

The great irony of geopolitical writing, however, is that it was always a highly ideological and deeply politicized form of analysis. Geopolitical theory from

Ratzel to Mackinder, Haushofer to Bowman, Spykman to Kissinger was never an objective and disinterested activity but an organic part of the political philosophy and ambitions of these very public intellectuals. While the forms of geopolitical writing have varied among these and other authors, the practice of producing geopolitical theory has a common theme: the production of knowledge to aid the practice of statecraft and further the power of the state.

Within political geography, the geopolitical tradition has long been opposed by a tradition of resistance to such reasoning. A central problem that has dogged such resistance is its lack of a coherent and comprehensive theory of geopolitical writing and its relationship to the broader spatial practices that characterize the operation of international politics. This paper proposes such a theory by re-conceptualizing the conventional meaning of geopolitics using the concept of discourse. Our foundational premise is the contention that geography is a social and historical discourse which is always intimately bound up with questions of politics and ideology (Ó Tuathail, 1989). Geography is never a natural, non-discursive phenomenon which is separate from ideology and outside politics. Rather, geography as a discourse is a form of power/knowledge itself.

Geopolitics, we wish to suggest, should be critically re-conceptualized as a discursive practice by which intellectuals of statecraft "spatialize" international politics in such a way as to represent it as a "world" characterized by particular types of places, peoples and dramas. In our understanding, the study of geopolitics is the study of the spatialization of international politics by core powers and hegemonic states. This definition needs careful explication.

The notion of discourse has become an important object of investigation in contemporary critical social science, particularly that which draws inspiration from the writings of the French philosopher Michel Foucault (1980). Within the discipline of international relations, there has been a series of attempts to incorporate the notion of discourse into the study of the practices of international politics. [. . .] Discourses are best conceptualized as sets of capabilities people have, as sets of socio-cultural resources used by people in the construction of meaning about their world and their activities. It is NOT simply speech or written statements but the rules by which verbal speech and written statements are made meaningful. Discourses enable one to write, speak, listen and act meaningfully. They are a set of capabilities, an ensemble of rules by which readers/listeners and speakers/audiences are able to take what they hear and read and construct it into an organized, meaningful whole. [. . .]

Discourses, like grammars, have a virtual and not an actual existence. They are not overarching constructs in the way that "structures" are sometimes represented. Rather, they are real sets of capabilities whose existence we infer from their realizations in activities, texts and speeches. Neither are they absolutely deterministic. Discourses enable. One can view these capabilities or rules as permitting a certain bounded field of possibilities and reasoning as the process by which certain possibilities are actualized. [. . .] Discourses are never static but are constantly mutating and being modified by human practice. The study of geopolitics in discursive terms, therefore, is the study of the sociocultural resources and rules by which geographies of international politics get written.

The notion of "intellectuals of statecraft" refers to a whole community of state bureaucrats, leaders, foreign-policy experts and advisors throughout the world who comment upon, influence and conduct the activities of statecraft. Ever since the development of the modern state system in the sixteenth century there has been a community of intellectuals of statecraft. Up until the twentieth century this community was rather small and restricted, with most intellectuals also being practitioners of statecraft. In the twentieth century, however, this community has become quite extensive and internally specialized. Within the larger states at least, one can differentiate between types of intellectuals of statecraft on the basis of their institutional setting and style of reasoning. Within civil society there are "defense intellectuals" associated with particular defense contractors and weapons systems. There is also a specialized community of security intellectuals in various public think-tanks. [. . .] One finds a different form of intellectualizing from public intellectuals of statecraft such as Henry Kissinger or Zbigniew Brzezinski who, as former top governmental officials, command a wide audience for their opinions in national newspapers and foreign-policy journals. Within political society itself there are different gradations amongst the foreign-policy community from those who design, articulate and order foreign policy from the top to those actually charged with implementing particular foreign policies and practicing statecraft (whether diplomatic or military) on a daily basis. [. . .]

We wish to propose four theses which follow from our preliminary observations on reasoning processes and intellectuals of statecraft. The first of these is that the study of geopolitics as we have defined it involves the comprehensive study of statecraft as a set of social practices. Geopolitics is not a discrete and relatively contained activity confined only to a small group of "wise men" who speak in the language of classical geopolitics. Simply to describe a foreign-policy problem is to engage in geopolitics, for one is implicitly and tacitly normalizing a particular world. One could describe geopolitical reasoning as the creation of the backdrop or setting upon which "international politics" takes place, but such would be a simplistic view. The creation of such a setting is itself part of world politics. This setting itself is more than a single backdrop but an active component part of the drama of world politics. To designate a place is not simply to define a location or setting. It is to open up a field of possible taxonomies and trigger a series of narratives, subjects and appropriate foreign-policy responses. Merely to designate an area as Islamic is to designate an implicit foreign policy (Said, 1981). Simply to describe a different or indeed the same place as "Western" (e.g. Egypt) is silently to operationalize a competing set of foreign-policy operators. Geopolitical reasoning begins at a very simple level and is a pervasive part of the practice of international politics. It is an innately political process of representation by which the intellectuals of statecraft designate a world and "fill" it with certain dramas, subjects, histories and dilemmas. All statespersons engage in the practice; it is one of the norms of the world political community.

Our second thesis is that most geopolitical reasoning in world politics is of a practical and not a formal type. Practical geopolitical reasoning is reasoning by means of consensual and unremarkable assumptions about places and their particular identities. This is the reasoning of practitioners of statecraft, of statespersons, politicians and military commanders. This is to be contrasted with the formal geopolitical reasoning of strategic thinkers and public intellectuals (such as those founding the "geopolitical tradition") who work in civil society and produce a highly codified system of ideas and principles to guide the conduct of statecraft. The latter forms of knowledge tend to have highly formalized rules of statement, description and debate. By contrast, practical geopolitical reasoning tends to be of a common-sense type which relies on the narratives and binary distinctions found in societal mythologies.

In the case of colonial discourse, there are contrasts between white and non-white, civilized and backward, Western and non-Western, adult and child. [. . .]

Our third thesis is that the study of geopolitical reasoning necessitates studying the production of geographical knowledge within a particular state and throughout the modern world system. Geographical knowledge is produced at a multiplicity of different sites throughout not only the nation-state, but the world political community. From the classroom to the living-room, the newspaper office to the film studio, the pulpit to the presidential office, geographical knowledge about a world is being produced, reproduced and modified. The challenge for the student of geopolitics is to understand how geographical knowledge is transformed into the reductive geopolitical reasoning of intellectuals of statecraft. How are places reduced to security commodities, to geographical abstractions which need to be "domesticated", controlled, invaded or bombed rather than understood in their complex reality? How, for example, did Truman metamorphose the situation in Greece in March 1947 – it was the site of a complex civil war at the time – into the Manichean terms of the Truman Doctrine? The answer we suspect is rather ironic given the common-sense meaning of geography as "place facts": geopolitical reasoning works by the active suppression of the complex geographical reality of places in favor of controllable geopolitical abstractions.

Our fourth thesis concerns the operation of geopolitical reasoning within the context of the modern world-system. Throughout the history of the modern world-system, intellectuals of statecraft from core states – particularly those states which are competing for hegemony – have disproportionate influence and power over how international political space is represented. A hegemonic world power, such as the United States in the immediate post-war period, is by definition a "rule-writer" for the world community. Concomitant with its material power is the power to represent world politics in certain ways. Those in power within the institutions of the hegemonic state become the deans of world politics, the administrators, regulators and geographers of international affairs. Their power is a power to constitute the terms of geopolitical world order, an ordering of international space which defines the central drama of international politics in particularistic ways. Thus not only can they represent in their own terms particular regional conflicts, whose causes may be quite localized (e.g. the

Greek civil war), but they can help create conditions whereby peripheral and semi-peripheral states actively adopt and use the geopolitical reasoning of the hegemon. [. . .]

PRACTICAL GEOPOLITICAL REASONING IN AMERICAN FOREIGN POLICY

Given our re-conceptualization of geopolitics, any analysis of American geopolitics must necessarily be more than an analysis of the formal geopolitical reasoning of a series of "wise men" of strategy (Mahan, Spykman, Kissinger and others). American geopolitics involves the study of the different historical means by which US intellectuals of statecraft have spatialized international politics and represented it as a "world" characterized by particular types of places, peoples and dramas. Such is obviously a vast undertaking and we wish to make but three general observations on the contours of American geopolitical reasoning. Before doing so, however, it is important to note two factors about the American case. First, we must acknowledge the key role the Presidency plays in the assemblage of meaning about international politics within the United States (and internationally since the US became a world power). In ethnographic terms, the US President is the chief bricoleur of American political life, a combination of storyteller and tribal shaman. One of the great powers of the Presidency, invested by the sanctity, history and rituals associated with the institution – the fact that the media take their primary discursive cues from the White House – is the power to describe, represent, interpret and appropriate. It is a formidable power but not an absolute power, for the art of description and appropriation (e.g. President Reagan's representation of the Nicaraguan contras as the "moral equivalents of the founding fathers") must have resonances with the Congress, the established media and the American public. The generation of such resonances often requires the repetition and re-cycling of certain themes and images even though the socio-historical context of their use may have changed dramatically. One has the attempted production of continuity by the incorporation of "strategic terms", "key metaphors" and "key symbols" into geopolitical reasoning. Behind all of these is the assumption of a power of appropriateness in the use of certain relatively fixed terms and phrases.

Secondly, we must recognize that American involvement with world politics has followed a distinctive cultural logic or set of presuppositions and orientations, what Gramsci called "Americanismo" (Gramsci, 1971). In particular, economic freedom – in the form of "free" business activity and the political conditions necessary for this – has been a central element in American culture. This has given rise to an attempt to reconstruct foreign places in an American image. US foreign-policy experiences with Mexico, China, Central America, the Caribbean and the Philippines all bear witness to this fundamental feature of US foreign policy (Agnew, 1983).

The first of our three observations on practical geopolitical reasoning in American foreign policy is that representations of "America" as a place are pervasively mythological. "America" is a place which is at once real, material and bounded (a territory with quiddity) yet also a mythological, imaginary and universal ideal with no specific spatial bounds. Ever since early modern times, North America and the Caribbean have had the transgressive aura of a place "beyond the line" [. . .] where might made right and the European treaties did not apply. [. . .]

Secondly, there is a tension between a universal omnipresent image of "America" and a different spatially-bounded image of the place. On one hand, American discourse consistently plays upon the unique geographical location of "America" yet simultaneously asserts that the principles of this "New World" are universal and not spatially confined there. The geography evoked in the American Declaration of Independence was not continental or hemispheral but universal. Its concern was with "the earth", the "Laws of Nature and of Nature's God", and all of "mankind". In this universalist vision, "America" is positioned as being equivalent with the strivings of a universal human nature. "The cause of America", Paine (1969: 23) proclaimed, "is in a great measure the cause of all mankind". The freedoms it struggles for are, in Reagan's terms, the freedoms desired by "all the peoples of the world". "America" is at once a territorially-defined state and a universal ideal, a place on the North American continent and a mythical homeland of freedom.

For the late eighteenth and most of the nineteenth century, the spatially-bounded sense of "America" was the one that predominated in US foreign-policy rhetoric. Even though the United States had closer economic, cultural and political ties with Europe than any other place, its foreign-policy rhetoric defined it as a separate and distinct sphere. "Europe", George Washington observed in his farewell address (1796), "has a set of primary interests which to us have none or a very remote relation. Hence she must be engaged in frequent controversies, the causes of which are essentially foreign to our concerns" (Richardson, 1905, vol. 1: 214). Washington's geopolitical reasoning was largely a negative one which defined the American sphere as extra-European (like Persia and Turkey) rather than a system complete and to itself. [. . .] The unilateral declaration of what later became known as the Monroe Doctrine affirmed such a position, stating that the political system of the European powers is different from that of America. Therefore, the United States would "consider any attempt on their part to extend their system to any portion of this hemisphere as dangerous to our peace and safety". An "American hemisphere", of course, was an arbitrary social construct – for the United States can be located in many different hemispheres, depending on where one decides to center them (e.g. a Northern hemisphere, a so-called Western hemisphere or a predominantly land hemisphere: see Boggs, 1945). Such geopolitical reasoning was imaginary and the putative bonds of affinity between the Latin republics of South America and the white Anglo-Saxon republic of the North equally imaginary.

By the late nineteenth century, the increasing wealth and power of the US state, together with the scramble for colonies among the European powers, produced a foreign policy which subordinated the hemispheral identity of the United States to universalist themes and identities concerning race, civilization and Christianity. [. . .] The United States was beginning to consider itself a "world power" with "principles" that were no longer qualified as contingently applicable to the "American hemisphere". McKinley and Theodore Roosevelt's racial script was followed by Woodrow Wilson's crusade for what he and US political culture took to be democracy. That Wilsonian internationalism did not succeed was partly due to the re-invigoration of the mythology that an isolationist "America" is the true and pure "America". Yet while the United States in the 1930s steered clear of political alliances with the rest of the world, its business enterprises continued their long-standing economic expansionism overseas. By the time of the Truman Doctrine, the US no longer conceptualized itself as *a* world power but as *the* world power. The geopolitical reasoning of Truman [. . .] was abstract and universal. Containment had no clearly conceptualized geographical limitations. Its genuine space was the abstract universal isotropic plane wherein right does perpetual battle with wrong, liberty with totalitarianism and Americanism with the forces of un-Americanism.

A third feature of American discourse is the strong lines it draws between the space of the "Self" and the space of the "Other" (Dalby, 1988). Like the cultural maps of many nations, American political discourse is given shape by a frontier which separates civilization from savagery in Turner's (1920) terms or an "Iron Curtain" marking the free world from the "evil empire". Robertson (1980: 92) notes:

> Frontiers and lines are powerful symbols for Americans. The moving frontier was never only a geographical line: it was a palpable barrier which separated the wilderness from civilization. It distinguished Americans, with their beliefs and their ideals, from savages and strangers, those "others" who could not be predicted or trusted. It divided the American nation from other nations, and marked its independence.

While such a point is valid, one can overstate the uniquely American character of this practice. Early European experiences, particularly the Iberian reconquista against the "infidel" and the English colonial experience with "heathens" in Ireland, were factors in the formation of imperialism as a "way of life" in the United States (Meinig, 1986; Williams, 1980). European discourses on colonialism, we have already noted, found their way into US foreign-policy practice not only in Theodore Roosevelt's time but even in determining the shape of the postwar world.

The processes of geopolitical world ordering in US foreign policy in the late 1940s are worthy of some detailed examination. [. . .] Let us consider the case of the two most famous American texts of that period, the "Long Telegram" and "Mr X" texts of George Kennan. The figure of George Kennan looms large in the annals of American foreign policy for it was Kennan who helped codify and constitute central elements of what became Cold War discourse. Kennan himself was, as Stephanson (1989: 157) observes, a man of the North, one to whom the vast heterogeneous area of the Third World was "a foreign space, wholly lacking in allure and best left to its own no doubt tragic fate". The crucial division in the world for Kennan and

the many others who made up the Atlanticist security community was that between the West and the East, between the world of maritime trading democracies and the Oriental world of xenophobic modern despotism. Trained at Princeton and in Germany and Estonia, Kennan developed something of an Old World *Weltanschauung* and brought this to bear in his early analyses of the USSR and world politics when working at the US Embassy in Moscow and later as Head of the Policy Planning Staff in Washington DC. In Kennan's two texts one can find at least three different strategies by which the USSR is represented. Each is worth exploring in detail.

The USSR as Oriental

Orientalism is premised, as Said (1978: 12) notes, on a primitive geopolitical awareness of the globe as composed of two unequal worlds, the Orient and the Occident. For Kennan and the Cold War discourse he helped codify, the USSR is part of the "Other" world, the Oriental world. In his famous "Long Telegram" Kennan describes the Soviet government as pervaded by an atmosphere of Oriental secretiveness and conspiracy. In the "Mr X" article published in *Foreign Affairs* in July 1947 he expounds on his thesis that the "political personality of Soviet power" is "the product of ideology and circumstances", the latter being the stamp of Russia's history and geography:

> The very teachings of Lenin himself require great caution and flexibility in the pursuit of communist purposes. Again, these precepts are fortified by the lessons of Russian history: of centuries of obscure battles between nomadic forces over the stretches of a vast unfortified plain. Here caution, circumspection, flexibility and deception are the valuable qualities, and their value finds natural appreciation in the Russian or oriental mind. (Kennan, 1947: 574).

In an earlier passage, Kennan had noted the paranoia of Soviet leaders. "Their particular brand of fanaticism", he noted, "was too fierce and too jealous to envisage any permanent sharing of power". In a revealing sentence he then noted: "From the Russian-Asiatic world out of which they had emerged they carried with them a skepticism as to the possibilities of permanent and peaceful coexistence of rival forces"

(Kennan, 1947: 570). Pietz (1988) notes that the Cold War discourse Kennan helped shape was "post-colonialist" in the sense that it drew upon and was assembled from many familiar and pervasive colonial discourses such as Orientalism and the putative primitiveness of non-Western regions and spaces. Totalitarianism, the theoretical anchor of Cold War discourse, came to be known as "nothing other than traditional Oriental despotism plus modern police technology" (Pietz, 1988: 58).

The USSR as potential rapist

Another pre-existent source from which Cold War discourse and representations of the USSR were assembled was patriarchal mythology – particularly that concerning fables of female vulnerability, rape and guardianship. In the descriptions being constructed around the USSR and Communism at this time the image of penetration was frequently evoked. The leaders of the USSR were a "frustrated" and "discontented" lot who "found in Marxist theory a highly convenient rationalization for their own instinctive desires" (Kennan, 1947: 569). Marxism was only a "fig leaf of moral and intellectual responsibility which cloaked essentially naked instinctive desires". These instinctive desires produced Soviet "aggressiveness" (another favorite Cold War description of the USSR) and "fluid and constant pressure to extend the limits of Russian police power which are together the natural and instinctive urges of Russian rulers" (Kennan, 1946: 54).

In the face of this instinctive behavior, the US needed to be aware that the USSR "cannot be charmed or talked out of existence" (Kennan, 1947: 576). The USSR was a wily and flexible power that would employ a variety of different "tactical maneuvers" (e.g. peaceful co-existence) to woo the West, particularly a vulnerable and psychologically-weakened Western Europe which was disposed to wishful thinking. Given this situation, the policy of the United States needed to be "that of a long-term, patient but firm and vigilant containment of Russian expansive tendencies" (Kennan, 1947: 575). The United States needed to act as the tough masculine guardian of Western Europe. If the policy of "adroit and vigilant application of counter-force at a series of constantly shifting geographical and political points, corresponding to the shirts and maneuvers of Soviet policy" was patiently

followed by the United States, then the weaknesses of the Soviet Union itself would become apparent. Turning the sexual grid of intelligibility on the USSR itself, Kennan (1947: 578) wrote that as long as the deficiencies that characterize Soviet society are not corrected,

> Russia will remain economically a vulnerable, and in a certain sense an impotent, nation, capable of exporting its enthusiasms and of radiating the strange charm of its primitive political vitality but unable to back up those articles of export by the real evidence of material power and prosperity. [. . .]

The Red flood

In tandem with the patriarchal mythology described above, one also had the recurring representation of Soviet foreign policy and Communism as a flood. The image of the Red flood was a particularly powerful element in fascist mythology during the inter-war period where, as Theweleit (1987: 230) chronicles in Weimar Germany, the powerful metaphor "engenders a clearly ambivalent state of excitement. It is threatening but also attractive [. . .]". Many different elements are at play here: situations and boundaries are fluid, solid ground becomes soft and swampy, barriers are breached, repressed instincts come bursting forth – water and sea as symbolic of the unconscious, the undisciplined id – and conditions are unrestrained, anarchic and dangerous. The response of the Freikorps, in Theweleit's account, is to act as firm, erect dams against this anarchic degeneration of society. With both feet securely planted on solid ground, they contained the Red flood and brought death to all that flowed. The very foundations of society, after all, were under attack. Switching to Kennan's Mr X article, we find the following graphic passage which defines the very nature of the Soviet threat to Western Europe:

> It's [the USSR's] political action is a fluid stream which moves constantly, wherever it is permitted to move, towards a given goal. Its main concern is to make sure that it has filled every nook and cranny available to it in the basin of world power. But if it finds unassailable barriers in its path, it accepts these philosophically and accommodates itself to them.

The main thing is that there should always be pressure, unceasing constant pressure, towards the desired goal. (Kennan, 1947: 575)

The image of the flood, which has also a sexual dimension (unrestrained, gushing desire, etc.), is critical, for it is by this means that the geography of containment becomes constituted. If the Soviet threat has the characteristics of a flood then one needs firm and vigilant containment along all of the Soviet border. Containment is thus constituted as a virtually global and not singularly Western European task. Effective containment in Western Europe, so the scenario goes, will lead to increasing Soviet pressure on the Middle East and Asia which eventually could result in the USSR spilling out into one or more of these regions. Such an image is easily reinforced by appropriate cartographic visuals featuring bleeding red maps of the USSR spreading outwards, or menacingly penetrating arrows busily trying to break out. The explanation of why US security managers instinctively read the North Korean invasion of South Korea as an act of Soviet expansionism certainly must address the power of such pre-existent images and scenarios. The formal geopolitical reasoning found in the different strategies of containment (Gaddis, 1982) rested, we suspect, on the flimsy foundations of widely shared practical geopolitical preconceptions.

CONCLUSION

[. . .] The irony of practical geopolitical representations of place is that in order to succeed they actually necessitate the abrogation of genuine geographical knowledge about the diversity and complexity of places as social entities. Describing the USSR then (or Iraq today) as Orientalist is a work of geographical abstractionism. A complex, diverse and heterogeneous social mosaic of places is hypostatized into a singular over-determined and predictable actor. As a consequence therefore the United States was put in the ironic situation of being simultaneously tremendously geographically ignorant of the USSR and [Saddam Hussein's] Iraq yet preoccupied with that state and its influence in world politics. [. . .] Contemporary geography, in deconstructing its own vocabulary and critically exploring the forms of practical geopolitical reasoning that circulate within states, can [. . .] help create descriptions of the world based not on reductive geopolitical reasoning but on critical geographical knowledge.

REFERENCES

Agnew, J. A. (1983) An excess of "national exceptionalism": towards a new political geography of American foreign policy. *Political Geography Quarterly* 2, 151–366.

Boggs, S. (1945) This hemisphere. *Department of State Bulletin* 6 May.

Dalby, S. (1988) Geopolitical discourse. The Soviet Union as Other. *Alternatives* 13, 415–422.

Foucault, M. (1980) *Power/Knowledge*. New York: Pantheon.

Gaddis, J. L. (1982) *Strategies of Containment*. New York: Oxford University Press.

Gramsci, A. (1971) *Selections from the Prison Notebooks*. New York: International Publishers.

Gray, C. (1988) *The Geopolitics of Superpower*. Lexington: University of Kentucky Press.

Kaldor, M. (1990) After the Cold War. *New Left Review* 80, 25–37.

Kennan, G. (1946) The "Long Telegram". In *Containment: Documents on American Foreign Policy and Strategy, 1945–1950* (T. Etzol and J. Gaddis eds), pp. 50–63. New York: Columbia University Press.

Kennan, G. ["Mr X"] (1947) The sources of Soviet conduct. *Foreign Affairs* 25, 566–582.

Kennan, G. (1967) *Memoirs 1925–19 50*. Boston: Little Brown.

Lafeber, W. (1963) *The New Empire: An Interpretation of American Expansionism, 1860–1898*. Ithaca, NY: Cornell University Press.

Meinig, D. (1986) *The Shaping of America. Volume 1: Atlantic America, 1492–1800*. New Haven: Yale University Press.

Ó Tuathail, G. (1989) Critical geopolitics: the social construction of place and space in the practice of statecraft. Unpublished PhD thesis, Syracuse University.

Paine, T. (1969) *The Essential Thomas Paine*. New York: Mentor.

Pietz, W. (1988) The "post-colonialism" of Cold War discourse. *Social Text* 19/20, 55–75.

Reagan, R. (1988) Peace and Democracy for Nicaragua. Address to the nation on February 2, 1988. *Department of State Bulletin* 88(2133), 32–35.

Richardson, J. (1905) *A Compilation of Messages and Papers of the Presidents, 1789–1902*, 12 vols. Washington DC: Bureau of National Literature and Art.

Robertson, J. O. (1980) *American Myth, American Reality*. New York: Hill and Wang.

Said, E. (1978) *Orientalism*. New York: Vintage Books.

Said, E. (1981) *Covering Islam*. New York: Vintage Books.

Stephanson, A. (1989) *George Kennan and the Art of Foreign Policy*. Boston: Harvard University Press.

Theweleit, K. (1987) *Male Fantasies. Volume 1: Women, Floods, Bodies, History*. Minneapolis: University of Minnesota Press.

Turner, F. J. (1920) *The Frontier of American History*. New York: Henry Holt.

Whitaker, A. (1954) *The Western Hemisphere Idea: Its Rise and Decline*. Ithaca, NY: Cornell University Press.

Williams, W. A. (1980) *Empire as a Way of Life*. Oxford: Oxford University Press.

13

Common Sense and the Common Danger

Policy Statement of the Committee on the Present Danger

from *Alerting America: The Papers of the Committee on the Present Danger* (1984)

I

Our country is in a period of danger, and the danger is increasing. Unless decisive steps are taken to alert the nation, and to change the course of its policy, our economic and military capacity will become inadequate to assure peace with security.

The threats we face are more subtle and indirect than was once the case. As a result, awareness of danger has diminished in the United States, in the democratic countries with which we are naturally and necessarily allied, and in the developing world.

There is still time for effective action to ensure the security and prosperity of the nation in peace, through peaceful deterrence and concerted alliance diplomacy. A conscious effort of political will is needed to restore the strength and coherence of our foreign policy; to revive the solidarity of our alliances; to build constructive relations of cooperation with other nations whose interests parallel our own – and on that sound basis to seek reliable conditions of peace with the Soviet Union, rather than an illusory detente.

Only on such a footing can we and the other democratic industrialized nations, acting together, work with the developing nations to create a just and progressive world economy – the necessary condition of our own prosperity and that of the developing nations and Communist nations as well. In that framework, we shall be better able to promote human rights, and to help deal with the great and emerging problems of food, energy, population, and the environment.

II

The principal threat to our nation, to world peace, and to the cause of human freedom is the Soviet drive for dominance based upon an unparalleled military buildup.

The Soviet Union has not altered its long-held goal of a world dominated from a single center – Moscow. It continues, with notable persistence, to take advantage of every opportunity to expand its political and military influence throughout the world: in Europe; in the Middle East and Africa; in Asia; even in Latin America; in all the seas.

The scope and sophistication of the Soviet campaign have been increased in recent years, and its tempo quickened. It encourages every divisive tendency within and among the developed states and between the developed and the underdeveloped world. Simultaneously, the Soviet Union has been acquiring a network of positions including naval and air bases in the Southern Hemisphere which support its drive for dominance in the Middle East, the Indian Ocean, Africa, and the South Atlantic.

For more than a decade, the Soviet Union has been enlarging and improving both its strategic and its conventional military forces far more rapidly than the United States and its allies. Soviet military power and its rate of growth cannot be explained or justified by considerations of self-defense. The Soviet Union is consciously seeking what its spokesmen call "visible preponderance" for the Soviet sphere. Such preponderance, they explain, will permit the Soviet Union "to transform the conditions of world politics" and determine the direction of its development.

The process of Soviet expansion and the worldwide deployment of its military power threaten our interest in the political independence of our friends and allies, their and our fair access to raw materials, the freedom of the seas, and in avoiding a preponderance of adversary power.

These interests can be threatened not only by direct attack, but also by envelopment and indirect aggression. The defense of the Middle East, for example, is vital to the defense of Western Europe and Japan. In the Middle East the Soviet Union opposes those just settlements between Israel and its Arab neighbors which are critical to the future of the area. Similarly, we and much of the rest of the world are threatened by renewed coercion through a second round of Soviet-encouraged oil embargoes.

III

Soviet expansionism threatens to destroy the world balance of forces on which the survival of freedom depends. If we see the world as it is, and restore our will, our strength and our self-confidence, we shall find resources and friends enough to counter that threat. There is a crucial moral difference between the two superpowers in their character and objectives. The United States – imperfect as it is – is essential to the hopes of those countries which desire to develop their societies in their own ways, free of coercion.

To sustain an effective foreign policy, economic strength, military strength, and a commitment to leadership are essential. We must restore an allied defense posture capable of deterrence at each significant level and in those theaters vital to our interests. The goal of our strategic forces should be to prevent the use of, or the credible threat to use, strategic weapons in world politics; that of our conventional forces, to prevent other forms of aggression directed against our interests. Without a stable balance of forces in the world and policies of collective defense based upon it, no other objective of our foreign policy is attainable.

As a percentage of Gross National Product, US defense spending is lower than at any time in twenty-five years. For the United States to be free, secure and influential, higher levels of spending are now required for our ready land, sea, and air forces, our strategic deterrent, and, above all, the continuing modernization of those forces through research and development. The increased level of spending required is well within our means so long as we insist on all feasible efficiency in our defense spending. We must also expect our allies to bear their fail-share of the burden of defense.

From a strong foundation, we can pursue a positive and confident diplomacy, addressed to the full array of our economic, political and social interests in world politics. It is only on this basis that we can expect successfully to negotiate hardheaded and verifiable agreements to control and reduce armaments.

If we continue to drift, we shall become second best to the Soviet Union in overall military strength; our alliances will weaken; our promising rapprochement with China could be reversed. Then we could find ourselves isolated in a hostile world, facing the unremitting pressures of Soviet policy backed by an overwhelming preponderance of power. Our national survival itself would be in peril, and we should face, one after another, bitter choices between war and acquiescence under pressure.

IV

We are Independents, Republicans and Democrats who share the belief that foreign and national security policies should be based upon fundamental considerations of the nation's future and well being, not that of one faction or party.

We have faith in the maturity, good sense and fortitude of our people. But public opinion must be informed before it can reach considered judgments and make them effective in our democratic system. Time, weariness, and the tragic experience of Vietnam have weakened the bipartisan consensus which sustained our foreign policy between 1940 and the mid-1960s. We must build a fresh consensus to expand the opportunities and diminish the dangers of a world in flux.

We have therefore established the Committee on the Present Danger to help promote a better understanding of the main problems confronting our foreign policy, based on a disciplined effort to gather the facts and a sustained discussion of their significance for our national security and survival.

14

Appeal for European Nuclear Disarmament (END)

END Committee

from *Protest and Survive* (1980)

We are entering the most dangerous decade in human history. A third world war is not merely possible, but increasingly likely. Economic and social difficulties in advanced industrial countries, crisis, militarism and war in the third world compound the political tensions that fuel a demented arms race. In Europe, the main geographical stage for the East-West confrontation, new generations of ever more deadly nuclear weapons are appearing.

For at least twenty-five years, the forces of both the North Atlantic and the Warsaw alliance have each had sufficient nuclear weapons to annihilate their opponents, and at the same time to endanger the very basis of civilized life. But with each passing year, competition in nuclear armaments has multiplied their numbers, increasing the probability of some devastating accident or miscalculation.

As each side tries to prove its readiness to use nuclear weapons, in order to prevent their use by the other side, new, more "usable" nuclear weapons are designed and the idea of "limited" nuclear war is made to sound more and more plausible. So much so that this paradoxical process can logically only lead to the actual use of nuclear weapons.

Neither of the major powers is now in any moral position to influence smaller countries to forgo the acquisition of nuclear armament. The increasing spread of nuclear reactors and the growth of the industry that installs them, reinforce the likelihood of worldwide proliferation of nuclear weapons, thereby multiplying the risks of nuclear exchanges.

Over the years, public opinion has pressed for nuclear disarmament and detente between the con-tending military blocs. This pressure has failed. An increasing proportion of world resources is expended on weapons, even though mutual extermination is already amply guaranteed. This economic burden, in both East and West, contributes to growing social and political strain, setting in motion a vicious circle in which the arms race feeds upon the instability of the world economy and vice versa: a deathly dialectic.

We are now in great danger. Generations have been born beneath the shadow of nuclear war, and have become habituated to the threat. Concern has given way to apathy. Meanwhile, in a world living always under menace, fear extends through both halves of the European continent. The powers of the military and of internal security forces are enlarged, limita-tions are placed upon free exchanges of ideas and between persons, and civil rights of independent minded individuals are threatened, in the West as well as the East.

We do not wish to apportion guilt between the political and military leaders of East and West. Guilt lies squarely upon both parties. Both parties have adopted menacing postures and committed aggressive actions in different parts of the world.

The remedy lies in our own hands. We must act together to free the entire territory of Europe, from Poland to Portugal, from nuclear weapons, air and submarine bases, and from all institutions engaged in research into or manufacture of nuclear weapons. We ask the two super powers to withdraw all nuclear weapons from European territory. In particular, we ask the Soviet Union to halt production of SS 20 medium-range missiles and we ask the United States not to implement the decision to develop cruise missiles and Pershing II missiles for deployment in Western Europe. We also urge the ratification of the SALT II agreement, as a necessary step towards the renewal of effective negotiations on general and complete disarmament.

At the same time, we must defend and extend the right of all citizens, East or West, to take part in this common movement and to engage in every kind of exchange. We appeal to our friends in Europe, of every faith and persuasion, to consider urgently the ways in which we can work together for these common objectives. We envisage a European-wide campaign, in which every kind of exchange takes place; in which representatives of different nations and opinions confer and co-ordinate their activities; and in which less formal exchanges, between universities, churches, women's organizations, trade unions, youth organizations, professional groups and individuals, take place with the object of promoting a common object: to free all of Europe from nuclear weapons.

We must commence to act as if a united, neutral and pacific Europe already exists. We must learn to be loyal, not to "East" or "West", but to each other, and we must disregard the prohibitions and limitations imposed by any national state.

It will be the responsibility of the people of each nation to agitate for the expulsion of nuclear weapons and bases from European soil and territorial waters, and to decide upon its own means and strategy, concerning its own territory. These will differ from one country to another, and we do not suggest that any single strategy should be imposed. But this must be part of a trans-continental movement in which every kind of exchange takes place.

We must resist any attempt by the statesmen of East or West to manipulate this movement to their own advantage. We offer no advantage to either NATO or the Warsaw Alliance. Our objectives must be to free Europe from confrontation, to enforce detente between the United States and the Soviet Union, and, ultimately, to dissolve both great power alliances.

In appealing to fellow-Europeans, we are not turning our backs on the world. In working for the peace of Europe we are working for the peace of the world. Twice in this century Europe has disgraced its claims to civilization by engendering world war. This time we must repay our debts to the world by engendering peace.

This appeal will achieve nothing if it is not supported by determined and inventive action, to win more people to support it. We need to mount an irresistible pressure for a Europe free of nuclear weapons.

We do not wish to impose any uniformity on the movement nor to pre-empt the consultations and decisions of those many organizations already exercising their influence for disarmament and peace. But the situation is urgent. The dangers steadily advance. We invite your support for this common objective, and we shall welcome both your help and advice.

15

The End of History?

Francis Fukuyama
from *The National Interest* (1989)

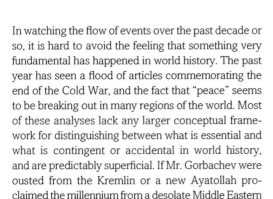

In watching the flow of events over the past decade or so, it is hard to avoid the feeling that something very fundamental has happened in world history. The past year has seen a flood of articles commemorating the end of the Cold War, and the fact that "peace" seems to be breaking out in many regions of the world. Most of these analyses lack any larger conceptual framework for distinguishing between what is essential and what is contingent or accidental in world history, and are predictably superficial. If Mr. Gorbachev were ousted from the Kremlin or a new Ayatollah proclaimed the millennium from a desolate Middle Eastern capital, these same commentators would scramble to announce the rebirth of a new era of conflict.

And yet, all of these people sense dimly that there is some larger process at work, a process that gives coherence and order to the daily headlines. The twentieth century saw the developed world descend into a paroxysm of ideological violence, as liberalism contended first with the remnants of absolutism, then bolshevism and fascism, and finally an updated Marxism that threatened to lead to the ultimate apocalypse of nuclear war. But the century that began full of self-confidence in the ultimate triumph of Western liberal democracy seems at its close to be returning full circle to where it started: not to an "end of ideology" or a convergence between capitalism and socialism, as earlier predicted, but to an unabashed victory of economic and political liberalism.

The triumph of the West, of the Western idea, is evident first of all in the total exhaustion of viable systematic alternatives to Western liberalism. In the past decade, there have been unmistakable changes in the intellectual climate of the world's two largest communist countries, and the beginnings of significant reform movements in both. But this phenomenon extends beyond high politics and it can be seen also in the ineluctable spread of consumerist Western culture in such diverse contexts as the peasants' markets and color television sets now omnipresent throughout China, the cooperative restaurants and clothing stores opened in the past year in Moscow, the Beethoven piped into Japanese department stores, and the rock music enjoyed alike in Prague, Rangoon, and Tehran.

What we may be witnessing is not just the end of the Cold War, or the passing of a particular period of postwar history, but the end of history as such: that is, the end point of mankind's ideological evolution and the universalization of Western liberal democracy as the final form of human government. This is not to say that there will no longer be events to fill the pages of *Foreign Affairs'* yearly summaries of international relations, for the victory of liberalism has occurred primarily in the realm of ideas or consciousness and is as yet incomplete in the real or material world. But there are powerful reasons for believing that it is the ideal that will govern the material world *in the long run*. To understand how this is so, we must first consider some theoretical issues concerning the nature of historical change.

The notion of the end of history is not an original one. Its best known propagator was Karl Marx, who believed that the direction of historical development was a purposeful one determined by the interplay of material forces, and would come to an end only with the achievement of a communist Utopia that would finally resolve all prior contradictions. But the concept of history as a dialectical process with a beginning, a middle, and an end was borrowed by Marx from his great German predecessor, Georg Wilhelm Friedrich Hegel.

For better or worse, much of Hegel's historicism has become part of our contemporary intellectual baggage. The notion that mankind has progressed through a series of primitive stages of consciousness on his path to the present, and that these stages corresponded to concrete forms of social organization, such as tribal, slave-owning, theocratic, and finally democratic-egalitarian societies, has become inseparable from the modern understanding of man. Hegel was the first philosopher to speak the language of modern social science, insofar as man for him was the product of his concrete historical and social environment and not, as earlier natural right theorists would have it, a collection of more or less fixed "natural" attributes. The mastery and transformation of man's natural environment through the application of science and technology was originally not a Marxist concept, but a Hegelian one. Unlike later historicists whose historical relativism degenerated into relativism *tout court*, however, Hegel believed that history culminated in an absolute moment – a moment in which a final, rational form of society and state became victorious.

It is Hegel's misfortune to be known now primarily as Marx's precursor, and it is our misfortune that few of us are familiar with Hegel's work from direct study, but only as it has been filtered through the distorting lens of Marxism. In France, however, there has been an effort to save Hegel from his Marxist interpreters and to resurrect him as the philosopher who most correctly speaks to our time. Among those modern French interpreters of Hegel, the greatest was certainly Alexandre Kojeve, a brilliant Russian emigre who taught a highly influential series of seminars in Paris in the 1930s at the *Ecole Practique des Hautes Etudes*. While largely unknown in the United States, Kojeve had a major impact on the intellectual life of the continent. Among his students ranged such future luminaries as Jean-Paul Sartre on the Left and Raymond Aron on the Right; postwar existentialism borrowed many of its basic categories from Hegel via Kojeve.

Kojeve sought to resurrect the Hegel of the *Phenomenology of Mind*, the Hegel who proclaimed history to be at an end in 1806. For as early as this Hegel saw in Napoleon's defeat of the Prussian monarchy at the Battle of Jena the victory of the ideals of the French Revolution, and the imminent universalization of the state incorporating the principles of liberty and equality. Kojeve, far from rejecting Hegel in

light of the turbulent events of the next century and a half, insisted that the latter had been essentially correct. The Battle of Jena marked the end of history because it was at that point that the *vanguard* of humanity (a term quite familiar to Marxists) actualized the principles of the French Revolution. While there was considerable work to be done after 1806 – abolishing slavery and the slave trade, extending the franchise to workers, women, blacks, and other racial minorities, etc. – the basic *principles* of the liberal democratic state could not be improved upon. The two world wars in this century and their attendant revolutions and upheavals simply had the effect of extending those principles spatially, such that the various provinces of human civilization were brought up to the level of its most advanced outposts, and of forcing those societies in Europe and North America at the vanguard of civilization to implement their liberalism more fully.

The state that emerges at the end of history is liberal insofar as it recognizes and protects through a system of law man's universal right to freedom, and democratic insofar as it exists only with the consent of the governed. For Kojeve, this so-called "universal homogenous state" found real-life embodiment in the countries of postwar Western Europe – precisely those flabby, prosperous, self-satisfied, inward-looking, weak-willed states whose grandest project was nothing more heroic than the creation of the Common Market. But this was only to be expected. For human history and the conflict that characterized it was based on the existence of "contradictions": primitive man's quest for mutual recognition, the dialectic of the master and slave, the transformation and mastery of nature, the struggle for the universal recognition of rights, and the dichotomy between proletarian and capitalist. But in the universal homogenous state, all prior contradictions are resolved and all human needs are satisfied. There is no struggle or conflict over "large" issues, and consequently no need for generals or statesmen; what remains is primarily economic activity. And indeed, Kojeve's life was consistent with his teaching. Believing that there was no more work for philosophers as well, since Hegel (correctly understood) had already achieved absolute knowledge, Kojeve left teaching after the war and spent the remainder of his life working as a bureaucrat in the European Economic Community, until his death in 1968.

To his contemporaries at mid-century, Kojeve's proclamation of the end of history must have seemed like the typical eccentric solipsism of a French intellectual, coming as it did on the heels of World War II and at the very height of the Cold War. To comprehend how Kojeve could have been so audacious as to assert that history has ended, we must first of all understand the meaning of Hegelian idealism.

For Hegel, the contradictions that drive history exist first of all in the realm of human consciousness, i.e. on the level of ideas – not the trivial election year proposals of American politicians, but ideas in the sense of large unifying world views that might best be understood under the rubric of ideology. Ideology in this sense is not restricted to the secular and explicit political doctrines we usually associate with the term, but can include religion, culture, and the complex of moral values underlying any society as well. Hegel's view of the relationship between the ideal and the real or material worlds was an extremely complicated one, beginning with the fact that for him the distinction between the two was only apparent. He did not believe that the real world conformed or could be made to conform to ideological preconceptions of philosophy professors in any simple minded way, or that the "material" world could not impinge on the ideal. Indeed, Hegel the professor was temporarily thrown out of work as a result of a very material event, the Battle of Jena. But while Hegel's writing and thinking could be stopped by a bullet from the material world, the hand on the trigger of the gun was motivated in turn by the ideas of liberty and equality that had driven the French Revolution.

For Hegel, all human behavior in the material world, and hence all human history, is rooted in a prior state of consciousness – an idea similar to the one expressed by John Maynard Keynes when he said that the views of men of affairs were usually derived from defunct economists and academic scribblers of earlier generations. This consciousness may not be explicit and self-aware, as are modern political doctrines, but may rather take the form of religion or simple cultural or moral habits. And yet this realm of consciousness *in the long run* necessarily becomes manifest in the material world, indeed creates the material world in its own image. Consciousness is cause and not effect, and can develop autonomously from the material world; hence the real subtext underlying the apparent jumble of current events is the history of ideology.

Failure to understand that the roots of economic behavior lie in the realm of consciousness and culture leads to the common mistake of attributing material causes to phenomena that are essentially ideal in nature. For example, it is common place in the West to interpret the reform movements first in China and most recently in the Soviet Union as the victory of the material over the ideal – that is, a recognition that ideological incentives could not replace material ones in stimulating a highly productive modern economy, and that if one wanted to prosper one had to appeal to baser forms of self-interest. But the deep defects of socialist economies were evident thirty or forty years ago to any one who chose to look. Why was it that these countries moved away from central planning only in the 1980s? The answer must be found in the consciousness of the elites and leaders ruling them, who decided to opt for the "Protestant" life of wealth and risk over the "Catholic" path of poverty and security. That change was in no way made inevitable by the material conditions in which either country found itself on the eve of the reform, but instead came about as the result of the victory of one idea over another.

For Kojeve, as for all good Hegelians, understanding the underlying processes of history requires understanding developments in the realm of consciousness or ideas, since consciousness will ultimately remake the material world in its own image. To say that history ended in 1806 meant that mankind's ideological evolution ended in the ideals of the French or American Revolutions: while particular regimes in the real world might not implement these ideals fully, their theoretical truth is absolute and could not be improved upon. Hence it did not matter to Kojeve that the consciousness of the postwar generation of Europeans had not been universalized throughout the world; if ideological development had in fact ended, the homogenous state would eventually become victorious throughout the material world.

I have neither the space nor, frankly, the ability to defend in depth Hegel's radical idealist perspective. The issue is not whether Hegel's system was right, but whether his perspective might uncover the problematic nature of many materialist explanations we often take for granted. This is not to deny the role of material factors as such. To a literal minded idealist, human society can be built around any arbitrary set of principles regardless of their relationship to the material world. And in fact men have proven themselves able to endure the most extreme material hardships in the name of ideas that exist in the realm of the spirit alone, be it the divinity of cows or the nature of the Holy Trinity.

But while man's very perception of the material world is shaped by his historical consciousness of it, the material world can clearly affect in return the viability of a particular state of consciousness. In particular, the spectacular abundance of advanced liberal economies and the infinitely diverse consumer culture made possible by them seem to both foster and preserve liberalism in the political sphere. I want to avoid the materialist determinism that says that liberal economics inevitably produces liberal politics, because I believe that both economics and politics presuppose an autonomous prior state of conscious-ness that makes them possible. But that state of consciousness that permits the growth of liberalism seems to stabilize in the way one would expect at the end of history if it is underwritten by the abundance of a modern free market economy. We might summarize the content of the universal homogenous state as liberal democracy in the political sphere combined with easy access to VCRs [video-cassette recorders] and stereos in the economic.

Have we in fact reached the end of history? Are there, in other words, any fundamental "contradictions" in human life that cannot be resolved in the context of modern liberalism, that would be resolvable by an alternative political-economic structure? If we accept the idealist premises laid out above, we must seek an answer to this question in the realm of ideology and consciousness. Our task is not to answer exhaus-tively the challenges to liberalism promoted by every crackpot messiah around the world, but only those that are embodied in important social or political forces and movements, and which are therefore part of world history. For our purposes, it matters very little what strange thoughts occur to people in Albania or Burkina Faso, for we are interested in what one could in some sense call the common ideological heritage of mankind.

In the past century, there have been two major challenges to liberalism, those of fascism and of communism. The former saw the political weakness, materialism, anomie, and lack of community of the West as fundamental contradictions in liberal societies that could only be resolved by a strong state that forged a new "people" on the basis of national exclusiveness. Fascism was destroyed as a living ideology by World War II. This was a defeat, of course, on a very material level, but it amounted to a defeat of the idea as well. What destroyed fascism as an idea was not universal moral revulsion against it, since plenty of people were

willing to endorse the idea as long as it seemed the wave of the future, but its lack of success. After the war, it seemed to most people that German fascism as well as its other European and Asian variants were bound to self-destruct. There was no material reason why new fascist movements could not have sprung up again after the war in other locales, but for the fact that expansionist ultranationalism, with its promise of unending conflict leading to disastrous military defeat, had completely lost its appeal. The ruins of the Reich chancellory as well as the atomic bombs dropped on Hiroshima and Nagasaki killed this ideology on the level of consciousness as well as materially, and all of the proto-fascist movements spawned by the German and Japanese examples like the Peronist movement in Argentina or Subhas Chandra Bose's Indian National Army withered after the war.

The ideological challenge mounted by the other great alternative to liberalism, communism, was far more serious. Marx, speaking Hegel's language, asserted that liberal society contained a fundamental contradiction that could not be resolved within its context, that between capital and labor, and this contradiction has constituted the chief accusation against liberalism ever since. But surely, the class issue has actually been successfully resolved in the West. As Kojeve (among others) noted, the egalitarianism of modern America represents the essential achieve-ment of the classless society envisioned by Marx. This is not to say that there are not rich people and poor people in the United States, or that the gap between them has not grown in recent years. But the root causes of economic inequality do not have to do with the underlying legal and social structure of our society, which remains fundamentally egalitarian and moderately redistributionist, so much as with the cultural and social characteristics of the groups that make it up, which are in turn the historical legacy of premodern conditions. Thus black poverty in the United States is not the inherent product of liberalism, but is rather the "legacy of slavery and racism" which persisted long after the formal abolition of slavery.

As a result of the receding of the class issue, the appeal of communism in the developed Western world, it is safe to say, is lower today than any time since the end of World War I. This can be measured in any number of ways: in the declining membership and electoral pull of the major European communist parties, and their overtly revisionist programs; in the corresponding electoral success of conservative parties from Britain and Germany to the United States and Japan, which are unabashedly pro-market and anti-statist; and in an intellectual climate whose most "advanced" members no longer believe that bourgeois society is something that ultimately needs to be overcome. This is not to say that the opinions of progressive intellectuals in Western countries are not deeply pathological in any number of ways. But those who believe that the future must inevitably be socialist tend to be very old, or very marginal to the real political discourse of their societies.

One may argue that the socialist alternative was never terribly plausible for the North Atlantic world, and was sustained for the last several decades primarily by its success outside of this region. But it is precisely in the non-European world that one is most struck by the occurrence of major ideological transformations. Surely the most remarkable changes have occurred in Asia. Due to the strength and adaptability of the indigenous cultures there Asia became a battleground for a variety of imported Western ideologies early in this century. Liberalism in Asia was a very weak reed in the period after World War I; it is easy today to forget how gloomy Asia's political future looked as recently as ten or fifteen years ago. It is easy to forget as well how momentous the outcome of Asian ideological struggles seemed for world political development as a whole.

The first Asian alternative to liberalism to be decisively defeated was the fascist one represented by Imperial Japan. Japanese fascism (like its German version) was defeated by the force of American arms in the Pacific war, and liberal democracy was imposed on Japan by a victorious United States. Western capitalism and political liberalism when transplanted to Japan were adapted and transformed by the Japanese in such a way as to be scarcely recognizable. Many Americans are now aware that Japanese industrial organization is very different from that prevailing in the United States or Europe, and it is questionable what relationship the factional maneuvering that takes place with the governing Liberal Democratic Party bears to democracy. Nonetheless, the very fact that the essential elements of economic and political liberalism have been so successfully grafted onto uniquely Japanese traditions and institutions guarantees their survival in the long run. More important is the contribution that Japan has made in turn to world history by following in the footsteps of the United States to create a truly universal consumer culture that has become both a symbol and an underpinning of the universal homogenous state. V.S. Naipaul travelling in Khomeini's Iran shortly after the revolution noted the omnipresent signs advertising the products of Sony, Hitachi, and JVC, whose appeal remained virtually irresistible and gave the lie to the regime's pretensions of restoring a state based on the rule of the *Shariah*. Desire for access to the consumer culture, created in large measure by Japan, has played a crucial role in fostering the spread of economic liberalism throughout Asia, and hence in promoting political liberalism as well.

The economic success of the other newly industrializing countries (NICs) in Asia following on the example of Japan is by now a familiar story. What is important from a Hegelian standpoint is that political liberalism has been following economic liberalism, more slowly than many had hoped but with seeming inevitability. Here again we see the victory of the idea of the universal homogenous state. South Korea had developed into a modern, urbanized society with an increasingly large and well-educated middle class that could not possibly be isolated from the larger democratic trends around them. Under these circumstances it seemed intolerable to a large part of this population that it should be ruled by an anachronistic military regime while Japan, only a decade or so ahead in economic terms, had parliamentary institutions for over forty years. Even the former socialist regime in Burma, which for so many decades existed in dismal isolation from the larger trends dominating Asia, was buffeted in the past year by pressures to liberalize both its economy and political system. It is said that unhappiness with strongman Ne Win began when a senior Burmese officer went to Singapore for medical treatment and broke down crying when he saw how far socialist Burma had been left behind by its ASEAN neighbors.

But the power of the liberal idea would seem much less impressive if it had not infected the largest and oldest culture in Asia, China. The simple existence of communist China created an alternative pole of ideological attraction, and as such constituted a threat to liberalism. But the past fifteen years have seen an almost total discrediting of Marxism-Leninism as an economic system. Beginning with the famous third plenum of the Tenth Central Committee in 1978, the Chinese Communist party set about de-collectivizing agriculture for the 800 million Chinese who still lived in the countryside. The role of the state in agriculture was reduced to that of a tax collector, while production of consumer goods was sharply increased in order to give peasants a taste of the universal homogenous state and thereby an incentive to work. The reform doubled Chinese grain output in only five years, and in the process created for Deng Xiao-ping a solid political base from which he was able to extend the reform to other parts of the economy. Economic statistics do not begin to describe the dynamism, initiative, and openness evident in China since the reform began. China could not now be described in any way as a liberal democracy. At present, no more than 20 per cent of its economy has been marketized, and most importantly it continues to be ruled by a self-appointed Communist party which has given no hint of wanting to devolve power. Deng has made none of Gorbachev's promises regarding democratization of the political system and there is no Chinese equivalent of *glasnost*. The Chinese leadership has in fact been much more circumspect in criticizing Mao and Maoism than Gorbachev with respect to Brezhnev and Stalin, and the regime continues to pay lip service to Marxism-Leninism as its ideological underpinning. But anyone familiar with the outlook and behavior of the new technocratic elite now governing China knows that Marxism and ideological principle have become virtually irrelevant as guides to policy, and that bourgeois consumerism has a real meaning in that country for the first time since the revolution. The various slow-downs in the pace of reform, the campaigns against "spiritual pollution" and crackdowns on political dissent are more properly seen as tactical adjustments made in the process of managing what is an extraordinarily difficult political transition. By ducking the question of political reform while putting the economy on a new footing, Deng has managed to avoid the breakdown of authority that has accompanied Gorbachev's *perestroika*. Yet the pull of the liberal idea continues to be very strong as economic power devolves and the economy becomes more open to the outside world. There are currently over 20,000 Chinese students studying in the US and other Western countries, almost all of them the children of the Chinese elite. It is hard to believe that when they return home to run the country they will be content for China to be the only country in Asia unaffected by the larger democratizing trend. The student demonstrations in Beijing that broke out first in December 1986 and recurred recently on the occasion of Hu Yao-bang's death were only the beginning of what will inevitably be mounting pressure for change in the political system as well.

What is important about China from the standpoint of world history is not the present state of the reform or even its future prospects. The central issue is the fact that the People's Republic of China can no longer act as a beacon for illiberal forces around the world, whether they be guerrillas in some Asian jungle or middle class students in Paris. Maoism, rather than being the pattern for Asia's future, became an anachronism, and it was the mainland Chinese who in fact were decisively influenced by the prosperity and dynamism of their overseas co-ethnics – the ironic ultimate victory of Taiwan. Important as these changes in China have been, however, it is developments in the Soviet Union – the original "homeland of the world proletariat" – that have put the final nail in the coffin of the Marxist-Leninist alternative to liberal democracy.

What has happened in the four years since Gorbachev's coming to power is a revolutionary assault on the most fundamental institutions and principles of Stalinism, and their replacement by other principles which do not amount to liberalism per se but whose only connecting thread is liberalism. This is most evident in the economic sphere, where the reform economists around Gorbachev have become steadily more radical in their support for free markets, to the point where some like Nikolai Shmelev do not mind being compared in public to Milton Friedman.

[. . .]

The Soviet Union could in no way be described as a liberal or democratic country now, nor do I think that it is terribly likely that *perestroika* will succeed such that the label will be thinkable any time in the near future. But at the end of history it is not necessary that all societies become successful liberal societies, merely that they end their ideological pretensions of representing different and higher forms of human society. And in this respect I believe that something very important has happened in the Soviet Union in the past few years: the criticisms of the Soviet system sanctioned by Gorbachev have been so thorough and devastating that there is very little chance of going back to either Stalinism or Brezhnevism in any simple way. Gorbachev has finally permitted people to say what they had privately understood for many years, namely, that the magical incantations of Marxism-Leninism were nonsense, that Soviet socialism was not superior to the West in any respect but was in fact a monumental failure. The conservative opposition in the USSR, consisting both of simple workers afraid of unemployment and inflation and of party officials fearful of losing their jobs and privileges, is outspoken and may be strong enough to force Gorbachev's ouster in the next few years. But what both groups desire is tradition, order, and authority; they manifest no deep commitment to Marxism-Leninism, except insofar as they have invested much of their own lives in it. For authority to be restored in the Soviet Union after Gorbachev's demolition work, it must be on the basis of some new and vigorous ideology which has not yet appeared on the horizon.

If we admit for the moment that the fascist and communist challenges to liberalism are dead, are there any other ideological competitors left? Or put another way, are there contradictions in liberal society beyond that of class that are not resolvable? Two possibilities suggest themselves, those of religion and nationalism.

The rise of religious fundamentalism in recent years within the Christian, Jewish, and Muslim traditions has been widely noted. One is inclined to say that the revival of religion in some way attests to a broad un-happiness with the impersonality and spiritual vacuity of liberal consumerist societies. Yet while the empti-ness at the core of liberalism is most certainly a defect in the ideology – indeed, a flaw that one does not need the perspective of religion to recognize – it is not at all clear that it is remediable through politics. Modern liberalism itself was historically a consequence of the weakness of religiously-based societies which, failing to agree on the nature of the good life, could not provide even the minimal preconditions of peace and stability. In the contemporary world only Islam has offered a theocratic state as a political alternative to both liberalism and communism. But the doctrine has little appeal for non-Muslims, and it is hard to believe that the movement will take on any universal significance. Other less organized religious impulses have been successfully satisfied within the sphere of personal life that is permitted in liberal societies.

The other major "contradiction" potentially un-resolvable by liberalism is the one posed by nationalism and other forms of racial and ethnic consciousness. It is certainly true that a very large degree of conflict since the Battle of Jena has had its roots in nationalism. Two cataclysmic world wars in this century have been spawned by the nationalism of the developed world in various guises, and if those passions have been muted to a certain extent in postwar Europe, they are still extremely powerful in the Third World. Nationalism has been a threat to liberalism historically in Germany,

and continues to be one in isolated parts of "post-historical" Europe like Northern Ireland.

But it is not clear that nationalism represents an irreconcilable contradiction in the heart of liberalism. In the first place, nationalism is not one single phe-nomenon but several, ranging from mild cultural nostalgia to the highly organized and elaborately articulated doctrine of National Socialism. Only systematic nationalisms of the latter sort can qualify as a formal ideology on the level of liberalism or com-munism. The vast majority of the world's nationalist movements do not have a political program beyond the negative desire of independence from some other group or people, and do not offer anything like a comprehensive agenda for socio-economic organiza-tion. As such, they are compatible with doctrines and ideologies that do offer such agendas. While they may constitute a source of conflict for liberal societies, this conflict does not arise from liberalism itself so much as from the fact that the liberalism in question is incomplete. Certainly a great deal of the world's ethnic and nationalist tension can be explained in terms of peoples who are forced to live in unrepresentative political systems that they have not chosen.

While it is impossible to rule out the sudden appearance of new ideologies or previously unrecog-nized contradictions in liberal societies, then, the present world seems to confirm that the fundamental principles of socio-political organization have not advanced terribly far since 1806. Many of the wars and revolutions fought since that time have been undertaken in the name of ideologies which claimed to be more advanced than liberalism, but whose pre-tensions were ultimately unmasked by history. In the meantime, they have helped to spread the universal homogenous state to the point where it could have a significant effect on the overall character of international relations.

What are the implications of the end of history for international relations? Clearly, the vast bulk of the Third World remains very much mired in history, and will be a terrain of conflict for many years to come. But let us focus for the time being on the larger and more developed states of the world who, after all, account for the greater part of world politics.

[. . .]

The passing of Marxism-Leninism first from China and then from the Soviet Union will mean its death as a living ideology of world historical significance. For while there may be some isolated true believers left in places like Managua, Pyongyang, or Cambridge, Massachusetts, the fact that there is not a single large state in which it is a going concern undermines completely its pretensions to being in the vanguard of human history. And the death of this ideology means the growing "Common Marketization" of international relations, and the diminution of the likelihood of large-scale conflict between states.

This does not by any means imply the end of international conflict *per se*. For the world at that point would be divided between a part that was historical and a part that was post-historical. Conflict between states still in history, and between those states and those at the end of history, would still be possible. There would be still be a high and perhaps rising level of ethnic and nationalist violence, since those are impulses completely played out, even in parts of the post-historical world. Palestinians and Kurds, Sikhs and Tamils, Irish Catholics and Walloons, Armenians and Azeris, will continue to have their unresolved grievances. This implies that terrorism and wars of national liberation will continue to be an important item on the national agenda. But large-scale conflict must involve large states still caught in the grip of history, and they are what appear to be passing from the scene.

The end of history will be a very sad time. The struggle for recognition, the willingness to risk one's own life for a purely ideological struggle that called forth daring, courage, imagination, and idealism, will be replaced by economic calculation, the endless solving of technical problems, environmental concerns, and the satisfaction of sophisticated consumer demands. In the post-historical period there will be neither art nor philosophy, just the perpetual caretaking of the museum of human history. I can feel in myself, and see in others around me, a powerful nostalgia for the time when history existed. Such nostalgia, in fact, will continue to fuel competition and conflict even in the post-historical world for some time to come. Even though I recognize its inevitability, I have the most ambivalent feelings for the civilization that has been created in Europe since 1945, with its North Atlantic and Asian offshoots. Perhaps this very prospect of centuries of boredom at the end of history will serve to get history started once again.

Cartoon 4 Workers of the world

Turning an old communist slogan on its head, Matt Wuerker represents the 1989 revolutions in Eastern Europe as revolts against communist icons in favour of Western consumer icons. The workers of the world have become consumers of world commodities. The pot-belly on Ronald McDonald, however, indicates that the utopia of mass consumption is not what it seems but has ugly consequences, namely unhealthy and environmentally wasteful habits of consumption.

Source: M. Wuerker

PART THREE

Twenty-first Century Geopolitics

INTRODUCTION TO PART THREE

Gearóid Ó Tuathail

On May 9, 2005, world leaders gathered in Moscow's Red Square to commemorate the sixtieth anniversary of the end of World War II. US President Bush stood next to President Putin of Russia as a military parade commemorated the greatest military achievement of the Soviet Union, the repulsion of Nazi invasion and capture of Berlin by the Red Army. In a national address two weeks prior, Putin declared that "the biggest geopolitical catastrophe of the [20th] century" was the collapse of the Soviet Union (Kuchins, 2005). The sentiment was widely shared by many former citizens of the Soviet Union, though not by the Baltic states, which joined the European Union in 2004, nor by the victims of Stalinism. For many, the end of the Soviet Union was the beginning of a dramatic decline in their standard of living. Once the vanguard of modernity, Soviet infrastructure is now crumbling while its education, health and military systems are in poor condition across most former Soviet states. In Russia, retirees saw their pensions dwindle while state owned enterprises were sold off to well connected oligarchs, like Roman Abramovich (the owner of Chelsea football club), who accumulated personal fortunes from natural resources that had once belonged to all (Hoffman, 2002). Russia's life expectancy rate has dropped to 58 years for men (lower than Bangladesh) while emigration, declining births, and rising HIV/AIDS and tuberculosis rates threaten to diminish the state's population (149 million in 1991) to below 100 million by 2050 (Specter, 2004). Confronting the harsh realities of the diminished state capacity, income inequalities induced by oligarchic capitalism, and poor health, many older former Soviet citizens understandably yearn for the imagined strength and "greatness" of the Soviet Union.

While many former Soviet citizens suffered, citizens of the United States enjoyed the longest economic expansion in peacetime in the 1990s. "Globalization" was pronounced as the new defining drama in world politics as geoeconomic stories eclipsed geopolitical ones in public discourse (Friedman, 2000). With Cold War bureaucracies and an "iron triangle" entrenched at the heart of its political system, the United States recast itself as the "sole remaining superpower" in world affairs and continued spending money on its military-industrial complex, though it was not clear what "superpower" was to be used for in a world where the United States had no military competitors. Two interrelated questions faced US geopoliticans in the post-Cold War period: first, what were the new strategic challenges facing the United States and its allies and, second, how should the United States adapt its Cold War institutions to deal with them? The first set of answers to both questions was provided during the presidency of Bill Clinton (1993–2001), the second set during the presidency of George W. Bush (2001–2009). Due to a series of conjunctural factors, from Bush's controversial appointment as US president in 2000 to the 9/11 terrorist attacks, US foreign policy in the early twenty-first century veered dramatically under Bush's leadership from multilateralism to unilateralism, from engagement with a broad array of global dangers to narrow preoccupation with the threat of terrorism and war in Iraq (Daalder and Lindsey, 2003). The transformation of the United States from a stabilizing superpower to an aggressive one – repudiating international treaties, bypassing international institutions, circumventing international law, invading sovereign states, launching preemptive war and openly threatening other states – has propelled the conduct of US foreign policy under George W. Bush to the forefront of twenty-first century geopolitics. How and why has the United States become such a "rogue nation" (Prestowitz, 2003)? Are the US actions evidence of a "new imperialism" (Hardt and Negri, 2001; Harvey, 2003)? Many salient dramas characterize the geopolitics of the twenty-first century – global health pandemics, the economic rise of China, rising energy prices, climatic instability, state failure, genocide, and long-standing regional confrontations over Kashmir, Israel/Palestine and Taiwan – but the transformation of the United States into a "rogue superpower" in world affairs is a central feature of early twenty-first century geopolitics and the one we focus on in this part of the book.

THE CLINTON ADMINISTRATION AND NEOCONSERVATIVISM

In November 1992, the former governor of Arkansas, Bill Clinton, was elected president of the United States. Clinton's presidential campaign was organized around the slogan "It's the economy, stupid" and his victory, ending twelve years of Republican rule in the White House, marked the triumph of domestic economic issues over international geopolitical concerns. The Clinton administration conducted its foreign policy in a manner similar to the elder Bush administration it replaced, continuing its multilateral gradualism and emphasis on free trade initiatives (Clinton signed the NAFTA, the North American Free Trade Agreement into law). Like the elder Bush administration, it accepted European leadership of the response to the Bosnian war and refrained from intervening with force to bring the war to an end. Unlike the Bush administration, however, the Clinton administration supported funding for international family planning organizations that discussed and accepted abortion, overturning a "gag rule" imposed since Reagan. Seeking a successor to "containment" as an organizing strategy for US foreign policy, the Clinton administration proposed "enlargement" and pursued policies that expanded NATO in Central Europe, and emphasized "rule of law" and "good governance" in international institutions (Albright, 2003; Clinton, 2004; Mallaby, 2004).

From the moment of his election (with 43 percent of the vote in a three way contest), a certain segment of the American population was unreconciled to Clinton as president. Clinton was a southern Christian yet also liberal and pro-choice. He was committed to diversity in government and promoted "third way" policies like a progressive tax system, diversity in school curricula and secular values in public life. He also had a reputation for womanizing. These factors made Clinton an object of hate, particularly among "born again" Christians, hard-core conservative activists and chauvinistic nationalists in the military and beyond. This sentiment was organized and channeled by wealthy conservative donors into a concerted campaign against the Clinton presidency. By the middle of his second term, the Clinton presidency had become enmeshed in a sex scandal that was used by a Special Prosecutor to support an impeachment campaign against him. The impeachment vote did not succeed but the campaign polarized the American electorate with significant implications for the post-Clinton years.

Geopolitics was not immune to the raging "cultural wars" of the 1990s and two geopolitical discourses that emerged during the Clinton years were influential visions of twenty-first century geopolitics. Both were variants of neoconservativism, one pessimistic the other expansively ambitious, which rearticulated long-standing tendencies in US geopolitical culture in different ways. As noted in Part Two, the term neoconservative was used to describe those on the right who rejected Nixon–Kissinger's policy of détente. Unlike traditional conservativism, which was for limited government and suspicious of foreign entanglements, neoconservativism was a Wilsonianism of the right, embracing big (defense spending) government and a messianic mission to spread "freedom" worldwide, but not

Wilson's commitment to international institutions and multilateral means. In fact, neoconservatives regarded the United Nations with contempt, viewing it as a constraining and enfeebling influence on US "sovereignty." Eschewing traditional conservativism's respect for limits, neoconservatives advocated an activist imperial role for the United States as "the sole remaining superpower" that should be willing to confront "tyranny" abroad and unapologetically promote American interests and values. Unlike political realists, neoconservatives believed that moral values – their political version of them which they project as "universal values" – are central to world politics. Neoconservativism, in sum, was an imperial vision of the US role in world affairs, a "Wilsonianism in [marching] boots" (Ash, 2004: 116).

The first neoconservative discourse that gained prominence during the Clinton presidency was Samuel P. Huntington's "Clash of Civilizations" story (Reading 16). Published in *Foreign Affairs*, the journal of the Council on Foreign Relations, Huntington claimed that the great divisions among humankind and the dominating source of conflict in the future will be cultural. Nation states will remain the most powerful actors in world politics, but the principal conflicts in global politics will occur between nations and groups of different civilizations. The clash of civilizations, he argued, will dominate world politics (Huntington 1993: 22). Despite uncertainty about the exact number ("seven or eight"), Huntington claims that civilizations are primordial and "stretch back deep into history." Identity – "What are you?" – is a cultural given, he claims, that cannot be changed. Civilizational identities and conflicts between them, he adds, lead to "civilization rallying" as groups and states belonging to one civilization that become involved in a war with peoples from another civilization rally to the support of their own "kin and country."

The most fundamental of all civilizational clashes, the "central axis of world politics," for Huntington is the conflict between "the West and the Rest." Some state leaders are trying to join the West but their countries are torn between joining "Western civilization" and adhering to non-Western alternatives (Islam for Turkey, Slavic Orthodoxy for Russia and Ibero-American Catholicism for Mexico). Others are trying to challenge the West's primacy and dominance. While asserting the existence of primordial civilizations, Huntington invents a new hybrid cross-civilizational network which he terms "the Confucian–Islamic connection." This is a network of "weapon states" that are a dangerous Otherness against which "the West" must mobilize and act because, according to Huntington, they do not share Western concepts and goals like arms control and non-proliferation. Echoing Cold War style discourses of danger, Huntington argues that this new enemy is relentlessly building military power and threatening "Western interests, values and power."

Huntington's article is a classic example of the geopolitics genre. A central organizing plot of world politics is asserted as primordial (underlying all events) by the all-seeing geopolitician. The cultural complexity and geographical diversity of the planet is reduced into a few macro categories, in this case "civilizations" (the earth labeling game). Global space is stripped of its plurality and world politics of its multiplicity. The categorization scheme then becomes an explanatory system, with metaphors of "blocs," "faultlines," "clashes" and "torn countries" serving as supposed analytics of explanation. Threats and dangers are identified to the identity/space/interest championed by the geopolitician, and an "us" versus "them" narrative generates policy recommendations involving vigilance and military spending.

Huntington's narrative has many problems, and these reveal his neoconservative commitments. First, Huntington's concept of civilizations is fraught with contradictions. Civilizations are analytical artifacts not on-the-ground realities. Read Huntington closely to grasp the qualifiers he constructs around his notion of "civilizations": "Civilizations obviously blend and overlap" and "they rise and fall; they divide and merge." He wants to claim that "while the lines between them are seldom sharp, they are real." But if the lines are "seldom sharp," then this undermines his claims about fault *lines*, clashes and torn countries. The whole plot about clashing essentialized opposites is undermined if civilizations cannot be distinguished from each other (remember, he is not even certain how many there are). Rather than acknowledge interconnections, interdependence, cross-cultural learning and hybridity, Huntington asserts essentialized clashing civilizations ("they are real"), in particular the notion that "the West" is in a struggle against "the Rest." This is a standard neoconservative preconception, a vision of world politics as a "cultural war" between "the West" and Otherness, with Huntington's version sharing similarities to other fatalistic and "declinist" visions.

Second, Huntington's conception of identity is similarly contradictory. Again, read the article closely. "People can and do redefine their identities," he writes. Yet, a few pages later he writes: "In conflicts between civilizations, the question is 'What are you?' That is a given that cannot be changed." Huntington is correct that people can and do change their identities: they are socially constructed and dependent upon logics of personal and social ascription. Yet, as a neoconservative, Huntington is committed to a primordialist conception of identity, to the notion that there are certain "essential types" of identities. This discourse is a recycled version of the racial hierarchy discourse we discussed in Part One, with religion, language and culture serving as thinly disguised surrogates for race. Huntington holds that multiple identities, particularly religious identities, in one place create conflicts: heterogeneity of belief itself is the explanation for conflict. This is misleading – the Yugoslav wars in the 1990s, for example, were caused by elite manipulations amidst economic crisis, not religious or ethnicity diversity per se – but a standard neoconservative prejudice. Elsewhere in Huntington's writings, most notably in the book version of *The Clash of Civilizations* (Huntington, 1996) and his book *Who Are We?* (Huntington, 2004), his political commitments become quite clear. Huntington considers "Anglo-Protestant culture" to be the bedrock of the United States/West and sees this as being undermined by "multiculturalism" and, in particular, the Catholic Hispanic population of the United States. His "West versus the Rest" struggle is a culture war not only against potential enemies with different moral values overseas but also against potential threats at home from communities that refuse to learn English and assimilate into mainstream Anglo-Protestant culture. Huntington's antipathy towards Hispanics is matched by his antipathy towards Islam and Chinese culture. Some consider his 1993 claim – "A central focus of conflict for the immediate future will be between the West and several Islamic-Confucian states" – as prophetic, without realizing that Huntington's discourse is a frame-working of conflict in those terms.

The second example of neoconservative geopolitics is the "Statement of Principles" from a group called "Project for a New American Century" (Reading 17; for further documents see the website http:// newamericancentury.org). Like the Committee on the Present Danger, this is a grouping of neoconservative intellectuals (including Fuyukama), media pundits and out-of-power neoconservative policy makers. Organized by William Kristol, son of Irving Kristol and former aide to Vice-President Dan Quayle, the project is designed as a "brain trust" for Republican candidates running for office, particularly the presidency. Though they call themselves "conservative," the project is thoroughly neoconservative, the terminological choice signaling an ambition to remake conservativism around neoconservative convictions. The philosophical underpinnings are Straussian: liberals are held to be wishy-washy moral relativists whereas conservatives are best when they are firm moral absolutists, when they "stand for what is right" (Norton, 2004). This philosophical contrast, besides being simplifying and misleading, is deeply gendered. Neoconservatives have a masculinity fetish and celebrate "strength" in the face of "evil." Force, will, character and firmness are desirable masculine virtues; empathy, appeasement, pragmatism and flexibility are ultimately feminized vices. Theodore Roosevelt, Winston Churchill and Ronald Reagan are masculine "strong leaders" whereas Jimmy Carter and Bill Clinton are constructed as feminized "weak leaders."

The strategic goal of the group is reasserting American supremacy in world affairs. The very name announces the goal. Just as the twentieth century was, in Henry Luce's terms, the "American century," the twenty-first century should also be a "new American century." The goal is not that the United States coordinates its affairs with its allies in an increasingly multilateral world of global dangers (the Clinton foreign policy). Rather, the goal is American preeminence; the United States must be so strong that none of its allies can combine to check its power. Preeminent military supremacy is essential to forging a Pax Americana, which is why the document, predictably, calls for greater levels of defense spending. Neoconservatives had four great causes in the late 1990s: unequivocal support for missile defense, unambiguous support for Taiwan in its conflict with mainland China, overthrowing Saddam Hussein, and unconditional support for Israel (Mann, 2004).

THE BUSH ADMINISTRATION AND SEPTEMBER 11 TERRORIST ATTACKS

George W. Bush was elected president of the United States after the US Supreme Court ordered the termination of vote recounts in the state of Florida in December 2000 (if all the votes had been counted, later analysis showed, the Democratic candidate Al Gore would have won Florida and thus the presidency). His victory was a triumph for the Clinton-hating movement and for neoconservative intellectuals, many of whom received top positions in the new Bush administration. Donald Rumsfeld and Paul Wolfowitz, both signatories of the "Statement of Principles," were now running the Pentagon, John Bolton was at the State Department (though under non-signatory Secretary of State Colin Powell) while Richard Cheney, another signatory, was Vice-President. During the presidential campaign George W. Bush promised to conduct a "humble" foreign policy and opposed American efforts at "nation-building" across the world. In office, however, Bush's foreign policy thinking was shaped by neoconservatism more than traditional Republican pragmatism or Kissingerian realism. The abortion "gag rule" was reimposed while the administration announced that it would not abide by the Kyoto Accords or support the establishment of an International Criminal Court.

An overriding priority for the new administration was the construction of a multibillion dollar ballistic missile defense system. Before becoming Secretary of Defense, Rumsfeld had rescued this idea from a professional intelligence estimate that it was not immediately needed, heading a commission modeled after "Team B" that concluded that the danger to the United States from a missile attack was greater than the CIA and other intelligence agencies were reporting (Mann, 2004: 241). The scenario of a nuclear missile launch by North Korea was the basis for peddling the program but an undeclared motivation for its construction, beyond lucrative contracts to defense corporations, was fear of China's emergent nuclear program. Many in the new Bush administration were ideologically hostile to China. Fortunately, more moderate pragmatic voices prevailed in April 2001 when the Chinese government captured a US military spy plane after it collided with a Chinese aircraft, killing the young pilot.

Some within the new administration were planning an invasion of Iraq (Suskind, 2004). While this represented "unfinished business" from the time of the elder Bush, George W. Bush did not follow in his father's footsteps. He pursued a markedly unilateralist foreign policy agenda which reflected the influence of both the chauvinistic nationalism of the Jacksonian tradition in American foreign policy, and the crusading Wilsonianism of the right found in neoconservative thinking (Mead, 2002). Though son of a president and a multimillionaire from "front work" for groups of investors in Texas (including rich Saudis), Bush was unlikely presidential material. He did not possess a valid passport, had never traveled to Europe or Asia, and was woefully ignorant of world politics and leaders. Throughout his life, he demonstrated strong anti-intellectual tendencies and frequently fell back upon reasoning that downscaled complex questions into simpleminded categories and contrasts. His battle with alcoholism made him a "born again" Christian with attendant moral righteousness (Ivans, 2000). Bush was a formidable political campaigner but amidst the world culture of educated professionals that surrounded the US presidency, his cognitive and communications shortcomings were apparent. This weakness at the center of executive power has had significant consequences for the United States. Initially, for example, combating international terrorism was a neglected presidential priority. Former counter-terrorism chief, Richard Clarke, documents in detail how the Bush administration did not take the threat seriously until it was too late (Clarke, 2004).

The terrorist attacks of September 11, 2001 were an unprecedented attack on two symbols of globalization and American power. Citizens from eighty-six different states were horribly killed in the attacks, and the chaos they caused in New York and Washington DC reverberated for months (offices in both cities were the target of anthrax-in-the-mail on the heels of the attacks). The outpouring of sympathy for the United States in the days after the attacks was extraordinary. President Putin of Russia was the first to offer sympathy and support. The American anthem was played at Buckingham Palace, London. The French newspaper, *Le Monde*, declared in a banner headline that "we are all Americans." NATO members invoked Article 5 of its charter for the first time, where an armed attack against one is considered an attack against all. The UN Security Council passed a resolution requiring all states to deny terrorists safe harbor and suppress terrorist financing. Even states like Iraq, Iran and Syria condemned the attacks. International states lined up to support the US ultimatum to the Taliban in Afghanistan to turn over the Al-Qaeda leaders responsible for organizing the criminal attacks.

The attacks were immediately made meaningful by government officials and the media as "a declaration of war" against the United States (this was the banner headline in a number of newspapers). Bush officials instinctively Americanized the event and the dead, analogizing it to Pearl Harbor. Describing them as "terrorism" seemed inadequate. President Bush pronouncing that the

> deadly attacks which were carried out yesterday against our country were more than acts of terror. They were acts of war . . . Freedom and democracy are under attack. This will be a monumental struggle of good versus evil. But good will prevail.
>
> (Bush, September 12, 2001; see http://www.whitehouse.gov to read the full remarks)

These initial descriptions and classification had implications. The United States owned the event and would dictate the response: "we will rally the world," Bush declared. The attacks were war not terrorism. The United States was attacked but so also were the universal concepts the United States was automatically assumed to represent (the standard confusion of America with universal values that characterizes US geopolitical culture). It was a struggle of good versus evil. On the basic description of the attacks as "war," Juergensmeyer (2002) writes:

> war is an enticing conceptual construct. . . . It points to a dichotomous opposition on an absolute scale. War suggests an all-or-nothing struggle against an enemy who is determined to destroy. No compromise is deemed possible. The very existence of the opponent is a threat, and until the enemy is either crushed or contained, one's own existence cannot be secure. What is striking about a martial attitude is the certainty of one's position and the willingness to defend it, or impose it on others, to the end.
>
> (Juergensmeyer 2002: 31)

Interpreting the 9/11 terrorist attacks as a "declaration of war" by "evil-doers" (a popular Bush construct that reveals his downscaled worldview) against a United States representing universal values produced a geopolitics dominated by unilateralism, martial attitudes, simplifying polarizations and moral absolutism. At a memorial Bush declared that "our responsibility to history is clear: to answer these attacks and rid the world of evil" (Bush, September 14, 2001; see http://whitehouse.gov to read the full address). The first formal statement of this terror geopolitics was when President Bush addressed a joint session of Congress nine days after the attacks. In it Bush echoes the language of the Truman Doctrine declaring:

> Every nation, in every region, now has a decision to make. Either you are with us, or you are with the terrorists. From this day forward, any nation that continues to harbor or support terrorism will be regarded by the United States as a hostile regime.

Manichean moralistic discourse is, as we have seen, a persistent feature of US geopolitical culture and it was apparent that Bush was much more comfortable constructing a downscaled geopolitical world of good versus evil than acknowledging the complex geopolitics of actual regions, in particular the United States' "special relationship" with Saudi Arabia where fifteen of the nineteen hijackers originated. The other persistent feature of American geopolitical culture – the conceit that the United States' interests were the interests of all – led Bush to pronounce the global war on terror as not just "America's fight." "This is the world's fight," he declared. "This is civilization's fight. This is the fight of all who believe in progress and pluralism, tolerance and freedom" (Bush, September 20, 2001; see http://whitehouse.gov). The conceit was an imperial one yet, because the US war to topple the Taliban regime in Afghanistan was widely supported, the hubris it expressed would not become exposed until the Bush administration unilaterally decided that Iraq was next.

There were many other interpretations and possible responses to the 9/11 attacks than launching a global war on terror (GWOT). As many commentators pointed out, one cannot wage war against a tactic of violence. Also, the meaning of "terrorism" was not universally shared, with some even charging that the US government sponsored "terrorism" by funding anti-Castro Cuban exiles, the Nicaraguan *contras*, right wing paramilitaries in Latin America or "selective assassinations" by the Israel government. Others saw the 9/11 attacks as a vindication of Huntington's ideas and even painted visions of apocalyptic showdown between Christianity and Islam in the Middle East (Moyers, 2005). A critical reaction to popular readings of 9/11 is provided by the secular Palestinian intellectual, Edward Said, a professor at Columbia University in Manhattan, a few miles from the site of the first attack (Reading 18). Said makes three important observations. First, because of the spectacular horror of the 9/11 attacks, media pundits and politicians reach for a "vocabulary of gigantism and apocalypse" to constitute, categorize and mobilize. Big violent events seem to demand big polarized categories but, as Said points out, these categories are hopeless at capturing the messiness of our interdependent lived world and "the interconnectedness of innumerable lives." Second, refusing the simpleminded "West versus Islam" dichotomy, he points out the modernity of the methods of the terrorist hijackers. Gray (2003: 25) deepens this point, arguing that the intellectual roots of radical Islam are in the European Counter-Enlightenment and that "the Romantic belief that the world can be reshaped by an act of will is as much a part of the modern world as the Enlightenment ideal of a universal civilization based on reason." Both notions, he added, are myths. Third, noting the upsurge in Islamophobia across the United States in the wake of September 11, Said offers a different set of categories by which the post-9/11 world can be understood: power and powerlessness, reason and ignorance, justice and injustice. While he wrote, however, the most powerful military in the world was bombing one of the poorest countries in the world. US public opinion was enflamed and chauvinistic nationalism fueled sentiments of revenge. The head of the CIA's counterterrorism center articulated the prevailing conception of justice: by the time the United States military was finished with them, "they [the terrorists] will have flies on their eyeballs" (Woodward, 2002).

THE PREEMPTIVE WAR STRATEGY AND THE INVASION OF IRAQ

The September 11 attacks did not change the worldview of American neoconservatives; it merely reinforced it and made them even more aggressive in promoting their agenda. Missile defense was lavishly funded and construction on the initial stage of its deployment begun. The administration repudiated the Anti-Ballistic Missile Treaty in December 2001, a cornerstone of arms control during the Cold War. And preparations for the public relations campaign to justify the invasion of Iraq began. The campaign was launched with the publication of a new National Security Strategy in September 2002. This publication is a formalized statement of the strategic goals and philosophy of the administration, with various government agencies writing sections under the supervision of White House officials. The 2002 document does not bluntly speak of a strategy of military supremacy. Instead one has phrases like the United States will seek "to create a balance of power that favors human freedom" and "the United States will use this moment of opportunity to extend the benefits of freedom across the globe" (pp. i, ii). The document articulates the imperial conceit that is the abiding alibi for US interventionism and state violence across the planet: "the United States must defend liberty and justice because these principles are right and true for all people everywhere" (p. 3).

The most significant aspect of the document, in beginning the buildup to the invasion of Iraq, is its articulation of a doctrine of preemptive war against rogue states. Whereas previously defense and intelligence analysis had considered the capabilities, intentions and immediate preparations for war by adversaries (e.g. moving troops to the frontier or loading missiles), the new position of the US government was that the malevolent identity of adversaries alone is justification for a preemptive attack by the United States. The document states it thus:

> It has taken almost a decade for us to comprehend the true nature of this threat [potential use of weapons of mass destruction]. Given the goals of rogue states and terrorists, the United States can no longer solely rely on a reactive posture as we have in the past. The inability to deter a potential attacker, the immediacy of today's threats, and the magnitude of potential harm that could be caused by our adversaries' choice of weapons, do not permit that option. We cannot let our enemies strike first.
>
> (White House, 2004: 15; see http://www.whitehouse.gov to read the full document)

While the United States had always reserved the right to act preemptively in the past, the formal codification of a doctrine of attack based merely on the perceived malevolent identity of an adversary was a qualitative change in US strategic thinking. It appeared to be the geopolitical formalization of Bush's apparent hyperbolic claim about the need to "rid the world of evil." This use of theological categories to make sense of the world political map reached a notable apogee in Bush's 2002 State of the Union address when he condemned an "axis of evil, arming to threaten the peace of the world," comprising Iraq, Iran and North Korea, three states that had minimal relations with each other, two of them neighbors that had fought a bloody war, one despotic, the other a theocracy, and the third a relic communist dictatorship (Bush, January 29, 2002; http://www. whitehouse.gov). That they were in alliance was absurd. That they were "evil" was a theological truth to the neoconservatives running the Bush administration.

While the Bush administration had concluded that Saddam Hussein's Iraq was "evil" and, therefore, was the first candidate for preemptive war, most other leaders and professional diplomats in the international community held to the principle that evidence of WMD capabilities and of immanent intention to use them were required before the US campaign for "regime change" could be supported. What unfolded from September 2002 until the actual invasion of Iraq on March 19, 2003 by the United States, and some allied states like the United Kingdom and Poland, was a long saga over intelligence sources, credible evidence, reliability and professional judgment. The past was precedent for neoconservatives. Given their long-standing suspicion of professional intelligence bureaucracies, they revived the "Team B" work-around maneuver by creating special panels and teams to deliver them the intelligence they wanted. Ideological preconception trumped empirical evidence and professional judgment. The intelligence was made to fit the case for war (Packer, 2005).

There were many geopolitical justifications for the belligerence in US foreign and military strategy in the wake of September 11, 2001. Two are worthy of note (see also Frum and Pearle, 2003). The first was an article by Robert Kagan (2002) called 'Power and Weakness' in *Policy Review*, a journal published by the conservative Hoover Institution, that earned considerable press attention. Kagan was a friend of William Kristol and like him the son of a celebrated neoconservative, classics professor Donald Kagan, another of the signatories of the "Statement of Principles." Kristol and Kagan had edited a book together called *Present Dangers* that was a compendium of neoconservative thinking reiterating classic "present danger" discourse (Kagan and Kristol, 2000). Kagan's article sought to explain the growing divide between the United States and the European Union's leading members in terms of different contexts and philosophies. Following his fellow neoconservatives, Fuyukama and Huntington, Kagan's argument is overstated and sweeping. Typical of the geopolitical genre, there is a central polarity between two competing abstractions, the "United States" and "Europe" which, consistent with the masculinity fetish of neoconservatives, are deeply gendered. In the most mediagenic line of the article, Kagan pronounced that:

> on major strategic and international questions today, Americans are from Mars and Europeans are from Venus: They agree on little and understand one another less and less. And this state of affairs is not transitory – the product of one American election or one catastrophic event. The reasons for the transatlantic divide are deep, long in development, and likely to endure. When it comes to setting national priorities, determining threats, defining challenges, and fashioning and implementing foreign and defense policies, the United States and Europe have parted ways.
> (Kagan, 2002: 3; for the full article see http://www.policyreview.org/JUN02/kagan.html)

Kagan's thesis is simplistic (indeed, he concedes it is a "caricature"). Europe is turning away from power and is reluctant to use force in world affairs. It lives in a post-historical paradise of peace and relative prosperity, the European Union, and its policies develop from its fundamental military weakness. The United States "remains mired in history, exercising power in the anarchic Hobbesian world where international laws and rules are unreliable and where true security and the defense and promotion of a liberal order still depend on the possession and use of military might." As with Fukuyama and Huntington, Kagan's exaggerated claims earned him considerable press attention and a book contract, though the subsequent product was exceedingly slim (Kagan, 2003; Lindberg, 2005).

Simpleminded contracts between essentialized geographical abstractions is the stuff of conventional geopolitics and hardly new. But Kagan's claims dealt with the fracturing of the West itself and had the

THREE

good fortune of appearing at a time when the Bush administration was losing international support because of its fixation with overthrowing the regime of Saddam Hussein. Yet even the Bush administration did not accept Kagan's thesis, distinguishing between an "old Europe" (France and Germany principally) that was about appeasement (i.e. did not support the invasion of Iraq) and a "new Europe" that supported the US policy of creating "a balance of power that favors human freedom" (i.e. marching Wilsonianism). Contra Kagan, public opinion studies as well as popular opposition to the Iraqi war in the United States revealed a divide not between the "United States" and "Europe" but between American Republican Party voters (who tend to be bellicose Jacksonian nationalists) and mainstream social democratic voters in both Europe and North America (Ash, 2004: 67).

The second justification for an imperial American foreign policy and the war in Iraq appeared on the eve of the war in *Esquire* magazine (Reading 19). Thomas P.M. Barnett is a faculty member at the US Naval War College and a Pentagon aide to Rumsfeld on force transformation and strategic futures. He is one of a community of radical neoconservative defense intellectuals, like Frank Gaffney at the Center for Security Policy, who have long hyped a "revolution in military affairs" as a means of transforming Pentagon organization, strategy and appropriation expenditures. "The Pentagon New Map" follows the conventional geopolitical narrative formula for popular impact and a book contract. World politics revolves around a singular plot that Barnett reveals as globalization and the "rule sets" and connectivities it produces. The world political map is divided into two clashing spatial zones, a Functioning Core that has a stable rule set and thick connectivity, and a Non-Integrating Gap where rule sets are arbitrary, dysfunctional or non-existent. This is mapped in a graphic manner with a seamless line helpfully distinguishing the zones (though the overriding commitment to an unbroken line separating two zones of homogenized space leaves North Korea in the Functioning Core!). The implications of the thesis for the United States are then explained: America will fight its future wars in the Gap, not against large modern states like itself. Finally, the message is captured in a catchy slogan: "disconnectedness defines danger." Predictably, a book with a major publisher to expand the thesis and profit from its media reverberation followed (Barnett, 2004).

Barnett writes with plenty of techno-futurist discourse and attitude. Read the essay closely to grasp how his central metaphors are cybernetic: rule sets, on-line/off-line, system security/immunity/ perturbation/feedback, firewall, etc. World political space is like the operating space of a computer system: there are certain zones that are properly formatted and run well, and there are other bad sectors that need repair through intervention, reformatting and reconnection to the central operating system. That system is globalization, its operating system neoliberal rule sets and its chief "system administrator" the United States (Barnett, 2004: 315). Think critically about how these metaphors justify not simply the war in Iraq or the preemptive war strategy of the United States but an even more ambitious project of intervening across the planet. "If a country is either losing out to globalization or rejecting much of the content flows associated with its advance, there is a far greater chance that the US will end up sending forces at some point." The United States' imperial role is naturalized, and builds upon the dubious assumption that Bin Laden and Al-Qaeda "are pure products of the Gap" (a denial of the United States' long history of involvement with Saudi Arabia in general and Bin Laden in particular) that functioned only in the "ends of the earth." This then leads to the exceedingly trite and superficial conclusion: "A country's potential to warrant a US military response is inversely related to its globalization connectivity."

In a critical reading of Barnett's thesis, Roberts et al. (2003) argue it exemplifies a more widespread form of "neoliberal geopolitics" which they define as the combined action of both neoliberal idealism about the virtues of free markets, openness and global economic integration, and American military force (see also Dalby, 2005). This perhaps formalizes the contingencies of the Iraq war and Barnett's provocations too much for it is worth underscoring how remarkably unrealistic his vision is of future US military interventions. His is a fantasy map of US imperial domination (benignly spun as "system administration") that is a recipe for endless interventionism, a planetary US military war machine and unlimited resources for "repair and reconnection" (so-called "nation-building"). While articulating the imperial fantasies of neoconservativism, it also amply demonstrates its hubris, self-delusion and overreaching impracticality. The Pentagon's new map is an imperial dream. It encountered the real world when the United States invaded Iraq in March 2003 and its subsequent failure to stabilize this one relatively small country is indicative of how delusionary it is.

The war in Iraq was a dilemma for some liberal internationalists who, though opposed to the unilateralism of much of the Bush administration's foreign policy, nevertheless supported the overthrow of Saddam Hussein because of his brutal human rights record. This group can be described as "liberal imperialists." They share the vision of a proactive US promoting "democracy" and "liberty" across the globe, with an accent on human rights and the liberation of oppressed peoples. Canadian Professor Michael Ignatieff, Director of the Carr Center for Human Rights Policy at Harvard University, articulates the moral dilemmas of supporting an imperialist United States in "The Burden," published three months before the Iraqi invasion (Reading 20).

Ignatieff makes three arguments about confronting realities, as he sees them. First, the United States is an empire in denial, and in the post-9/11 world, this is a problem. The United States is a non-territorial empire which he terms an "empire lite, a global hegemony whose grace notes are free markets, human rights and democracy, enforced by the most awesome military power the world has ever known." Ignatieff suggests that human rights activists have to accept that the use of force in international affairs can be positive and that "many people owe their freedom to the exercise of American military power." There is ambivalence about imperial burdens in US geopolitical culture, however, with many arguing that "in becoming an empire it [the United States] risks losing its soul" (Agnew, 2003; Bacevich, 2002). Ignatieff argues that Iraq "lays bare the realities of the US role." "Virtuous disengagement," he declares, "is no longer a possibility" and it is reasonable for it to act forcefully given the dangers posed by weapons of mass destruction in the hands of Saddam Hussein or terrorist networks. Note that Ignatieff accepts multiple storylines promoted by the Bush administration: the justification of so-called "preemptive" warfare, the "intelligence" claims that Iraq has weapons of mass destruction, that weapons of mass destruction "would render Saddam the master of the region," even the claim that Iraq could intimidate and deter the United States.

Second, Ignatieff argues that if the United States takes on Iraq it will be taking on a long-term burden that involves the whole region. Ignatieff believes that the United States needs to address the oppression of the Palestinians by the Israeli state if it is to try to ameliorate regional rage and stop the production of future terrorists. Unaddressed in his account is the unqualified commitment of neoconservatives, many Jewish, to the repressive policies (including state-sanctioned terrorism and assassination by missile strike) of the state of Israel. US war planners, like Paul Wolfowitz and Douglas Feith, are committed Zionists and neoconservatives because, among other factors, of family experience with Nazi genocide (Boyer, 2004; Goldberg, 2005). Deeply skeptical about "appeasing tyrants," they are also deeply committed to Israel's fight against "terrorism." The larger context of that conflict is ignored. For Ignatieff, producing a "two-state solution" is a test of American leadership. It is one it continues to fail because of domestic power networks and ideological commitments, factors missing from Ignatieff's analysis.

Third, Ignatieff argues the burden of empire cannot be borne by the United States alone. It needs allies yet is alienating them by its militarism and unilateral methods. Ignatieff paraphrases Kagan's narrative on the United States and Europe, conceding rather than challenging its simpleminded polarities and militarism. He ends with a warning for the United States. It risks reproducing the classic failing of imperial states: hubris. This has many forms: the delusions induced by accepting one's own heroic self-image, the confusion of global military means with global military power, and failure to attend to the limits of empire. Many aspects of Ignatieff's essay are contestable – his misleading description of UN inaction, his acceptance of Bush administration narratives, and his lack of critical engagement with the political economy and culture of empire (see De Zengotita, 2003; Gregory, 2004) – but it remains relevant as a discussion of the dangers to the United States from its Iraqi venture.

AMERICAN NATIONALISM AND THE PROBLEMS OF A ROGUE SUPERPOWER

The final reading in Part Three is by Anatol Lieven, a British journalist and research scholar living in Washington DC (Reading 21). The extract is the introduction to his book *America, Right or Wrong* (Lieven, 2004). In it Lieven provides an analysis of why the United States has become a rogue superpower, not the conservative hegemon it should be, but, as he puts it, "an unsatisfied and even revolutionary power, kicking to pieces the hill on which it is the king." His explanation lies in the nature of American nationalism, the leadership of George W. Bush and the traumatizing effects of the terrorist attacks of September 11, 2001 on US political life and culture. American nationalism, he argues, is a complex synthesis made up of, on the one hand, a popular American Creed (thesis) that articulates standard civil nationalist narratives of equality, liberty and acceptance of diversity yet, on the other hand, is also composed of a more exclusivist ethnonationalism (antithesis) dominated by white Christian anxieties and preoccupations. The American Creed is the acceptable face of American identity and the basis of its universalist discourse about liberty, freedom and democracy in world affairs. Yet, the practice of American power is often animated and driven by the more chauvinistic attitudes and concerns of American ethnonationalism. Many, while accepting the rhetoric of the American Creed, hold a purist vision of it as the product of the virtue of a distinctive ethnoreligious community, namely white Anglo-Protestants (see Huntington's arguments). This tension within the heart of American popular self-identity became a polarized divide as a result of the "cultural wars" of the 1990s and the mobilization of populist ethnoreligious anti-elitism by the Republican Party against liberal civic nation principles to seize the White House in 2000 and keep it in 2004 (Frank, 2004).

The 9/11 attacks facilitated the production of a sense of righteous victimhood in US geopolitical culture. An innocent party was wronged by evil forces and now there would be payback. This classic narrative, the plot of so many American films, was the constructed justification for the unleashing of fulsome displays of American violence on Afghanistan and Iraq (Ó Tuathail, 2003, 2005). What Timothy Garton Ash (2004: 114) terms "the hubris of the wounded" led to the overreaching extension of the war against Al-Qaeda into a more generalized war against "terrorism" and "evil." Lieven argues the Bush administration's "will to empire" has propelled it into conflict with the world of multilateral institutions and governance structures American hegemony produced (the United Nations, the World Bank, etc.). Many Americans, he points out, "are in revolt against the world which America itself has made." Rather than patiently building consensus thus enhancing American power and legitimacy, the Bush administration has gone it alone and suffered the consequences as it now finds itself bogged down in Iraq. Like the European elites before 1914 (with Bush himself an American equivalent of the obdurate Wilhelm II of Germany), the Bush administration "has allowed its own national chauvinism and limitless ambition to compromise the security and stability of the world capitalist system of which America is the custodian and greatest beneficiary."

Lieven's incisive arguments point to the contradictions of rogue superpower: actions officially proclaimed as enhancing the state's security end up undermining it, and producing more enemies and potential terrorists not less (Mann, 2003). With no end to violence in Afghanistan and Iraq in sight, it is worth concluding by noting the failures and weaknesses of rogue superpower behavior. First, the United States' unilateralism has diminished not multiplied American power. Through its high-handed policies and hostility to international institutions and mechanisms, the United States has lost friends and pushed indifferent others into antagonism. Levels of anti-Americanism have risen dramatically since 2001, even in states historically sympathetic to the United States like Italy, the United Kingdom, Ireland, France and Germany. Second, the United States went to war when it did not need to do so. The smartest strategists avoid war when they can. The United States showed strategic impatience in its confrontation with Saddam Hussein and rushed to war without an international coalition or a plan on how to win the peace. It has been paying for these strategic blunders ever since. Also, the discourse of preventative war has established a dangerous precedent in world affairs. US intelligence has been thoroughly delegitimized by its Iraq intelligence failure and future collective action by the international community against North Korea and Iran are likely compromised by the divisive legacy of the Iraqi war.

Third, the United States has needlessly destroyed human lives and wasted financial resources in Iraq. The war on terror, combined with irresponsible tax policies at home, has created massive US budget deficits. As of January 2006, the number of civilian dead from the Iraqi war is estimated at between 28,000 and 32,000 (see http://www.iraqbodycount.net/). US military deaths are approaching 2500 with over 16,000 soldiers wounded. Financial expenditure by the United States in Iraq by 2006 will reach $250 billion, with much of this going to private security firms and other corporations with close ties to the White House. The Iraq war is calculated as costing the United States $5.6 billion a month, which is higher than the monthly cost of the Vietnam war adjusted for inflation (for statistical updates see the Institute for Policy Studies Iraq index page: http://www.ips-dc.org/iraq/index.htm). Today, the US state runs a huge budget deficit which is covered by foreign investors, from China, Japan, Europe and the Middle East, buying its treasury bonds (the US national debt is $7.8 trillion: http://see www.ustreas.gov). This situation is unsustainable and hardly the basis for continued US strength in world affairs.

Fourth, in letting its military strategy be driven by a transparent desire for revenge, the United States has sanctioned immoderate displays of American violence and illegal detention and torture tactics that have undermined its moral authority. While the bombing of Afghanistan resulted in the death of many innocents, it was legally sanctioned by a sympathetic international community. The Iraq war, however, was an illegal war of aggression under international law, without a triggering provocation save "intelligence reports" that later proved false. The prison camp at Guantánomo was established to evade the Geneva Conventions and American domestic law until US courts thwarted this tactic (Amnesty International, 2005). Both there and in the infamous Abu Ghraib prison in Baghdad, the United States has used torture against suspected terrorists and insurgents (Hersh, 2004). The United States has cooperated with allies to facilitate the export of suspects for torture abroad (Human Rights Watch, 2005). The means of fighting the global war on terror, in other words, are undermining the ostensible end. The United States has little credibility in the eyes of many nowadays as an upholder of "the rule of law" and a promoter of "freedom."

Fifth, the greatest weapon of terrorists is their ability to spread fear. As Benjamin Barber (2003: 50) has argued, it is not terrorism but fear that is the real enemy. How the Bush administration has chosen to fight its GWOT is through the institutionalization of a society built around fear, and through displays of violence designed to defeat terrorist fear with state fear. The end result, however, is merely the multiplication of fear throughout society. Throughout 2001–2004, the US government scared its own citizens with a color-coded system of "terrorist attack" warnings. Americans, consequently, are more afraid of terrorism than genuine threats to their safety from hand-gun violence or vehicle accidents. Their government has become more intrusive and their civil liberties are more compromised, with wire tapping and surveillance of American citizens sanctioned by an imperial White House brazenly flaunting the law. As Barber argues, the strategy of defeating the fear by amplifying fear has created an "empire of fear." In such circumstances only demagogues and extremists thrive.

Finally, the Bush administration's war against terror is premised on a Manichaeism that has induced dangerous polarization in world affairs. While it is overly trite and simplistic to argue that Bush versus Bin Laden is a "clash of fundamentalisms" (Ali, 2003), there is a strong case to be made that the Bush administration's response to 9/11 has swelled the ranks of terrorist organizations, made rogue states more determined to acquire nuclear weapons, and congealed a harmful divide between "Islam" and "the West" that has negative consequences across the world. For example, the 2005 London bombers cited US and British policies in the Middle East as explanation for their horrific acts of terrorism. While talk of an essential antagonism between the "Islamic world" and "the West" abounds, the simplifying logic of these geopolitical labels should be rejected by all. It is often forgotten that three fourths of the world's Muslim population lives outside the Middle East, in states like India, Bangladesh, Turkey and Indonesia. Islam has a rich internal complexity and its geographic extent is global. Europe has an estimated 15 million Muslims, the United States, 3 to 6 million (Aslan, 2005). As Said notes, "Islam is inside from the start," its foundational texts entwined with those of Judaism and Christianity. There is no homogeneous "Islamic terrorism" (or Jewish or Christian terrorism), no innate "clash of civilizations" or "war" between "Islam" and "the West." The moral absolutism and brutal violence of terror geopolitics makes it difficult to avoid these seductive simplifications but that is the task of critical geopolitics today, to think otherwise so the interconnections and hybridity as well as the culturally contested meaning of twenty first century geopolitics are always remembered.

REFERENCES AND FURTHER READING

On Russia and American empire

Daalder, I. and Lindsey, J. (2003) *America Unbound: The Bush Revolution in Foreign Policy.* Washington, DC: Brookings Institution Press.

Friedman, T. (2000) *The Lexus and the Olive Tree: Understanding Globalization.* New York: Anchor.

Gardner, L. and Young, M. (2005) *The New American Empire.* New York: New Press.

Hardt, M. and Negri, A. (2001) *Empire.* Cambridge, MA: Harvard University Press.

Harvey, D. (2003) *The New Imperialism.* Oxford: Oxford University Press.

Hoffman, D. (2002) *The Oligarchs: Wealth and Power in the New Russia.* New York: Public Affairs.

Johnson, C. (2004) *The Sorrows of Empire: Militarism, Secrecy and the End of the Republic.* New York: Henry Holt.

Kuchins, A. (2005) "Europe's Last Geopolitician?" *Profil*, May 9.

Prestowitz, C. (2003) *Rogue Nation: American Unilateralism and the Failure of Good Intentions.* New York: Basic Books.

Specter, M. (2004) "The Devastation: Russia and AIDS," *New Yorker*, October 11, 58 ff.

On the Clinton administration and neoconservativism

Albright, M. (2003) *Madame Secretary: A Memoir*. New York: Miramax Books.
Clinton, B. (2004) *My Life*. New York: Knopf.
Huntington, S.P. (1993) "The Clash of Civilizations," *Foreign Affairs*, 72(3): 22–28.
—— (1996) *The Clash of Civilizations and the Remaking of World Order*. New York: Simon & Schuster.
—— (2004) *Who Are We? The Challenges to America's Identity*. New York: Simon & Schuster.
Mallaby, S. (2004) *The World's Banker*. New York: Penguin.
Norton, A. (2004) *Leo Strauss and the Politics of the American Empire*. New Haven, CT: Yale University Press.

On the Bush administration and the September 11 attacks

Barber, B. (2003) *Fear's Empire: War, Terrorism, and Democracy*. New York: Norton.
Clarke, R. (2004) *Against All Enemies: Inside America's War on Terror*. New York: Free Press.
Gray, J. (2003) *Al Qaeda and What it Means to be Modern*. London: Faber & Faber.
Ivans, M. (2000) *Shrub: The Short but Happy Political Life of George W. Bush*. New York: Vintage.
Juergensmeyer, M. (2002) "Religious Terror and Global War," in *Understanding September 11*, eds. C. Calhoun, P. Price and A. Timmer. New York: New Press.
Mann, J. (2004) *Rise of the Vulcans: The History of Bush's War Cabinet*. New York: Viking.
Mead, W.R. (2002) *Special Providence: American Foreign Policy and How it Changed the World*. New York: Routledge.
Moyers, B. (2005) "Welcome to Doomsday," *New York Review of Books*, 52(5), March 24.
Suskind, R. (2004) *The Price of Loyalty*. New York: Simon & Schuster.
Woodward, B. (2002) *Bush at War*. New York: Simon & Schuster.

On preemptive war and the invasion of Iraq

Barnett, T.P.M. (2004) *The Pentagon's New Map: War and Peace in the Twenty First Century*. New York: Putnam.
Boyer, P. (2004) "The Believer: Paul Wolfowitz Defends his War," *New Yorker*, November 1, 46.
Dalby, S. (2005) "The Pentagon's New Imperial Cartography: Tabloid Realism and the War on Terror," in *Spaces of Political Violence*, eds. D. Gregory and A. Pred. New York: Routledge.
De Zengotita, T. (2004) "The Romance of Empire and the Politics of Self-Love," *Harpers*, July.
Frum, D. and Pearle, R. (2003) *An End to Evil: How to Win the War on Terror*. New York: Random House.
Goldberg, J. (2005) "A Little Learning: What Douglas Feith Knew and When He Knew It," *New Yorker*, May 9, 36.
Kagan, R. (2002) "Power and Weakness," *Policy Review*, 113.
—— (2003) *Of Paradise and Power: America and Europe in the New World Order*. New York: Knopf.
Kagan, R. and Kristol, W. (2000) *Present Dangers: Crisis and Opportunity in American Foreign and Defense Policy*. San Francisco, CA: Encounter Books.
Lindberg, T. (2005) *Beyond Paradise and Power: Europe, America and the Future of a Troubled Partnership*. New York: Routledge.
Ó Tuathail, G. (2003) "'Just Out Looking for a Fight:' American Affect and the Invasion of Iraq," *Antipode*, 35(5): 856–870.
Packer, G. (2005) *The Assassins Gate: America in Iraq*. New York: Farrar, Straus and Giroux.

Roberts, S., Secor, A. and Sparke, M. (2003) "Neoliberal Geopolitics," *Antipode*, 35(5): 886–897.

Woodward, B. (2004) *Plan of Attack*. New York: Simon & Schuster.

On American nationalism and the problems of rogue superpower

Agnew, J. (2003) "American Hegemony into American Empire," *Antipode*, 35(5): 871–885.

Ali, T. (2003) *The Clash of Fundamentalisms*. London: Verso.

Amnesty International (2005) *2005 Annual Report*. Available at http://www.amnestyusa.org/annualreport/annualreport.html [last accessed June 12, 2005].

Ash, T.G. (2004) *Free World: America, Europe and the Surprising Future of the West*. New York: Random House.

Aslan, R. (2005) *No God but God: The Origins, Evolution, and Future of Islam*. New York: Random House.

Bacevich, A. (2002) *American Empire: The Realities and Consequences of US Diplomacy*. Cambridge, MA: Harvard University Press.

Frank, T. (2004) *What's the Matter with Kansas? How Conservatives Won the Heart of America*. New York: Metropolitan Books.

Gregory, D. (2004) *The Colonial Present*. Oxford: Blackwell.

Hersh, S. (2004) *Chain of Command: The Road from 9/11 to Abu Ghraib*. New York: HarperCollins.

Human Rights Watch (2005) *Black Hole: The Fate of Islamicists Rendered to Egypt*. New York. Available at http://www.hrw.org/reports/2005/egypt0505/ [last accessed June 12, 2005].

Lieven, A. (2004) *America Right or Wrong: An Anatomy of American Nationalism*. Oxford: Oxford University Press.

Mann, M. (2003) *Incoherent Empire*. London: Verso.

Ó Tuathail, G. (2005) "The Frustrations of Geopolitics and the Pleasures of War: *Behind Enemy Lines* and American Geopolitical Culture," *Geopolitics*, 10(2): 356–377.

Cartoon 5 Happy Coincidence!
This Wuerker cartoon points out that there is, in actuality, a structural relationship between US "defense" and "environmental" policies and the material interests of particular groups bankrolling the political career of George W. Bush. Money is the lifeblood of US politics. Without campaign donations from wealthy interests, politicians do not get elected to public office.

Source: M. Wuerker

The Clash of Civilizations?

Samuel P. Huntington
from *Foreign Affairs* (1993)

THE NEXT PATTERN OF CONFLICT

World politics is entering a new phase, and intellectuals have not hesitated to proliferate visions of what it will be – the end of history, the return of traditional rivalries between nation states, and the decline of the nation state from the conflicting pulls of tribalism and globalism, among others. Each of these visions catches aspects of the emerging reality. Yet they all miss a crucial, indeed a central, aspect of what global politics is likely to be in the coming years.

It is my hypothesis that the fundamental source of conflict in this new world will not be primarily ideological or primarily economic. The great divisions among humankind and the dominating source of conflict will be cultural. Nation states will remain the most powerful actors in world affairs, but the principal conflicts of global politics will occur between nations and groups of different civilizations. The clash of civilizations will dominate global politics. The fault lines between civilizations will be the battle lines of the future.

Conflict between civilizations will be the latest phase in the evolution of conflict in the modern world. For a century and a half after the emergence of the modern international system with the Peace of Westphalia, the conflicts of the Western world were largely among princes – emperors, absolute monarchs and constitutional monarchs attempting to expand their bureaucracies, their armies, their mercantilist economic strength and, most important, the territory they ruled. In the process they created nation states, and beginning with the French Revolution the principal lines of conflict were between nations rather than princes. In 1793, as R. R. Palmer put it, "The wars of kings were over; the wars of peoples had begun." This nineteenth-century pattern lasted until the end of World War I. Then, as a result of the Russian Revolution and the reaction against it, the conflict of nations yielded to the conflict of ideologies, first among communism, fascism-Nazism and liberal democracy, and then between communism and liberal democracy. During the Cold War, this latter conflict became embodied in the struggle between the two superpowers, neither of which was a nation state in the classical European sense and each of which defined its identity in terms of its ideology.

These conflicts between princes, nation states and ideologies were primarily conflicts within Western civilization, "Western civil wars," as William Lind has labeled them. This was as true of the Cold War as it was of the world wars and the earlier wars of the seventeenth, eighteenth and nineteenth centuries. With the end of the Cold War, international politics moves out of its Western phase, and its center piece becomes the interaction between the West and non-Western civilizations and among non-Western civilizations. In the politics of civilizations, the peoples and governments of non-Western civilizations no longer remain the objects of history as targets of Western colonialism but join the West as movers and shapers of history.

THE NATURE OF CIVILIZATIONS

During the Cold War the world was divided into the First, Second and Third Worlds. Those divisions are no longer relevant. It is far more meaningful now to group countries not in terms of their political or economic systems or in terms of their level of economic development but rather in terms of their culture and civilization.

What do we mean when we talk of a civilization? A civilization is a cultural entity. Villages, regions, ethnic groups, nationalities, religious groups, all have distinct cultures at different levels of cultural heterogeneity. The culture of a village in southern Italy may be different from that of a village in northern Italy, but both will share in a common Italian culture that distinguishes them from German villages. European communities, in turn, will share cultural features that distinguish them from Arab or Chinese communities. Arabs, Chinese and Westerners, however, are not part of any broader cultural entity. They constitute civilizations. A civilization is thus the highest cultural grouping of people and the broadest level of cultural identity people have short of that which distinguishes humans from other species. It is defined both by common objective elements, such as language, history, religion, customs, institutions, and by the subjective self-identification of people. People have levels of identity; a resident of Rome may define himself with varying degrees of intensity as a Roman, an Italian, a Catholic, a Christian, a European, a Westerner. The civilization to which he belongs is the broadest level of identification with which he intensely identifies. People can and do redefine their identities and, as a result, the composition and boundaries of civilizations change.

Civilizations may involve a large number of people, as with China ("a civilization pretending to be a state," as Lucian Pye put it), or a very small number of people, such as the Anglophone Caribbean. A civilization may include several nation states, as is the case with Western, Latin American and Arab civilizations, or only one, as is the case with Japanese civilization. Civilizations obviously blend and overlap, and may include subcivilizations. Western civilization has two major variants, European and North American, and Islam has its Arab, Turkic and Malay subdivisions. Civilizations are nonetheless meaningful entities, and while the lines between them are seldom sharp, they are real. Civilizations are dynamic; they rise and fall; they divide and merge. And, as any student of history knows, civilizations disappear and are buried in the sands of time.

Westerners tend to think of nation states as the principal actors in global affairs. They have been that, however, for only a few centuries. The broader reaches of human history have been the history of civilizations.

In *A Study of History*, Arnold Toynbee identified 21 major civilizations; only six of them exist in the contemporary world.

WHY CIVILIZATIONS WILL CLASH

Civilization identity will be increasingly important in the future, and the world will be shaped in large measure by the interactions among seven or eight major civilizations. These include Western, Confucian, Japanese, Islamic, Hindu, Slavic-Orthodox, Latin American and possibly African civilizations. The most important conflicts of the future will occur along the cultural fault lines separating these civilizations from one another.

Why will this be the case?

First, differences among civilizations are not only real; they are basic. Civilizations are differentiated from each other by history, language, culture, tradition and, most important, religion. The people of different civilization have different views on the relations between God and man, the individual and the group, the citizen and the state, parents and children, husband and wife, as well as differing views on the relative importance of rights and responsibilities, liberty and authority, equality and hierarchy. These differences are the product of centuries. They will not soon disappear. They are far more fundamental than differences among political ideologies and political regimes. Differences do not necessarily mean conflict, and conflict does not necessarily mean violence. Over the centuries, however, differences among civilizations have generated the most prolonged and the most violent conflicts.

Second, the world is becoming a smaller place. The interactions between peoples of different civilizations are increasing; these increasing interactions intensify civilization consciousness and awareness of differences between civilizations and commonalities within civilizations. North African immigration to France generates hostility among Frenchmen and at the same time increased receptivity to immigration by "good" European Catholic Poles. Americans react far more negatively to Japanese investment than to larger investments from Canada and European countries. Similarly, as Donald Horowitz has pointed out, "An Ibo may be . . . an Owerri Ibo or an Onitsha Ibo in what was the Eastern region of Nigeria. In Lagos, he is simply an Ibo. In London, he is a Nigerian. In New York, he is an African." The interactions among peoples of different civilizations enhance the civilization-consciousness of people that, in turn, invigorates differences and animosities stretching or thought to stretch back deep into history.

Third, the processes of economic modernization and social change throughout the world are separating people from longstanding local identities. They also weaken the nation state as a source of identity. In much of the world religion has moved in to fill this gap, often in the form of movements that are labeled "fundamentalist." Such movements are found in Western Christianity, Judaism, Buddhism and Hinduism, as well as in Islam. In most countries and most religions the people active in fundamentalist movements are young, college-educated, middle-class technicians, professionals and business persons. The "unsecularization of the world," George Weigel has remarked, "is one of the dominant social facts of life in the late twentieth century." The revival of religion, "la revanche de Dieu," as Gilles Kepel labeled it, provides a basis for identity and commitment that transcends national boundaries and unites civilizations.

Fourth, the growth of civilization-consciousness is enhanced by the dual role of the West. On the one hand, the West is at a peak of power. At the same time, however, and perhaps as a result, a return to the roots phenomenon is occurring among non-Western civilizations. Increasingly one hears references to trends toward a turning inward and "Asianization" in Japan, the end of the Nehru legacy and the "Hinduization" of India, the failure of Western ideas of socialism and nationalism and hence "re-Islamization" of the Middle East, and now a debate over Westernization versus Russianization in Boris Yeltsin's country. A West at the peak of its power confronts non-Wests that increasingly have the desire, the will and the resources to shape the world in non-Western ways.

In the past, the elites of non-Western societies were usually the people who were most involved with the West, had been educated at Oxford, the Sorbonne or Sandhurst, and had absorbed Western attitudes and values. At the same time, the populace in non-Western countries often remained deeply imbued with the indigenous culture. Now, however, these relationships are being reversed. A de-Westernization and indigenization of elites is occurring in many non-Western countries at the same time that Western, usually American, cultures, styles and habits become more popular among the mass of the people.

Fifth, cultural characteristics and differences are less mutable and hence less easily compromised and resolved than political and economic ones. In the former Soviet Union, communists can become democrats, the rich can become poor and the poor rich, but Russians cannot become Estonians and Azeris cannot become Armenians. In class and ideological conflicts, the key question was "Which side are you on?" and people could and did choose sides and change sides. In conflicts between civilizations, the question is "What are you?" That is a given that cannot be changed. And as we know, from Bosnia to the Caucasus to the Sudan, the wrong answer to that question can mean a bullet in the head. Even more than ethnicity, religion discriminates sharply and exclusively among people. A person can be half-French and half-Arab and simultaneously even a citizen of two countries. It is more difficult to be half-Catholic and half-Muslim.

Finally, economic regionalism is increasing. The proportions of total trade that were infra-regional rose between 1980 and 1989 from 51 per cent to 59 per cent in Europe, 33 per cent to 37 per cent in East Asia, and 32 per cent to 36 per cent in North America. The importance of regional economic blocs is likely to continue to increase in the future. On the one hand, successful economic regionalism will reinforce civilization-consciousness. On the other hand, economic regionalism may succeed only when it is rooted in a common civilization. The European Community rests on the shared foundation of European culture and Western Christianity. The success of the North American Free Trade Area depends on the convergence now underway of Mexican, Canadian and American cultures. Japan, in contrast, faces difficulties in creating a comparable economic entity in East Asia because Japan is a society and civilization unique to itself. However strong the trade and investment links Japan may develop with other East Asian countries, its cultural differences with those countries inhibit and perhaps preclude its promoting regional economic integration like that in Europe and North America.

[. . .]

As people define their identity in ethnic and religious terms, they are likely to see an "us" versus "them" relation existing between themselves and people of different ethnicity or religion. The end of ideologically defined states in Eastern Europe and the former Soviet Union permits traditional ethnic identities and animosities to come to the fore. Differences in culture and religion create differences over policy issues, ranging from human rights to immigration to trade and commerce to the environment. Geographical propinquity gives rise to conflicting territorial claims from Bosnia to Mindanao. Most important, the efforts of the West to promote its values of democracy and liberalism as universal values, to maintain its military predominance and to advance its economic interests engender countering responses from other civilizations. Decreasingly able to mobilize support and form coalitions on the basis of ideology, governments and groups will increasingly attempt to mobilize support by appealing to common religion and civilization identity.

The clash of civilizations thus occurs at two levels. At the micro-level, adjacent groups along the fault lines between civilizations struggle, often violently, over the control of territory and each other. At the macro-level, states from different civilizations compete for relative military and economic power, struggle over the control of international institutions and third parties, and competitively promote their particular political and religious values. [. . .]

THE FAULT LINES BETWEEN CIVILIZATIONS

The fault lines between civilizations are replacing the political and ideological boundaries of the Cold War as the flash points for crisis and bloodshed. The Cold War began when the Iron Curtain divided Europe politically and ideologically. The Cold War ended with the end of the Iron Curtain. As the ideological division of Europe has disappeared, the cultural division of Europe between Western Christianity, on the one hand, and Orthodox Christianity and Islam, on the other, has reemerged. The most significant dividing line in Europe, as William Wallace has suggested, may well be the eastern boundary of Western Christianity in the year 1500. This line runs along what are now the boundaries between Finland and Russia and between the Baltic states and Russia, cuts through Belarus and Ukraine separating the more Catholic western Ukraine from Orthodox eastern Ukraine, swings westward separating Transylvania from the rest of Romania, and then goes through Yugoslavia almost exactly along the line now separating Croatia and Slovenia from the rest of Yugoslavia. In the Balkans this line, of course,

coincides with the historic boundary between the Hapsburg and Ottoman empires. The peoples to the north and west of this line are Protestant or Catholic; they shared the common experiences of European history – feudalism, the Renaissance, the Reformation, the Enlightenment, the French Revolution, the Industrial Revolution; they are generally economically better off than the peoples to the east; and they may now look forward to increasing involvement in a common European economy and to the consolidation of democratic political systems. The peoples to the east and south of this line are Orthodox or Muslim; they historically belonged to the Ottoman or Tsarist empires and were only lightly touched by the shaping events in the rest of Europe; they are generally less advanced economically; they seem much less likely to develop stable democratic political systems. The Velvet Curtain of culture has replaced the Iron Curtain of ideology as the most significant dividing line in Europe. As the events in Yugoslavia show, it is not only a line of difference; it is also at times a line of bloody conflict.

Conflict along the fault line between Western and Islamic civilizations has been going on for 1300 years. After the founding of Islam, the Arab and Moorish surge west and north only ended at Tours in 732. From the eleventh to the thirteenth century the Crusaders attempted with temporary success to bring Christianity and Christian rule to the Holy Land. From the fourteenth to the seventeenth century, the Ottoman Turks reversed the balance, extended their sway over the Middle East and the Balkans, captured Constantinople, and twice laid siege to Vienna. In the nineteenth and early twentieth centuries as Ottoman power declined, Britain, France and Italy established Western control over most of North Africa and the Middle East.

After World War II, the West, in turn, began to retreat; the colonial empires disappeared; first Arab nationalism and then Islamic fundamentalism manifested themselves; the West became heavily dependent on the Persian Gulf countries for its energy; the oil-rich Muslim countries became money-rich and, when they wished to, weapons-rich. Several wars occurred between Arabs and Israel (created by the West). France fought a bloody and ruthless war in Algeria for most of the 1950s; British and French forces invaded Egypt in 1956; American forces went into Lebanon in 1958; subsequently American forces returned to Lebanon, attacked Libya, and engaged in various military encounters with Iran; Arab and Islamic terrorists, supported by at least three Middle Eastern governments, employed the weapon of the weak and bombed Western planes and installations and seized Western hostages. This warfare between Arabs and the West culminated in 1990, when the United States sent a massive army to the Persian Gulf to defend some Arab countries against aggression by another. In its aftermath NATO planning is increasingly directed to potential threats and instability along its "southern tier."

This centuries-old military interaction between the West and Islam is unlikely to decline. It could become more virulent. The Gulf War left some Arabs feeling proud that Saddam Hussein had attacked Israel and stood up to the West. It also left many feeling humiliated and resentful of the West's military presence in the Persian Gulf, the West's overwhelming military dominance, and their apparent inability to shape their own destiny. [. . .]

CIVILIZATION RALLYING: THE KIN-COUNTRY SYNDROME

Groups or states belonging to one civilization that become involved in war with people from a different civilization naturally try to rally support from other members of their own civilization. As the post-Cold War world evolves, civilization commonality, what H. D. S. Greenway has termed the "kin-country" syndrome, is replacing political ideology and traditional balance of power considerations as the principal basis for cooperation and coalitions. It can be seen gradually emerging in the post-Cold War conflicts in the Persian Gulf, the Caucasus and Bosnia. None of these was a full-scale war between civilizations, but each involved some elements of civilizational rallying, which seemed to become more important as the conflict continued and which may provide a foretaste of the future. [. . .]

Civilization rallying to date has been limited, but it has been growing, and it clearly has the potential to spread much further. As the conflicts in the Persian Gulf, the Caucasus and Bosnia continued, the positions of nations and the cleavages between them increasingly were along civilizational lines. Populist politicians, religious leaders and the media have found it a potent means of arousing mass support and of pressuring hesitant governments. In the coming years, the local conflicts most likely to escalate into major wars will be those, as in Bosnia and the Caucasus, along the fault lines between civilizations. The next world war, if there is one, will be a war between civilizations.

THE WEST VERSUS THE REST

The West is now at an extraordinary peak of power in relation to other civilizations. Its superpower opponent has disappeared from the map. Military conflict among Western states is unthinkable, and Western military power is unrivaled. Apart from Japan, the West faces no economic challenge. It dominates international political and security institutions and with Japan international economic institutions. Global political and security issues are effectively settled by a directorate of the United States, Britain and France, world economic issues by a directorate of the United States, Germany and Japan, all of which maintain extraordinarily close relations with each other to the exclusion of lesser and largely non-Western countries. Decisions made at the UN Security Council or in the International Monetary Fund that reflect the interests of the West are presented to the world as reflecting the desires of the world community. The very phrase "the world community" has become the euphemistic collective noun (replacing "the Free World") to give global legitimacy to actions reflecting the interests of the United States and other Western powers. Through the IMF and other international economic institutions, the West promotes its economic interests and imposes on other nations the economic policies it thinks appropriate. [. . .]

Western domination of the UN Security Council and its decisions, tempered only by occasional abstention by China, produced UN legitimation of the West's use of force to drive Iraq out of Kuwait and its elimination of Iraq's sophisticated weapons and capacity to produce such weapons. It also produced the quite unprecedented action by the United States, Britain and France in getting the Security Council to demand that Libya hand over the Pan Am 103 bombing suspects and then to impose sanctions when Libya refused. After defeating the largest Arab army, the West did not hesitate to throw its weight around in the Arab world. The West in effect is using international institutions, military power and economic resources to run the world in ways that will maintain Western predominance, protect Western interests and promote Western political and economic values.

That at least is the way in which non-Westerners see the new world, and there is a significant element of truth in their view. Differences in power and struggles for military, economic and institutional power are thus one source of conflict between the West and other civilizations. Differences in culture, that is basic values and beliefs, are a second source of conflict. V. S. Naipaul has argued that Western civilization is the "universal civilization" that "fits all men." At a superficial level much of Western culture has indeed permeated the rest of the world. At a more basic level, however, Western concepts differ fundamentally from those prevalent in other civilizations. Western ideas of individualism, liberalism, constitutionalism, human rights, equality, liberty, the rule of law, democracy, free markets, the separation of church and state, often have little resonance in Islamic, Confucian, Japanese, Hindu, Buddhist or Orthodox cultures. Western efforts to propagate such ideas produce instead a reaction against "human rights imperialism" and a reaffirmation of indigenous values, as can be seen in the support for religious fundamentalism by the younger generation in non-Western cultures. The very notion that there could be a "universal civilization" is a Western idea, directly at odds with the particularism of most Asian societies and their emphasis on what distinguishes one people from another. Indeed, the author of a review of 100 comparative studies of values in different societies concluded that "the values that are most important in the West are least important worldwide." In the political realm, of course, these differences are most manifest in the efforts of the United States and other Western powers to induce other peoples to adopt Western ideas concerning democracy and human rights. Modern democratic government originated in the West. When it has developed in non-Western societies it has usually been the product of Western colonialism or imposition.

The central axis of world politics in the future is likely to be, in Kishore Mahbubani's phrase, the conflict between "the West and the Rest" and the responses of non-Western civilizations to Western power and values. Those responses generally take one or a combination of three forms. At one extreme, non-Western states can, like Burma and North Korea, attempt to pursue a course of isolation, to insulate their societies from penetration or "corruption" by the West, and, in effect, to opt out of participation in the Western-dominated global community. The costs of this course, however, are high, and few states have pursued it exclusively. A second alternative, the equivalent of "bandwagoning" in international relations theory, is to attempt to join the West and accept its values and institutions. The third alternative is to attempt to "balance" the West by developing economic and military power and cooperating with other non-

Western societies against the West, while preserving indigenous values and institutions; in short, to modernize but not to Westernize.

THE TORN COUNTRIES

In the future, as people differentiate themselves by civilization, countries with large numbers of peoples of different civilizations, such as the Soviet Union and Yugoslavia, are candidates for dismemberment. Some other countries have a fair degree of cultural homogeneity but are divided over whether their society belongs to one civilization or another. These are torn countries. Their leaders typically wish to pursue a bandwagoning strategy and to make their countries members of the West, but the history, culture and traditions of their countries are non-Western. The most obvious and prototypical torn country is Turkey. The late twentieth-century leaders of Turkey have followed in the Attaturk tradition and defined Turkey as a modern, secular, Western nation state. They allied Turkey with the West in NATO and in the Gulf War; they applied for membership in the European Community. At the same time, however, elements in Turkish society have supported an Islamic revival and have argued that Turkey is basically a Middle Eastern Muslim society. In addition, while the elite of Turkey has defined Turkey as a Western society, the elite of the West refuses to accept Turkey as such. Turkey will not become a member of the European Community, and the real reason, as President Ozal said, "is that we are Muslim and they are Christian and they don't say that." Having rejected Mecca, and then being rejected by Brussels, where does Turkey look? Tashkent may be the answer. The end of the Soviet Union gives Turkey the opportunity to become the leader of a revived Turkic civilization involving seven countries from the borders of Greece to those of China. Encouraged by the West, Turkey is making strenuous efforts to carve out this new identity for itself.

During the past decade Mexico has assumed a position somewhat similar to that of Turkey. Just as Turkey abandoned its historic opposition to Europe and attempted to join Europe, Mexico has stopped defining itself by its opposition to the United States and is instead attempting to imitate the United States and to join it in the North American Free Trade Area. Mexican leaders are engaged in the great task of redefining Mexican identity and have introduced

fundamental economic reforms that eventually will lead to fundamental political change. In 1991 a top adviser to President Carlos Salinas de Gortari described at length to me all the changes the Salinas government was making. When he finished, I remarked: "That's most impressive. It seems to me that basically you want to change Mexico from a Latin American country into a North American country." He looked at me with surprise and exclaimed: "Exactly! That's precisely what we are trying to do, but of course we could never say so publicly." As his remark indicates, in Mexico as in Turkey, significant elements in society resist the redefinition of their country's identity. In Turkey, European-oriented leaders have to make gestures to Islam (Ozal's pilgrimage to Mecca); so also Mexico's North American-oriented leaders have to make gestures to those who hold Mexico to be a Latin American country (Salinas' Ibero-American Guadalajara summit).

Historically Turkey has been the most profoundly torn country. For the United States, Mexico is the most immediate torn country. Globally the most important torn country is Russia. The question of whether Russia is part of the West or the leader of a distinct Slavic Orthodox civilization has been a recurring one in Russian history. That issue was obscured by the communist victory in Russia, which imported a Western ideology, adapted it to Russian conditions and then challenged the West in the name of that ideology. The dominance of communism shut off the historic debate over Westernization versus Russification. With communism discredited, Russians once again face that question. President Yeltsin is adopting Western principles and goals and seeking to make Russia a "normal" country and a part of the West. Yet both the Russian elite and the Russian public are divided on this issue. Among the more moderate dissenters, Sergei Stankevich argues that Russia should reject the "Atlanticist" course, which would lead it to become European, to become a part of the world economy in rapid and organized fashion, to become the eighth member of the [Group of] Seven, and to put particular emphasis on Germany and the United States as the two dominant members of the Atlantic alliance.

While also rejecting an exclusively Eurasian policy, Stankevich nonetheless argues that Russia should give priority to the protection of Russians in other countries, emphasize its Turkic and Muslim connections, and promote "an appreciable redistribution of our resources, our options, our ties, and our interests

in favor of Asia, of the eastern direction." People of this persuasion criticize Yeltsin for subordinating Russia's interests to those of the West, for reducing Russian military strength, for failing to support traditional friends such as Serbia, and for pushing economic and political reform in ways injurious to the Russian people. [. . .]

To redefine its civilization identity, a torn country must meet three requirements. First, its political and economic elite has to be generally supportive of and enthusiastic about this move. Second, its public has to be willing to acquiesce in the redefinition. Third, the dominant groups in the recipient civilization have to be willing to embrace the convert. All three requirements in large part exist with respect to Mexico. The first two in large part exist with respect to Turkey. It is not clear that any of them exist with respect to Russia's joining the West. The conflict between liberal democracy and Marxism-Leninism was between ideologies which, despite their major differences, ostensibly shared ultimate goals of freedom, equality and prosperity. A traditional, authoritarian, nationalist Russia could have quite different goals. A Western democrat could carry on an intellectual debate with a Soviet Marxist. It would be virtually impossible for him to do that with a Russian traditionalist. If, as the Russians stop behaving like Marxists, they reject liberal democracy and begin behaving like Russians but not like Westerners, the relations between Russia and the West could again become distant and conflictual.

THE CONFUCIAN-ISLAMIC CONNECTION

The obstacles to non-Western countries joining the West vary considerably. They are least for Latin American and East European countries. They are greater for the Orthodox countries of the former Soviet Union. They are still greater for Muslim, Confucian, Hindu and Buddhist societies. Japan has established a unique position for itself as an associate member of the West: it is in the West in some respects but clearly not of the West in important dimensions. Those countries that for reason of culture and power do not wish to, or can not, join the West compete with the West by developing their own economic, military and political power. They do this by promoting their internal development and by cooperating with other non-Western countries. The most prominent form of this cooperation is the Confucian-Islamic connection that has emerged to challenge Western interests, values and power.

Almost without exception, Western countries are reducing their military power; under Yeltsin's leadership so also is Russia. China, North Korea and several Middle Eastern states, however, are significantly expanding their military capabilities. They are doing this by the import of arms from Western and non-Western sources and by the development of indigenous arms industries. One result is the emergence of what Charles Krauthammer has called "Weapon States," and the Weapon States are not Western states. Another result is the redefinition of arms control, which is a Western concept and a Western goal. During the Cold War the primary purpose of arms control was to establish a stable military balance between the United States and its allies and the Soviet Union and its allies. In the post-Cold War world the primary objective of arms control is to prevent the development by non-Western societies of military capabilities that could threaten Western interests. The West attempts to do this through international agreements, economic pressure and controls on the transfer of arms and weapons technologies.

The conflict between the West and the Confucian-Islamic states focuses largely, although not exclusively, on nuclear, chemical and biological weapons, ballistic missiles and other sophisticated means for delivering them, and the guidance, intelligence and other electronic capabilities for achieving that goal. The West promotes nonproliferation as a universal norm and nonproliferation treaties and inspections as means of realizing that norm. It also threatens a variety of sanctions against those who promote the spread of sophisticated weapons and proposes some benefits for those who do not. The attention of the West focuses, naturally, on nations that are actually or potentially hostile to the West.

The non-Western nations, on the other hand, assert their right to acquire and to deploy whatever weapons they think necessary for their security. They also have absorbed, to the full, the truth of the response of the Indian defense minister when asked what lesson he learned from the Gulf War: "Don't fight the United States unless you have nuclear weapons."

Nuclear weapons, chemical weapons and missiles are viewed, probably erroneously, as the potential equalizer of superior Western conventional power. China, of course, already has nuclear weapons;

Pakistan and India have the capability to deploy them. North Korea, Iran, Iraq, Libya and Algeria appear to be attempting to acquire them. A top Iranian official has declared that all Muslim states should acquire nuclear weapons, and in 1988 the president of Iran reportedly issued a directive calling for development of "offensive and defensive chemical, biological and radiological weapons."

Centrally important to the development of counter-West military capabilities is the sustained expansion of China's military power and its means to create military power. Buoyed by spectacular economic development, China is rapidly increasing its military spending and vigorously moving forward with the modernization of its armed forces. It is purchasing weapons from the former Soviet states; it is developing long-range missiles; in 1992 it tested a one-megaton nuclear device. It is developing power-projection capabilities, acquiring aerial refueling technology, and trying to purchase an aircraft carrier. Its military build-up and assertion of sovereignty over the South China Sea are provoking a multilateral regional arms race in East Asia. China is also a major exporter of arms and weapons technology. [. . .]

A Confucian-Islamic military connection has thus come into being, designed to promote acquisition by its members of the weapons and weapons technologies needed to counter the military power of the West. It may or may not last. At present, however, it is, as Dave McCurdy has said, "a renegades' mutual support pact, run by the proliferators and their backers." A new form of arms competition is thus occurring between Islamic-Confucian states and the West. In an old-fashioned arms race, each side developed its own arms to balance or to achieve superiority against the other side. In this new form of arms competition, one side is developing its arms and the other side is attempting not to balance but to limit and prevent that arms build-up while at the same time reducing its own military capability.

IMPLICATIONS FOR THE WEST

This article does not argue that civilization identities will replace all other identities, that nation states will disappear, that each civilization will become a single coherent political entity, that groups within a civilization will not conflict with and even fight each other. This paper does set forth the hypotheses that differences between civilizations are real and important; civilization-consciousness is increasing; conflict between civilizations will supplant ideological and other forms of conflict as the dominant global form of conflict; international relations, historically a game played out within Western civilization, will increasingly be de-Westernized and become a game in which non-Western civilizations are actors and not simply objects; successful political, security and economic international institutions are more likely to develop within civilizations than across civilizations; conflicts between groups in different civilizations will be more frequent, more sustained and more violent than conflicts between groups in the same civilization; violent conflicts between groups in different civilizations are the most likely and most dangerous source of escalation that could lead to global wars; the paramount axis of world politics will be the relations between the "West and the Rest"; the elites in some torn non-Western countries will try to make their countries part of the West, but in most cases face major obstacles to accomplishing this; a central focus of conflict for the immediate future will be between the West and several Islamic-Confucian states.

This is not to advocate the desirability of conflicts between civilizations. It is to set forth descriptive hypotheses as to what the future may be like. If these are plausible hypotheses, however, it is necessary to consider their implications for Western policy. These implications should be divided between short-term advantage and long-term accommodation. In the short term it is clearly in the interest of the West to promote greater cooperation and unity within its own civilization, particularly between its European and North American components; to incorporate into the West societies in Eastern Europe and Latin America whose cultures are close to those of the West; to promote and maintain cooperative relations with Russia and Japan; to prevent escalation of local inter-civilization conflicts into major inter-civilization wars; to limit the expansion of the military strength of Confucian and Islamic states; to moderate the reduction of Western military capabilities and maintain military superiority in East and Southwest Asia; to exploit differences and conflicts among Confucian and Islamic states; to support in other civilizations groups sympathetic to Western values and interests; to strengthen international institutions that reflect and legitimate Western interests and values and to promote the involvement of non-Western states in those institutions.

In the longer term other measures would be called for. Western civilization is both Western and modern. Non-Western civilizations have attempted to become modern without becoming Western. To date only Japan has fully succeeded in this quest. Non-Western civilizations will continue to attempt to acquire the wealth, technology, skills, machines and weapons that are part of being modern. They will also attempt to reconcile this modernity with their traditional culture and values. Their economic and military strength relative to the West will increase. Hence the West will increasingly have to accommodate these non-Western modern civilizations whose power approaches that of the West but whose values and interests differ significantly from those of the West. This will require the West to maintain the economic and military power necessary to protect its interests in relation to these civilizations. It will also, however, require the West to develop a more profound understanding of the basic religious and philosophical assumptions underlying other civilizations and the ways in which people in those civilizations see their interests. It will require an effort to identify elements of commonality between Western and other civilizations. For the relevant future, there will be no universal civilization, but instead a world of different civilizations, each of which will have to learn to coexist with the others.

Statement of Principles

Project for a New American Century
from http://newamericancentury.org (1997)

American foreign and defense policy is adrift. Conservatives have criticized the incoherent policies of the Clinton Administration. They have also resisted isolationist impulses from within their own ranks. But conservatives have not confidently advanced a strategic vision of America's role in the world. They have not set forth guiding principles for American foreign policy. They have allowed differences over tactics to obscure potential agreement on strategic objectives. And they have not fought for a defense budget that would maintain American security and advance American interests in the new century.

We aim to change this. We aim to make the case and rally support for American global leadership.

As the 20th century draws to a close, the United States stands as the world's preeminent power. Having led the West to victory in the Cold War, America faces an opportunity and a challenge: Does the United States have the vision to build upon the achievements of past decades? Does the United States have the resolve to shape a new century favorable to American principles and interests?

We are in danger of squandering the opportunity and failing the challenge. We are living off the capital – both the military investments and the foreign policy achievements – built up by past administrations. Cuts in foreign affairs and defense spending, inattention to the tools of statecraft, and inconstant leadership are making it increasingly difficult to sustain American influence around the world. And the promise of short-term commercial benefits threatens to override strategic considerations. As a consequence, we are jeopardizing the nation's ability to meet present threats and to deal with potentially greater challenges that lie ahead.

We seem to have forgotten the essential elements of the Reagan Administration's success: a military that is strong and ready to meet both present and future challenges; a foreign policy that boldly and purposefully promotes American principles abroad; and national leadership that accepts the United States' global responsibilities.

Of course, the United States must be prudent in how it exercises its power. But we cannot safely avoid the responsibilities of global leadership or the costs that are associated with its exercise. America has a vital role in maintaining peace and security in Europe, Asia, and the Middle East. If we shirk our responsibilities, we invite challenges to our fundamental interests. The history of the 20th century should have taught us that it is important to shape circumstances before crises emerge, and to meet threats before they become dire. The history of this century should have taught us to embrace the cause of American leadership.

Our aim is to remind Americans of these lessons and to draw their consequences for today. Here are four consequences:

- we need to increase defense spending significantly if we are to carry out our global responsibilities today and modernize our armed forces for the future;
- we need to strengthen our ties to democratic allies and to challenge regimes hostile to our interests and values;
- we need to promote the cause of political and economic freedom abroad;
- we need to accept responsibility for America's unique role in preserving and extending an international order friendly to our security, our prosperity, and our principles.

Such a Reaganite policy of military strength and moral clarity may not be fashionable today. But it is necessary if the United States is to build on the successes of this past century and to ensure our security and our greatness in the next.

The Clash of Ignorance

Edward W. Said

from *The Nation* (2001)

Samuel Huntington's article "The Clash of Civilizations?" appeared in the Summer 1993 issue of *Foreign Affairs*, where it immediately attracted a surprising amount of attention and reaction. Because the article was intended to supply Americans with an original thesis about "a new phase" in world politics after the end of the cold war, Huntington's terms of argument seemed compellingly large, bold, even visionary. He very clearly had his eye on rivals in the policy-making ranks, theorists such as Francis Fukuyama and his "end of history" ideas, as well as the legions who had celebrated the onset of globalism, tribalism and the dissipation of the state. But they, he allowed, had understood only some aspects of this new period. He was about to announce the "crucial, indeed a central, aspect" of what "global politics is likely to be in the coming years." Unhesitatingly he pressed on:

> It is my hypothesis that the fundamental source of conflict in this new world will not be primarily ideological or primarily economic. The great divisions among humankind and the dominating source of conflict will be cultural. Nation states will remain the most powerful actors in world affairs, but the principal conflicts of global politics will occur between nations and groups of different civilizations. The clash of civilizations will dominate global politics. The fault lines between civilizations will be the battle lines of the future.

Most of the argument in the pages that followed relied on a vague notion of something Huntington called "civilization identity" and "the interactions among seven or eight [*sic*] major civilizations," of which the conflict between two of them, Islam and the West, gets

the lion's share of his attention. In this belligerent kind of thought, he relies heavily on a 1990 article by the veteran Orientalist Bernard Lewis, whose ideological colors are manifest in its title, "The Roots of Muslim Rage." In both articles, the personification of enormous entities called "the West" and "Islam" is recklessly affirmed, as if hugely complicated matters like identity and culture existed in a cartoonlike world where Popeye and Bluto bash each other mercilessly, with one always more virtuous pugilist getting the upper hand over his adversary. Certainly neither Huntington nor Lewis has much time to spare for the internal dynamics and plurality of every civilization, or for the fact that the major contest in most modern cultures concerns the definition or interpretation of each culture, or for the unattractive possibility that a great deal of demagogy and downright ignorance is involved in presuming to speak for a whole religion or civilization. No, the West is the West, and Islam Islam.

The challenge for Western policy-makers, says Huntington, is to make sure that the West gets stronger and fends off all the others, Islam in particular. More troubling is Huntington's assumption that his perspective, which is to survey the entire world from a perch outside all ordinary attachments and hidden loyalties, is the correct one, as if everyone else were scurrying around looking for the answers that he has already found. In fact, Huntington is an ideologist, someone who wants to make "civilizations" and "identities" into what they are not: shut-down, sealed-off entities that have been purged of the myriad currents and counter-currents that animate human history, and that over centuries have made it possible for that history not only to contain wars of religion and imperial conquest but also to be one of exchange, cross-fertilization and sharing. This far less visible history is ignored in the rush to highlight the ludicrously compressed and constricted warfare that "the clash of civilizations" argues is the reality. When he published his book by the same title in 1996, Huntington tried to give his argument a little more subtlety and many, many more footnotes; all he did, however, was confuse himself and demonstrate what a clumsy writer and inelegant thinker he was.

The basic paradigm of West versus the rest (the cold war opposition reformulated) remained untouched, and this is what has persisted, often insidiously and implicitly, in discussion since the terrible events of September 11. The carefully planned and horrendous, pathologically motivated suicide attack and mass slaughter by a small group of deranged militants has been turned into proof of Huntington's thesis. Instead of seeing it for what it is – the capture of big ideas (I use the word loosely) by a tiny band of crazed fanatics for criminal purposes – international luminaries from former Pakistani Prime Minister Benazir Bhutto to Italian Prime Minister Silvio Berlusconi have pontificated about Islam's troubles, and in the latter's case have used Huntington's ideas to rant on about the West's superiority, how "we" have Mozart and Michelangelo and they don't. (Berlusconi has since made a halfhearted apology for his insult to "Islam.")

But why not instead see parallels, admittedly less spectacular in their destructiveness, for Osama bin Laden and his followers in cults like the Branch Davidians or the disciples of the Rev. Jim Jones at Guyana or the Japanese Aum Shinrikyo? Even the normally sober British weekly *The Economist*, in its issue of September 22–28, can't resist reaching for the vast generalization, praising Huntington extravagantly for his "cruel and sweeping, but nonetheless acute" observations about Islam. "Today," the journal says with unseemly solemnity, Huntington writes that "the world's billion or so Muslims are 'convinced of the superiority of their culture, and obsessed with the inferiority of their power.'" Did he canvas 100 Indonesians, 200 Moroccans, 500 Egyptians and fifty Bosnians? Even if he did, what sort of sample is that?

Uncountable are the editorials in every American and European newspaper and magazine of note adding to this vocabulary of gigantism and apocalypse, each use of which is plainly designed not to edify but to inflame the reader's indignant passion as a member of the "West," and what we need to do. Churchillian rhetoric is used inappropriately by self-appointed combatants in the West's, and especially America's,

war against its haters, despoilers, destroyers, with scant attention to complex histories that defy such reductiveness and have seeped from one territory into another, in the process overriding the boundaries that are supposed to separate us all into divided armed camps.

This is the problem with unedifying labels like Islam and the West: They mislead and confuse the mind, which is trying to make sense of a disorderly reality that won't be pigeonholed or strapped down as easily as all that. I remember interrupting a man who, after a lecture I had given at a West Bank university in 1994, rose from the audience and started to attack my ideas as "Western," as opposed to the strict Islamic ones he espoused. "Why are you wearing a suit and tie?" was the first retort that came to mind. "They're Western too." He sat down with an embarrassed smile on his face, but I recalled the incident when information on the September 11 terrorists started to come in: how they had mastered all the technical details required to inflict their homicidal evil on the World Trade Center, the Pentagon and the aircraft they had commandeered. Where does one draw the line between "Western" technology and, as Berlusconi declared, "Islam's" inability to be a part of "modernity"?

One cannot easily do so, of course. How finally inadequate are the labels, generalizations and cultural assertions. At some level, for instance, primitive passions and sophisticated knowhow converge in ways that give the lie to a fortified boundary not only between "West" and "Islam" but also between past and present, us and them, to say nothing of the very concepts of identity and nationality about which there is unending disagreement and debate. A unilateral decision made to draw lines in the sand, to undertake crusades, to oppose their evil with our good, to extirpate terrorism and, in Paul Wolfowitz's nihilistic vocabulary, to end nations entirely, doesn't make the supposed entities any easier to see; rather, it speaks to how much simpler it is to make bellicose statements for the purpose of mobilizing collective passions than to reflect, examine, sort out what it is we are dealing with in reality, the interconnectedness of innumerable lives, "ours" as well as "theirs."

In a remarkable series of three articles published between January and March 1999 in *Dawn*, Pakistan's most respected weekly, the late Eqbal Ahmad, writing for a Muslim audience, analyzed what he called the roots of the religious right, coming down very harshly on the mutilations of Islam by absolutists and fanatical tyrants whose obsession with regulating personal behavior promotes "an Islamic order reduced to a penal code, stripped of its humanism, aesthetics, intellectual quests, and spiritual devotion." And this "entails an absolute assertion of one, generally decontextualized, aspect of religion and a total disregard of another. The phenomenon distorts religion, debases tradition, and twists the political process wherever it unfolds." As a timely instance of this debasement, Ahmad proceeds first to present the rich, complex, pluralist meaning of the word *jihad* and then goes on to show that in the word's current confinement to indiscriminate war against presumed enemies, it is impossible "to recognize the Islamic – religion, society, culture, history or politics – as lived and experienced by Muslims through the ages." The modern Islamists, Ahmad concludes, are "concerned with power, not with the soul; with the mobilization of people for political purposes rather than with sharing and alleviating their sufferings and aspirations. Theirs is a very limited and time-bound political agenda." What has made matters worse is that similar distortions and zealotry occur in the "Jewish" and "Christian" universes of discourse.

It was Conrad, more powerfully than any of his readers at the end of the nineteenth century could have imagined, who understood that the distinctions between civilized London and "the heart of darkness" quickly collapsed in extreme situations, and that the heights of European civilization could instantaneously fall into the most barbarous practices without preparation or transition. And it was Conrad also, in *The Secret Agent* (1907), who described terrorism's affinity for abstractions like "pure science" (and by extension for "Islam" or "the West"), as well as the terrorist's ultimate moral degradation.

For there are closer ties between apparently warring civilizations than most of us would like to believe; both Freud and Nietzsche showed how the traffic across carefully maintained, even policed boundaries moves with often terrifying ease. But then such fluid ideas, full of ambiguity and skepticism about notions that we hold on to, scarcely furnish us with suitable, practical guidelines for situations such as the one we face now. Hence the altogether more reassuring battle orders (a crusade, good versus evil, freedom against fear, etc.) drawn out of Huntington's alleged opposition between Islam and the West, from which official discourse drew its vocabulary in the first days after the September 11 attacks. There's since been a noticeable de-escalation in that discourse, but to judge from the steady amount

of hate speech and actions, plus reports of law enforcement efforts directed against Arabs, Muslims and Indians all over the country, the paradigm stays on.

One further reason for its persistence is the increased presence of Muslims all over Europe and the United States. Think of the populations today of France, Italy, Germany, Spain, Britain, America, even Sweden, and you must concede that Islam is no longer on the fringes of the West but at its center. But what is so threatening about that presence? Buried in the collective culture are memories of the first great Arab-Islamic conquests, which began in the seventh century and which, as the celebrated Belgian historian Henri Pirenne wrote in his landmark book *Mohammed and Charlemagne* (1939), shattered once and for all the ancient unity of the Mediterranean, destroyed the Christian-Roman synthesis and gave rise to a new civilization dominated by northern powers (Germany and Carolingian France) whose mission, he seemed to be saying, is to resume defense of the "West" against its historical-cultural enemies. What Pirenne left out, alas, is that in the creation of this new line of defense the West drew on the humanism, science, philosophy, sociology and historiography of Islam, which had already interposed itself between Charlemagne's world and classical antiquity. Islam is inside from the start, as even Dante, great enemy of Mohammed, had to concede when he placed the Prophet at the very heart of his *Inferno*.

Then there is the persisting legacy of monotheism itself, the Abrahamic religions, as Louis Massignon aptly called them. Beginning with Judaism and Christianity, each is a successor haunted by what came before; for Muslims, Islam fulfills and ends the line of prophecy. There is still no decent history or demystification of the many-sided contest among these three followers – not one of them by any means a monolithic, unified camp – of the most jealous of all gods, even though the bloody modern convergence on Palestine furnishes a rich secular instance of what has been so tragically irreconcilable about them. Not surprisingly, then, Muslims and Christians speak readily of crusades and *jihads*, both of them eliding the Judaic presence with often sublime insouciance. Such an agenda, says Eqbal Ahmad, is "very reassuring to the men and women who are stranded in the middle of the ford, between the deep waters of tradition and modernity."

But we are all swimming in those waters, Westerners and Muslims and others alike. And since the waters are part of the ocean of history, trying to plow or divide them with barriers is futile. These are tense times, but it is better to think in terms of powerful and powerless communities, the secular politics of reason and ignorance, and universal principles of justice and injustice, than to wander off in search of vast abstractions that may give momentary satisfaction but little self-knowledge or informed analysis. "The Clash of Civilizations" thesis is a gimmick like "The War of the Worlds," better for reinforcing defensive self-pride than for critical understanding of the bewildering interdependence of our time.

Cartoon 6 Fear Level Tangerine

In the age of terror geopolitics, governments and mass media feed off each other to amplify fear, reducing politics to a primordial choice between candidates vying to appear tougher than the other. In such a climate, demagogues, extremist discourse and militarism thrives.

Source: M. Wuerker

19

The Pentagon's New Map

Thomas P.M. Barnett
from *Esquire* (2003)

Let me tell you why military engagement with Saddam Hussein's regime in Baghdad is not only necessary and inevitable, but good. When the United States finally goes to war again in the Persian Gulf, it will not constitute a settling of old scores, or just an enforced disarmament of illegal weapons, or a distraction in the war on terror. Our next war in the Gulf will mark a historical tipping point – the moment when Washington takes real ownership of strategic security in the age of globalization. That is why the public debate about this war has been so important: It forces Americans to come to terms with I believe is the new security paradigm that shapes this age, namely, *Disconnectedness defines danger*. Saddam Hussein's outlaw regime is dangerously disconnected from the globalizing world, from its rule sets, its norms, and all the ties that bind countries together in mutually assured dependence.

The problem with most discussion of globalization is that too many experts treat it as a binary outcome: Either it is great and sweeping the planet, or it is horrid and failing humanity everywhere. Neither view really works, because globalization as a historical process is simply too big and too complex for such summary judgments. Instead, this new world must be defined by where globalization has truly taken root and where it has not.

Show me where globalization is thick with network connectivity, financial transactions, liberal media flows, and collective security, and I will show you regions featuring stable governments, rising standards of living, and more deaths by suicide than murder. These parts of the world I call the Functioning Core, or Core. But show me where globalization is thinning or just plain absent, and I will show you regions plagued by politically repressive regimes, widespread poverty and disease, routine mass murder, and – most important – the chronic conflicts that incubate the next generation of global terrorists. These parts of the world I call the Non-Integrating Gap, or Gap.

Globalization's "ozone hole" may have been out of sight and out of mind prior to September 11, 2001, but it has been hard to miss ever since. And measuring the reach of globalization is not an academic exercise to an eighteen-year-old marine sinking tent poles on its far side. So where do we schedule the U.S. military's next round of away games? The pattern that has emerged since the end of the cold war suggests a simple answer: in the Gap.

The reason I support going to war in Iraq is not simply that Saddam is a cutthroat Stalinist willing to kill anyone to stay in power, nor because that regime has clearly supported terrorist networks over the years. The real reason I support a war like this is that the resulting long-term military commitment will finally force America to deal with the entire Gap as a strategic threat environment.

For most countries, accommodating the emerging global rule set of democracy, transparency, and free trade is no mean feat, which is something most Americans find hard to understand. We tend to forget just how hard it has been to keep the United States together all these years, harmonizing our own, competing internal rule sets along the way – through a Civil War, a Great Depression, and the long struggles for racial and sexual equality that continue to this day. As far as most states are concerned, we are quite unrealistic in our expectation that they should adapt themselves quickly to globalization's very American-looking rule set.

But you have to be careful with that Darwinian pessimism, because it is a short jump from apologizing for globalization-as-forced-Americanization to insinuating – along racial or civilization lines – that "*those people will simply never be like us.*" Just ten years ago, most experts were willing to write off poor Russia, declaring Slavs, in effect, genetically unfit for democracy and capitalism. Similar arguments resonated in most China-bashing during the 1990's, and you hear them today in the debates about the feasibility of imposing democracy on a post-Saddam Iraq – a sort of Muslims-are-from-Mars argument.

So how do we distinguish between who is really making it in globalization's Core and who remains trapped in the Gap? And how permanent is this dividing line? Understanding that the line between the Core and Gap is constantly shifting, let me suggest that the direction of change is more critical than the degree. So, yes, Beijing is still ruled by a "Communist party" whose ideological formula is 30 percent Marxist-Leninist and 70 percent *Sopranos*, but China just signed on to the World Trade Organization, and over the long run, that is far more important in securing the country's permanent Core status. Why? Because it forces China to harmonize its internal rule set with that of globalization – banking, tariffs, copyright protection, environmental standards. Of course, working to adjust your internal rule sets to globalization's evolving rule set offers no guarantee of success. As Argentina and Brazil have recently found out, following the rules (in Argentina's case, *sort of* following) does not mean you are panicproof, or bubbleproof, or even recessionproof. Trying to adapt to globalization does not mean bad things will never happen to you. Nor does it mean all your poor will immediately morph into stable middle class. It just means your standard of living gets better over time. In sum, it is always possible to fall off this bandwagon called globalization. And when you do, bloodshed will follow. If you are lucky, so will American troops.

So what parts of the world can be considered functioning right now? North America, much of South America, the European Union, Putin's Russia, Japan and Asia's emerging economies (most notably China and India), Australia and New Zealand, and South Africa, which accounts for roughly four billion out of a global population of six billion. Whom does that leave in the Gap? It would be easy to say "everyone else," but I want to offer you more proof than that and, by doing so, argue why I think the Gap is a long-term threat to more than just your pocketbook or conscience.

If we map out U.S. military responses since the end of the cold war, (see below), we find an overwhelming concentration of activity in the regions of the world that are excluded from globalization's growing Core – namely the Caribbean Rim, virtually all of Africa, the Balkans, the Caucasus, Central Asia, the Middle East and Southwest Asia, and much of Southeast Asia. That is roughly the remaining two billion of the world's population. Most have demographics skewed very young, and most are labeled, "low income" or "low middle income" by the World Bank (i.e., less than $3,000 annual per capita).

If we draw a line around the majority of those military interventions, we have basically mapped the Non-Integrating Gap. Obviously, there are outliers excluded geographically by this simple approach, such as an Israel isolated in the Gap, a North Korea adrift within the Core, or a Philippines straddling the line. But looking at the data, it is hard to deny the essential logic of the picture: If a country is either losing out to globalization or rejecting much of the content flows associated with its advance, there is a far greater chance that the U.S. will end up sending forces at some point. Conversely, if a country is largely functioning within globalization, we tend not to have to send our forces there to restore order to eradicate threats.

Now, that may seem like a tautology – in effect defining any place that has not attracted U.S. military intervention in the last decade or so as "functioning within globalization" (and vice versa). But think about this larger point: Ever since the end of World War II, this country has assumed that the real threats to its security resided in countries of roughly similar size, development, and wealth – in other words, other great powers like ourselves. During the cold war, that other great power was the Soviet Union. When the big Red machine evaporated in the early 1990's, we flirted with concerns about a united Europe, a powerhouse Japan, and – most recently – a rising China.

What was interesting about all those scenarios is the assumption that only an advanced state can truly threaten us. The rest of the world? Those less-developed parts of the world have long been referred to in military plans as the "Lesser Includeds," meaning that if we built a military capable of handling a great power's military threat, it would always be sufficient for any minor scenarios we might have to engage in the less advanced world.

That assumption was shattered by September 11. After all, we were not attacked by a nation or even an army but by a group of – in Thomas Friedman's vernacular – Super Empowered Individuals willing to die for their cause. September 11 triggered a system perturbation that continues to reshape our government (the new Department of Homeland Security), our economy (the de facto security tax we all pay), and even our society (Wave to the camera!). Moreover, it launched the global war on terrorism, the prism through which our government now views every bilateral security relationship we have across the world. In many ways, the September 11 attacks did the U.S. national-security establishment a huge favor by pulling us back from the abstract planning of future high-tech wars against "near peers" into the here-and-now threats to global order. By doing so, the dividing lines between Core and Gap were highlighted, and more important, the nature of the threat environment was thrown into stark relief.

Think about it: Bin Laden and Al Qaeda are pure products of the Gap – in effect, its most violent feed-back to the Core. They tell us how we are doing in exporting security to these lawless areas (not very well) and which states they would like to take "off line" from globalization and return to some seventh-century definition of the good life (any Gap state with a sizable Muslim population, especially Saudi Arabia). If you take this message from Osama and combine it with our military-intervention record of the last decade, a simple security rule set emerges: A country's potential to warrant a U.S. military response is inversely related to its globalization connectivity. There is a good reason why Al Qaeda was based first in Sudan and then later in Afghanistan: These are two of the most dis-connected countries in the world. Look at the other places U.S. Special Operations Forces have recently zeroed in on: northwestern Pakistan, Somalia, Yemen. We are talking about the ends of the earth as far as globalization is concerned. But just as important as "getting them where they live" is stopping the ability of these terrorist networks to access the Core via the "seam states" that lie along the Gap's bloody boundaries. It is along this seam that the Core will seek to suppress bad things coming out of the Gap. Which are some of these classic seam states? Mexico, Brazil, South Africa, Morocco, Algeria, Greece, Turkey, Pakistan, Thailand, Malaysia, the Philippines, and Indonesia come readily to mind. But the U.S. will not be the only Core state working this issue. For example, Russia has its own war on terrorism in the Caucasus, China is working its western border with more vigor,

and Australia was recently energized (or was it cowed?) by the Bali bombing.

If we step back for a minute and consider the broader implications of this new global map, then U.S. national-security strategy would seem to be: 1) Increase the Core's immune system capabilities for responding to September 11-like system perturba-tions; 2) Work the seam states to firewall the Core from the Gap's worst exports, such as terror, drugs, and pandemics; and, most important, 3) Shrink the Gap. Notice I did not just say Mind the Gap. The knee-jerk reaction of many Americans to September 11 is to say, "Let's get off our dependency on foreign oil, and then we won't have to deal with those people." The most naïve assumption underlying that dream is that reducing what little connectivity the Gap has with the Core will render it less dangerous to us over the long haul. Turning the Middle East into Central Africa will not build a better world for my kids. We cannot simply will those people away.

The Middle East is the perfect place to start. Diplomacy cannot work in a region where the biggest sources of insecurity lie not between states but within them. What is most wrong about the Middle East is the lack of personal freedom and how that translates into dead-end lives for most of the population – especially for the young. Some states like Qatar and Jordan are ripe for perestroika-like leaps into better political futures, thanks to younger leaders who see the inevitability of such change. Iran is likewise waiting for the right Gorbachev to come along – if he has not already.

What stands in the path of this change? Fear. Fear of tradition unraveling. Fear of the mullah's dis-approval. Fear of being labeled a "bad" or "traitorous" Muslim state. Fear of becoming a target of radical groups and terrorist networks. But most of all, fear of being attacked from all sides for being different – the fear of becoming Israel. The Middle East has long been a neighborhood of bullies eager to pick on the weak. Israel is still around because it has become – sadly – one of the toughest bullies on the block. The only thing that will change that nasty environment and open the floodgates for change is if some external power steps in and plays Leviathan full-time. Taking down Saddam, the region's bully-in-chief, will force the U.S. into playing that role far more fully than it has over the past several decades, primarily because Iraq is the Yugoslavia of the Middle East – a crossroads of civilizations that has historically required a dictatorship to keep the peace. As baby-sitting jobs go, this one will be a doozy, making our lengthy efforts in postwar Germany and Japan look simple in retrospect.

But it is the right thing to do, and now is the right time to do it, and we are the only country that can. Freedom cannot blossom in the Middle East without security, and security is this country's most influential public-sector export. By that I do not mean arms exports, but basically the attention paid by our military forces to any region's potential for mass violence. We are the only nation on earth capable of exporting security in a sustained fashion, and we have a very good track record of doing it.

Show me a part of the world that is secure in its peace and I will show you strong or growing ties between local militaries and the U.S. military. Show me regions where major war is inconceivable and I will show you permanent U.S. military bases and long-term security alliances. Show me the strongest investment relationships in the global economy and I will show you two postwar military occupations that remade Europe and Japan following World War II. This country has successfully exported security to globalization's Old Core (Western Europe, Northeast Asia) for half a century and to its emerging New Core (Developing Asia) for a solid quarter century follow-ing our mis-handling of Vietnam. But our efforts in the Middle East have been inconsistent – in Africa, almost nonexistent. Until we begin the systematic, long-term export of security to the Gap, it will increasingly export its pain to the Core in the form of terrorism and other instabilities.

Naturally, it will take a whole lot more than the U.S. exporting security to shrink the Gap. Africa, for example, will need far more aid than the Core has offered in the past, and the integration of the Gap will ultimately depend more on private investment than anything the Core's public sector can offer. But it all has to begin with security, because free markets and democracy cannot flourish amid chronic conflict.

Making this effort means reshaping our military establishment to mirror-image the challenge that we face. Think about it. Global war is not in the offing, primarily because our huge nuclear stockpile renders such war unthinkable – for anyone. Meanwhile, classic state-on-state wars are becoming fairly rare. So if the United States is in the process of "transforming" its military to meet the threats of tomorrow, what should it end up looking like? In my mind, we fight fire with fire. If we live in a world increasingly populated by Super-Empowered Individuals, we field a military of Super-Empowered-Individuals.

This may sound like additional responsibility for an already overburdened military, but that is the wrong way of looking at it, for what we are dealing with here are problems of success – not failure. It is America's continued success in deterring global war and obso-lescing state-on-state war that allows us to stick our noses into the far more difficult subnational conflicts and the dangerous transnational actors they spawn. I know most Americans do not want to hear this, but the real battlegrounds in the global war on terrorism are still over there. If gated communities and rent-a-cops were enough, September 11 never would have happened. History is full of turning points like that terrible day, but no turning-back-points. We ignore the Gap's existence at our own peril, because it will not go away until we as a nation respond to the challenge of making globalization truly global.

20

The American Empire: The Burden

Michael Ignatieff

from *The New York Times* magazine (2003)

In a speech to graduating cadets at West Point in June [2002], President Bush declared, "America has no empire to extend or utopia to establish." When he spoke to veterans assembled at the White House in November, he said: America has "no territorial ambitions. We don't seek an empire. Our nation is committed to freedom for ourselves and for others."

Ever since George Washington warned his countrymen against foreign entanglements, empire abroad has been seen as the republic's permanent temptation and its potential nemesis. Yet what word but "empire" describes the awesome thing that America is becoming? It is the only nation that polices the world through five global military commands; maintains more than a million men and women at arms on four continents; deploys carrier battle groups on watch in every ocean; guarantees the survival of countries from Israel to South Korea; drives the wheels of global trade and commerce; and fills the hearts and minds of an entire planet with its dreams and desires. A historian once remarked that Britain acquired its empire in "a fit of absence of mind." If Americans have an empire, they have acquired it in a state of deep denial. But Sept. 11 was an awakening, a moment of reckoning with the extent of American power and the avenging hatreds it arouses. Americans may not have thought of the World Trade Center or the Pentagon as the symbolic headquarters of a world empire, but the men with the box cutters certainly did, and so do numberless millions who cheered their terrifying exercise in the propaganda of the deed.

Being an imperial power, however, is more than being the most powerful nation or just the most hated one. It means enforcing such order as there is in the world and doing so in the American interest. It means laying down the rules America wants (on everything from markets to weapons of mass destruction) while exempting itself from other rules (the Kyoto Protocol on climate change and the International Criminal Court) that go against its interest. It also means carrying out imperial functions in places America has inherited from the failed empires of the 20th century – Ottoman, British and Soviet. In the 21st century, America rules alone, struggling to manage the insurgent zones – Palestine and the northwest frontier of Pakistan, to name but two – that have proved to be the nemeses of empires past.

Iraq lays bare the realities of America's new role. Iraq itself is an imperial fiction, cobbled together at the Versailles Peace Conference in 1919 by the French and British and held together by force and violence since independence. Now an expansionist human rights violator [i.e. Saddam Hussein] holds it together with terror. The United Nations lay dozing like a dog before the fire, happy to ignore Saddam, until an American president seized it by the scruff of the neck and made it bark. Multilateral solutions to the world's problems are all very well, but they have no teeth unless America bares its fangs.

America's empire is not like empires of times past, built on colonies, conquest and the white man's burden. We are no longer in the era of the United Fruit Company, when American corporations needed the Marines to secure their investments overseas. The 21st century imperium is a new invention in the annals of political science, an empire lite, a global hegemony whose grace notes are free markets, human rights and democracy, enforced by the most awesome military power the world has ever known. It is the imperialism of a people who remember that their country secured its independence by revolt against an empire, and who like to think of themselves as the friend of freedom everywhere. It is an empire without consciousness of itself as such, constantly shocked that its good intentions arouse resentment abroad. But that does not make it any less of an empire, with a conviction that it alone, in Herman Melville's words, bears "the ark of the liberties of the world."

In this vein, the president's National Security Strategy, announced in September, commits America to lead other nations toward "the single sustainable model for national success," by which he meant free markets and liberal democracy. This is strange rhetoric for a Texas politician who ran for office opposing nation-building abroad and calling for a more humble America overseas. But Sept. 11 changed everyone, including a laconic and anti-rhetorical president. His messianic note may be new to him, but it is not new to his office. It has been present in the American vocabulary at least since Woodrow Wilson went to Versailles in 1919 and told the world that he wanted to make it safe for democracy.

Ever since Wilson, presidents have sounded the same redemptive note while "frantically avoiding recognition of the imperialism that we in fact exercise," as the theologian Reinhold Niebuhr said in 1960. Even now, as President Bush appears to be maneuvering the country toward war with Iraq, the deepest implication of what is happening has not been fully faced: that Iraq is an imperial operation that would commit a reluctant republic to become the guarantor of peace, stability, democratization and oil supplies in a combustible region of Islamic peoples stretching from Egypt to Afghanistan. A role once played by the Ottoman Empire, then by the French and the British, will now be played by a nation that has to ask whether in becoming an empire it risks losing its soul as a republic.

As the United States faces this moment of truth, John Quincy Adams's warning of 1821 remains stark and pertinent: if America were tempted to "become the dictatress of the world, she would be no longer the ruler of her own spirit." What empires lavish abroad, they cannot spend on good republican government at home: on hospitals or roads or schools. A distended military budget only aggravates America's continuing failure to keep its egalitarian promise to itself. And these are not the only costs of empire. Detaining two American citizens without charge or access to counsel

in military brigs, maintaining illegal combatants on a foreign island in a legal limbo, keeping lawful aliens under permanent surveillance while deporting others after secret hearings: these are not the actions of a republic that lives by the rule of law but of an imperial power reluctant to trust its own liberties. Such actions may still be a long way short of Roosevelt's internment of the Japanese, but that may mean only that the worst – following, say, another large attack on United States citizens that produces mass casualties – is yet to come.

The impending operation in Iraq is thus a defining moment in America's long debate with itself about whether its overseas role as an empire threatens or strengthens its existence as a republic. The American electorate, while still supporting the president, wonders whether his proclamation of a war without end against terrorists and tyrants may only increase its vulnerability while endangering its liberties and its economic health at home. A nation that rarely counts the cost of what it really values now must ask what the "liberation" of Iraq is worth. A republic that has paid a tiny burden to maintain its empire – no more than about 4 percent of its gross domestic product – now contemplates a bill that is altogether steeper. Even if victory is rapid, a war in Iraq and a postwar occupation may cost anywhere from $120 billion to $200 billion.

What every schoolchild also knows about empires is that they eventually face nemeses. To call America the new Rome is at once to recall Rome's glory and its eventual fate at the hands of the barbarians. A confident and carefree republic – the city on a hill, whose people have always believed they are immune from history's harms – now has to confront not just an unending imperial destiny but also a remote possibility that seems to haunt the history of empire: hubris followed by defeat.

II

Even at this late date, it is still possible to ask: Why should a republic take on the risks of empire? Won't it run a chance of endangering its identity as a free people? The problem is that this implies innocent options that in the case of Iraq may no longer exist. Iraq is not just about whether the United States can retain its republican virtue in a wicked world. Virtuous disengagement is no longer a possibility. Since Sept. 11, it has been about whether the republic can survive in safety at home without imperial policing abroad. Face to face with "evil empires" of the past, the republic reluctantly accepted a division of the world based on mutually assured destruction. But now it faces much less stable and reliable opponents – rogue states like Iraq and North Korea with the potential to supply weapons of mass destruction to a terrorist internationale. Iraq represents the first in a series of struggles to contain the proliferation of weapons of mass destruction, the first attempt to shut off the potential supply of lethal technologies to a global terrorist network.

Containment rather than war would be the better course, but the Bush administration seems to have concluded that containment has reached its limits – and the conclusion is not unreasonable. Containment is not designed to stop production of sarin, VX nerve gas, anthrax and nuclear weapons. Threatened retaliation might deter Saddam from using these weapons, but his continued development of them increases his capacity to intimidate and deter others, including the United States. Already his weapons have sharply raised the cost of any invasion, and as time goes by this could become prohibitive. The possibility that North Korea might quickly develop weapons of mass destruction makes regime change on the Korean peninsula all but unthinkable. Weapons of mass destruction would render Saddam the master of a region that, because it has so much of the world's proven oil reserves, makes it what a military strategist would call the empire's center of gravity.

Iraq may claim to have ceased manufacturing these weapons after 1991, but these claims remain unconvincing, because inspectors found evidence of activity after that date. So what to do? Efforts to embargo and sanction the regime have hurt only the Iraqi people. What is left? An inspections program, even a permanent one, might slow the dictator's weapons programs down, but inspections are easily evaded. That leaves us, but only as a reluctant last resort, with regime change.

Regime change is an imperial task par excellence, since it assumes that the empire's interest has a right to trump the sovereignty of a state. The Bush administration would ask, 'What moral authority rests with a sovereign who murders and ethnically cleanses his own people, has twice invaded neighboring countries and usurps his people's wealth in order to build palaces and lethal weapons'? And the administration is not alone. Not even Kofi Annan, the secretary general,

charged with defending the United Nations Charter, says that sovereignty confers impunity for such crimes, though he has made it clear he would prefer to leave a disarmed Saddam in power rather than risk the conflagration of war to unseat him.

Regime change also raises the difficult question for Americans of whether their own freedom entails a duty to defend the freedom of others beyond their borders. The precedents here are inconclusive. Just because Wilson and Roosevelt sent Americans to fight and die for freedom in Europe and Asia doesn't mean their successors are committed to this duty everywhere and forever. The war in Vietnam was sold to a skeptical American public as another battle for freedom, and it led the republic into defeat and disgrace.

Yet it remains a fact – as disagreeable to those left wingers who regard American imperialism as the root of all evil as it is to the right-wing isolationists, who believe that the world beyond our shores is none of our business – that there are many peoples who owe their freedom to an exercise of American military power. It's not just the Japanese and the Germans, who became democrats under the watchful eye of Generals MacArthur and Clay. There are the Bosnians, whose nation survived because American air power and diplomacy forced an end to a war the Europeans couldn't stop. There are the Kosovars, who would still be imprisoned in Serbia if not for Gen. Wesley Clark and the Air Force. The list of people whose freedom depends on American air and ground power also includes the Afghans and, most inconveniently of all, the Iraqis.

The moral evaluation of empire gets complicated when one of its benefits might be freedom for the oppressed. Iraqi exiles are adamant: even if the Iraqi people might be the immediate victims of an American attack, they would also be its ultimate beneficiaries. It would make the case for military intervention easier, of course, if the Iraqi exiles cut a more impressive figure. They feud and squabble and hate one another nearly as much as they hate Saddam. But what else is to be expected from a political culture pulverized by 40 years of state terror?

If only invasion, and not containment, can build democracy in Iraq, then the question becomes whether the Bush administration actually has any real intention of doing so. The exiles fear that a mere change of regime, a coup in which one Baathist thug replaces another, would suit American interests just as well, provided the thug complied with the interests of the

Pentagon and American oil companies. Whenever it has exerted power overseas, America has never been sure whether it values stability – which means not only political stability but also the steady, profitable flow of goods and raw materials – more than it values its own rhetoric about democracy. Where the two values have collided, American power has come down heavily on the side of stability, for example, toppling democratically elected leaders from Mossadegh in Iran to Allende in Chile. Iraq is yet another test of this choice. Next door in Iran, from the 1950's to the 1970's, America backed stability over democracy, propping up the autocratic rule of the shah, only to reap the whirlwind of an Islamic fundamentalist revolution in 1979 that delivered neither stability nor real democracy. Does the same fate await an American operation in Iraq?

International human rights groups, like Amnesty International, are dismayed at the way both the British government of Tony Blair and the Bush administration are citing the human rights abuses of Saddam to defend the idea of regime change. Certainly the British and the American governments maintained a complicit and dishonorable silence when Saddam gassed the Kurds in 1988. Yet now that the two governments are taking decisive action, human rights groups seem more outraged by the prospect of action than they are by the abuses they once denounced. The fact that states are both late and hypocritical in their adoption of human rights does not deprive them of the right to use force to defend them.

The disagreeable reality for those who believe in human rights is that there are some occasions – and Iraq may be one of them – when war is the only real remedy for regimes that live by terror. This does not mean the choice is morally unproblematic. The choice is one between two evils, between containing and leaving a tyrant in place and the targeted use of force, which will kill people but free a nation from the tyrant's grip.

III

Still, the claim that a free republic may sense a duty to help other people attain their freedom does not answer the prudential question of whether the republic should run such risks. For the risks are huge, and they are imperial. Order, let alone democracy, will take a decade to consolidate in Iraq. The Iraqi opposition's blueprints for a democratic and secular federation of Iraq's component peoples – Shiites, Sunnis, Kurds, Turkomans and others – are noble documents, but they are just paper unless American and then international troops, under United Nations mandate, remain to keep the peace until Iraqis trust one another sufficiently to police themselves. Like all imperial exercises in creating order, it will work only if the puppets the Americans install cease to be puppets and build independent political legitimacy of their own.

If America takes on Iraq, it takes on the reordering of the whole region. It will have to stick at it through many successive administrations. The burden of empire is of long duration, and democracies are impatient with long-lasting burdens – none more so than America. These burdens include opening up a dialogue with the Iranians, who appear to be in a political upsurge themselves, so that they do not feel threatened by a United States-led democracy on their border. The Turks will have to be reassured, and the Kurds will have to be instructed that the real aim of United States policy is not the creation of a Kurdish state that goes on to dismember Turkey. The Syrians will have to be coaxed into abandoning their claims against the Israelis and making peace. The Saudis, once democracy takes root next door in Iraq, will have to be coaxed into embracing democratic change themselves.

All this is possible, but there is a larger challenge still. Unseating an Arab government in Iraq while leaving the Palestinians to face Israeli tanks and helicopter gunships is a virtual guarantee of unending Islamic wrath against the United States. The chief danger in the whole Iraqi gamble lies here – in supposing that victory over Saddam, in the absence of a Palestinian-Israeli settlement, would leave the United States with a stable hegemony over the Middle East. Absent a Middle East peace, victory in Iraq would still leave the Palestinians face to face with the Israelis in a conflict in which they would destroy not only each other but American authority in the Islamic world as well.

The Americans have played imperial guarantor in the region since Roosevelt met with Ibn Saud in 1945 and Truman recognized Ben-Gurion's Israel in 1948. But it paid little or no price for its imperial pre-eminence until the rise of an armed Palestinian resistance after 1987. Now, with every day that American power appears complicit in Israeli attacks that kill civilians in the West Bank and in Gaza, and with the Arab nations giving their tacit support to Palestinian suicide bombers, the imperial guarantor finds itself dragged into a regional conflict that is one long hemorrhage of its diplomatic and military authority.

Properly understood, then, the operation in Iraq entails a commitment, so far unstated, to enforce a peace on the Palestinians and Israelis. Such a peace must, at a minimum, give the Palestinians a viable, contiguous state capable of providing land and employment for three million people. It must include a commitment to rebuild their shattered government infrastructure, possibly through a United Nations transitional administration, with U.N.-mandated peace-keepers to provide security for Israelis and Palestinians. This is an awesomely tall order, but if America cannot find the will to enforce this minimum of justice, neither it nor Israel will have any safety from terror. This remains true even if you accept that there are terrorists in the Arab world who will never be content unless Israel is driven into the sea. A successful American political strategy against terror depends on providing enough peace for both Israelis and Palestinians that extremists on either side begin to lose the support that keeps violence alive.

Paradoxically, reducing the size of the task does not reduce the risks. If an invasion of Iraq is delinked from Middle East peace, then all America will gain for victory in Iraq is more terror cells in the Muslim world. If America goes on to help the Palestinians achieve a state, the result will not win over those, like Osama bin Laden, who hate America for what it is. But at least it would address the rage of those who hate it for what it does.

This is finally what makes an invasion of Iraq an imperial act: for it to succeed, it will have to build freedom, not just for the Iraqis but also for the Palestinians, along with a greater sense of security for Israel. Again, the paradox of the Iraq operation is that half measures are more dangerous than whole measures. Imperial powers do not have the luxury of timidity, for timidity is not prudence; it is a confession of weakness.

IV

The question, then, is not whether America is too powerful but whether it is powerful enough. Does it have what it takes to be grandmaster of what Colin Powell has called the chessboard of the world's most inflammable region?

America has been more successful than most great powers in understanding its strengths as well as its limitations. It has become adept at using what is called soft power – influence, example and persuasion – in preference to hard power. Adepts of soft power understand that even the most powerful country in the world can't get its way all the time. Even client states have to be deferred to. When an ally like Saudi Arabia asks the United States to avoid flying over its country when bombing Afghanistan, America complies. When America seeks to use Turkey as a base for hostilities in Iraq, it must accept Turkish preconditions. Being an empire doesn't mean being omnipotent.

Nowhere is this clearer than in America's relations with Israel. America's ally is anything but a client state. Its prime minister has refused direct orders from the president of the United States in the past, and he can be counted on to do so again. An Iraq operation requires the United States not merely to prevent Israel from entering the fray but to make peace with a bitter enemy. Since 1948, American and Israeli security interests have been at one. But as the death struggle in Palestine continues, it exposes the United States to global hatreds that make it impossible for it to align its interests with those Israelis who are opposed to any settlement with the Palestinians that does not amount, in effect, to Palestinian capitulation. The issue is not whether the United States should continue to support the state of Israel, but which state, with which borders and which set of relations with its neighbors, it is willing to risk its imperial authority to secure. The apocalyptic violence of one side and the justified refusal to nego-tiate under fire on the other side leave precious little time to salvage a two-state solution for the Middle East. But this, even more than rescuing Iraq, is the supreme task – and test – of American leadership.

V

What assets does American leadership have at its disposal? At a time when an imperial peace in the Middle East requires diplomats, aid workers and civilians with all the skills in rebuilding shattered societies, American power projection in the area over-whelmingly wears a military uniform. "Every great power, whatever its ideology," Arthur Schlesinger Jr. once wrote, "has its warrior caste." Without realizing the consequences of what they were doing, successive American presidents have turned the projection of

American power to the warrior caste, according to the findings of research by Robert J. Lieber of Georgetown University. In President Kennedy's time, Lieber has found, the United States spent 1 percent of its G.D.P. on the nonmilitary aspects of promoting its influence overseas – State Department, foreign aid, the United Nations, information programs. Under Bush's presidency, the number has declined to just 0.2 percent.

Special Forces are more in evidence in the world's developing nations than Peace Corps volunteers and USAID food experts. As Dana Priest demonstrates in *The Mission*, a study of the American military, the Pentagon's regional commanders exercise more overseas diplomatic and political leverage than the State Department's ambassadors. Even if you accept that generals can make good diplomats and Special Forces captains can make friends for the United States, it still remains true that the American presence overseas is increasingly armed, in uniform and behind barbed wire and high walls. With every American Embassy now hardened against terrorist attack, the empire's overseas outposts look increasingly like Fort Apache. American power is visible to the world in carrier battle groups patrolling offshore and F-16's whistling overhead. In southern Afghanistan, it is the 82nd Airborne, bulked up in body armor, helmets and weapons, that Pashtun peasants see, not American aid workers and water engineers. Each month the United States spends an estimated $1 billion on military operations in Afghanistan and only $25 million on aid.

This sort of projection of power, hunkered down against attack, can earn the United States fear and respect, but not admiration and affection. America's very strength – in military power – cannot conceal its weakness in the areas that really matter: the elements of power that do not subdue by force of arms but inspire by force of example.

VI

It is unsurprising that force projection overseas should awaken resentment among America's enemies. More troubling is the hostility it arouses among friends, those whose security is guaranteed by American power. Nowhere is this more obvious than in Europe. At a moment when the costs of empire are mounting for America, her rich European allies matter financially. But in America's emerging global strategy, they have been demoted to reluctant junior partners. This makes

them resentful and unwilling allies, less and less able to understand the nation that liberated them in 1945.

For 50 years, Europe rebuilt itself economically while passing on the costs of its defense to the United States. This was a matter of more than just reducing its armed forces and the proportion of national income spent on the military. All Western European countries reduced the martial elements in their national identities. In the process, European identity (with the possible exception of Britain) became postmilitary and postnational. This opened a widening gap with the United States. It remained a nation in which flag, sacrifice and martial honor are central to national identity. Europeans who had once invented the idea of the martial nation-state now looked at American patriotism, the last example of the form, and no longer recognized it as anything but flag-waving extremism. The world's only empire was isolated, not just because it was the biggest power but also because it was the West's last military nation-state.

Sept. 11 rubbed in the lesson that global power is still measured by military capability. The Europeans discovered that they lacked the military instruments to be taken seriously and that their erstwhile defenders, the Americans, regarded them, in a moment of crisis, with suspicious contempt.

Yet the Americans cannot afford to create a global order all on their own. European participation in peacekeeping, nation-building and humanitarian reconstruction is so important that the Americans are required, even when they are unwilling to do so, to include Europeans in the governance of their evolving imperial project. The Americans essentially dictate Europe's place in this new grand design. The United States is multilateral when it wants to be, unilateral when it must be; and it enforces a new division of labor in which America does the fighting, the French, British and Germans do the police patrols in the border zones and the Dutch, Swiss and Scandinavians provide the humanitarian aid.

This is a very different picture of the world than the one entertained by liberal international lawyers and human rights activists who had hoped to see American power integrated into a transnational legal and economic order organized around the United Nations, the World Trade Organization, the International Criminal Court and other international human rights and environmental institutions and mechanisms. Successive American administrations have signed on to those pieces of the transnational legal order that

THE AMERICAN EMPIRE: THE BURDEN

suit their purposes (the World Trade Organization, for example) while ignoring or even sabotaging those parts (the International Criminal Court or the Kyoto Protocol) that do not. A new international order is emerging, but it is designed to suit American imperial objectives. America's allies want a multilateral order that will essentially constrain American power. But the empire will not be tied down like Gulliver with a thousand legal strings.

VII

On the new imperial frontier, in places like Afghanistan, Bosnia and Kosovo, American military power, together with European money and humanitarian motives, is producing a form of imperial rule for a postimperial age. If this sounds contradictory, it is because the impulses that have gone into this new exercise of power are contradictory. On the one hand, the semi-official ideology of the Western world – human rights – sustains the principle of self-determination, the right of each people to rule themselves free of outside interference. This was the ethical principle that inspired the decolonization of Asia and Africa after World War II. Now we are living through the collapse of many of these former colonial states. Into the resulting vacuum of chaos and massacre a new imperialism has reluctantly stepped – reluctantly because these places are dangerous and because they seemed, at least until Sept. 11, to be marginal to the interests of the powers concerned. But, gradually, this reluctance has been replaced by an understanding of why order needs to be brought to these places.

Nowhere, after all, could have been more distant than Afghanistan, yet that remote and desperate place was where the attacks of Sept. 11 were prepared. Terror has collapsed distance, and with this collapse has come a sharpened American focus on the necessity of bringing order to the frontier zones. Bringing order is the paradigmatic imperial task, but it is essential, for reasons of both economy and principle, to do so without denying local peoples their rights to some degree of self-determination.

The old European imperialism justified itself as a mission to civilize, to prepare tribes and so-called lesser breeds in the habits of self-discipline necessary for the exercise of self-rule. Self-rule did not necessarily have to happen soon – the imperial administrators hoped to enjoy the sunset as long as possible – but it

was held out as a distant incentive, and the incentive was crucial in co-opting local elites and preventing them from passing into open rebellion. In the new imperialism, this promise of self-rule cannot be kept so distant, for local elites are all creations of modern nationalism, and modern nationalism's primary ethical content is self-determination. If there is an invasion of Iraq, local elites must be "empowered" to take over as soon as the American imperial forces have restored order and the European humanitarians have rebuilt the roads, schools and houses. Nation-building seeks to reconcile imperial power and local self-determination through the medium of an exit strategy. This is imperialism in a hurry: to spend money, to get results, to turn the place back to the locals and get out. But it is similar to the old imperialism in the sense that real power in these zones – Kosovo, Bosnia, Afghanistan and soon, perhaps, Iraq – will remain in Washington.

VIII

At the beginning of the first volume of *The Decline and Fall of the Roman Empire*, published in 1776, Edward Gibbon remarked that empires endure only so long as their rulers take care not to overextend their borders. Augustus bequeathed his successors an empire "within those limits which nature seemed to have placed as its permanent bulwarks and boundaries: on the west the Atlantic Ocean; the Rhine and Danube on the north; the Euphrates on the east; and towards the south the sandy deserts of Arabia and Africa." Beyond these boundaries lay the barbarians. But the "vanity or ignorance" of the Romans, Gibbon went on, led them to "despise and sometimes to forget the outlying countries that had been left in the enjoyment of a barbarous independence." As a result, the proud Romans were lulled into making the fatal mistake of "confounding the Roman monarchy with the globe of the earth."

This characteristic delusion of imperial power is to confuse global power with global domination. The Americans may have the former, but they do not have the latter. They cannot rebuild each failed state or appease each anti-American hatred, and the more they try, the more they expose themselves to the overreach that eventually undermined the classical empires of old.

The secretary of defense may be right when he warns the North Koreans that America is capable of fighting on two fronts – in Korea and Iraq – simultaneously, but Americans at home cannot be overjoyed at such a prospect, and if two fronts are possible at once, a much larger number of fronts is not. If conflict in Iraq, North Korea or both becomes a possibility, Al Qaeda can be counted on to seek to strike a busy and overextended empire in the back. What this suggests is not just that overwhelming power never confers the security it promises but also that even the overwhelmingly powerful need friends and allies. In the cold war, the road to the North Korean capital, Pyongyang, led through Moscow and Beijing. Now America needs its old cold war adversaries more than ever to control the breakaway, bankrupt Communist rogue that is threatening America and her clients from Tokyo to Seoul.

Empires survive when they understand that diplomacy, backed by force, is always to be preferred to force alone. Looking into the still more distant future, say a generation ahead, resurgent Russia and China will demand recognition both as world powers and as regional hegemons. As the North Korean case shows, America needs to share the policing of nonproliferation and other threats with these powers, and if it tries, as the current National Security Strategy suggests, to prevent the emergence of any competitor to American global dominance, it risks everything that Gibbon predicted: overextension followed by defeat.

America will also remain vulnerable, despite its overwhelming military power, because its primary enemy, Iraq and North Korea notwithstanding, is not a state, susceptible to deterrence, influence and coercion, but a shadowy cell of fanatics who have proved that they cannot be deterred and coerced and who have hijacked a global ideology – Islam – that gives them a bottomless supply of recruits and allies in a war, a war not just against America but against her client regimes in the Islamic world. In many countries in that part of the world, America is caught in the middle of a civil war raging between incompetent and authoritarian regimes and the Islamic revolutionaries who want to return the Arab world to the time of the prophet. It is a civil war between the politics of pure reaction and the politics of the impossible, with America unfortunately aligned on the side of reaction. On Sept. 11, the American empire discovered that in the Middle East its local pillars were literally built on sand.

Until Sept. 11, successive United States administrations treated their Middle Eastern clients like gas stations. This was part of a larger pattern. After 1991 and the collapse of the Soviet empire, American presidents thought they could have imperial domination on the cheap, ruling the world without putting in place any new imperial architecture – new military alliances, new legal institutions, new international development organisms – for a postcolonial, post-Soviet world.

The Greeks taught the Romans to call this failure hubris. It was also, in the 1990's, a general failure of the historical imagination, an inability of the post-cold-war West to grasp that the emerging crisis of state order in so many overlapping zones of the world – from Egypt to Afghanistan – would eventually become a security threat at home. Radical Islam would never have succeeded in winning adherents if the Muslim countries that won independence from the European empires had been able to convert dreams of self-determination into the reality of competent, rule-abiding states. America has inherited this crisis of self-determination from the empires of the past. Its solution – to create democracy in Iraq, then hopefully roll out the same happy experiment throughout the Middle East – is both noble and dangerous: noble because, if successful, it will finally give these peoples the self-determination they vainly fought for against the empires of the past; dangerous because, if it fails, there will be nobody left to blame but the Americans.

The dual nemeses of empire in the 20th century were nationalism, the desire of peoples to rule themselves free of alien domination, and narcissism, the incurable delusion of imperial rulers that the "lesser breeds" aspired only to be versions of themselves. Both nationalism and narcissism have threatened the American reassertion of global power since Sept. 11.

IX

As the Iraqi operation looms, it is worth keeping Vietnam in mind. Vietnam was a titanic clash between two nation-building strategies, the Americans in support of the South Vietnamese versus the Communists in the north. Yet it proved impossible for foreigners to build stability in a divided country against resistance from a Communist elite fighting in the name of the Vietnamese nation. Vietnam is now one country, its civil war over and its long-term stability assured. An American operation in Iraq will not face a competing nationalist project, but across the Islamic world it will rouse the nationalist passions of people who want to rule themselves and worship as they please. As Vietnam shows, empire is no match, long-term, for nationalism.

America's success in the 20th century owed a great deal to the shrewd understanding that America's interest lay in aligning itself with freedom. Franklin Roosevelt, for example, told his advisers at Yalta in 1945, when he was dividing up the postwar world with Churchill and Stalin, that there were more than a billion "brown people" living in Asia, "ruled by a handful of whites." They resent it, the president mused aloud. America's goal, he said, "must be to help them achieve independence — 1,100,000,000 enemies are dangerous."

The core beliefs of our time are the creations of the anticolonial revolt against empire: the idea that all human beings are equal and that each human group has a right to rule itself free of foreign interference. It is at least ironic that American believers in these ideas have ended up supporting the creation of a new form of temporary colonial tutelage for Bosnians, Kosovars and Afghans — and could for Iraqis. The reason is simply that, however right these principles may be, the political form in which they are realized — the nationalist nation-building project — so often delivers liberated colonies straight to tyranny, as in the case of Baath Party rule in Iraq, or straight to chaos, as in Bosnia or Afghanistan. For every nationalist struggle that succeeds in giving its people self-determination and dignity, there are more that deliver their people only up to slaughter or terror or both. For every Vietnam brought about by nationalist struggle, there is a Palestinian struggle trapped in a downward spiral of terror and military oppression.

The age of empire ought to have been succeeded by an age of independent, equal and self-governing nation-states. But that has not come to pass. America has inherited a world scarred not just by the failures of empires past but also by the failure of nationalist movements to create and secure free states – and now, suddenly, by the desire of Islamists to build theocratic tyrannies on the ruins of failed nationalist dreams.

Those who want America to remain a republic rather than become an empire imagine rightly, but they have not factored in what tyranny or chaos can do to vital American interests. The case for empire is that it has become, in a place like Iraq, the last hope for democracy and stability alike. Even so, empires survive only by understanding their limits. Sept. 11 pitched the Islamic world into the beginning of a long and bloody struggle to determine how it will be ruled and by whom: the authoritarians, the Islamists or perhaps the democrats. America can help repress and contain the struggle, but even though its own security depends on the outcome, it cannot ultimately control it. Only a very deluded imperialist would believe otherwise.

Cartoon 7 Onward Into the Fog of War
This cartoon points out the irony of invading, bombing and killing for peace. Immodest displays of violence are likely to increase polarization and produce new generations of terrorists seeking revenge. The GWOT creates a self-sustaining cycle of violence.

Source: M. Wuerker

21

America, Right or Wrong

Anatol Lieven

from *America, Right or Wrong* (2004)

At first sight there is something surprising in this strange unrest of so many happy men, restless in the midst of abundance.

(Alexis de Tocqueville)

Traumatized by the terrorist attacks of September 11, 2001, Americans very naturally reacted by falling back on old patterns of belief and behavior. Among these patterns has been American nationalism. This nationalism embodies beliefs and principles of great and permanent value for America and the world, but it also contains very great dangers. Aspects of American nationalism imperil both the nation's global leadership and its success in the struggle against Islamist terrorism and revolution.

More than any other factor, it is the nature and extent of this nationalism which at the start of the twenty-first century divides the United States from a largely postnationalist Western Europe. Certain neoconservative and Realist writers have argued that American behavior in the world and American differences with Europe stem simply from the nation's possession of greater power and responsibility. It would be truer to say that this power enables America to do certain things. What it does, and how it reacts to the behavior of others, is dictated by America's political culture, of which different strands of nationalism form a critically important part.

Insofar as American nationalism has become mixed up with a chauvinist version of Israeli nationalism, it also plays an absolutely disastrous role in U.S. relations with the Muslim world and in fueling terrorism. One might say, therefore, while America keeps a splendid and welcoming house, it also keeps family demons

in its cellar. Usually kept under certain restraints, these demons were released by 9/11.

America enjoys more global power than any previous state. It dominates the world not only militarily, but also to a great extent culturally and economically, and derives immense national benefits from the current world system. Following the death of communism as an alternative version of modernization, American free market liberal democracy also enjoys ideological hegemony over the world. According to all precedents, therefore, the United States ought to be behaving as a conservative hegemon, defending the existing international order and spreading its values by example. After all, following World War II, the United States itself played the leading part in creating the institutions which between 2001 and 2003 the Bush administration sought to undermine.

Instead, under George W. Bush the nation was drawn toward the role of an unsatisfied and even revolutionary power, kicking to pieces the hill of which it is the king. In particular, many observers saw the idea of preventive war against potential threats (rather than preemptive war against imminent ones) as a decisive shift not only to unilateralism, but to a revolutionary, anti-status quo position in international affairs, a position reminiscent of Wilhelmine Germany before 1914 rather than Victorian Britain.

This book seeks to help explain why a country which after the terrorist attacks of September 11, 2001, had the chance to create a concert of all the world's major states – including Muslim ones – against Islamist revolutionary terrorism chose instead to pursue policies which divided the West, further alienated the Muslim world and exposed America itself to greatly increased danger. The most important reason why this has occurred is the character of American nationalism, which I analyze as a complex, multifaceted set of elements in the nation's political culture.

[...]

Nationalism has not been the usual prism through which American behavior has been viewed. Most Americans speak of their attachment to their country as patriotism or, in an extreme form, superpatriotism. Critics of the United States, at home and abroad, tend to focus on what has been called American imperialism. The United States today does harbor important forces which can be called imperialist in their outlook and aims. However, although large in influence, people holding these views are relatively few in number. They are to be found above all in overlapping sections of the intelligentsia and the foreign policy and security establishments, with a particular concentration among the so-called neoconservatives.

Unlike large numbers of Englishmen, Frenchmen and others at the time of their empires, the vast majority of ordinary Americans do not think of themselves as imperialist or as possessing an empire. As the aftermath of the Iraq War seems to be demonstrating, they are also not prepared to make the massive long-term commitments and sacrifices necessary to maintain a direct American empire in the Middle East and elsewhere.

Apart from the effects of modern culture on attitudes to military service and sacrifice, American culture historically has embodied a strong strain of isolationism. This isolationism is, however, a complex phenomenon which should not be understood simply as a desire to withdraw from the world. Rather, American isolationism forms another face of both American chauvinism and American messianism, in the form of a belief in America as a unique city on a hill. As a result, it is closely related to nationalist unilateralism in international affairs, since it forms part of a view that if the United States really has no choice at all but to involve itself with disgusting and inferior foreigners, it must absolutely control the process and must under no circumstances subject itself to foreign control or even advice.

Unlike previous empires, the U.S. national identity and what has been called American Creed are founded on adherence to democracy. However imperfectly democracy may be practiced at home and hypocritically preached abroad, democratic faith does set real limits to how far the United States can exert direct rule over other peoples. Therefore, since 1945 the United States has been an indirect empire resembling more closely the Dutch in the East Indies in the seventh and eighteenth centuries than the British in India.

As far as the mass of the American people is concerned, even an indirect American empire is still an empire in denial. In presenting its imperial plans to the American people, the Bush administration has been careful to package them as something else: on one hand, as part of a benevolent strategy of spreading American values of democracy and freedom; on the other, as an essential part of the defense not of an American empire, but of the nation itself.

A great many Americans are not only intensely nationalistic, but bellicose in response to any perceived attack or slight against the United States: "Don't Tread on Me!" as the rattlesnake on the American revolutionary flag declared. This attitude was summed up by that American nationalist icon, John Wayne, in his last role, as a dying gunfighter in the film *The Shootist* "I won't be wronged, I won't be insulted, and I won't be laid a hand on. I don't do these things to other people, and I require the same from them."

As an expression of pride, honor and a capacity for self-defense, these are sympathetic and indeed admirable words. However, in this context it is useful to remember an eighteenth-century expression, "to trail one's coat." This phrase means deliberately provoking a quarrel by allowing your coat to trail along the ground, that another man would step on it, thereby allowing you to challenge him to a duel. One might say that American imperialists trail America's coat across the whole world while most ordinary Americans are not looking and rely on those same Americans to react with "don't tread on me" nationalist fury when the coat is trodden on.

Coupled with an intense national solipsism and ignorance of the outside world among the American public, and with particular American prejudices against the religion of Islam, this bellicose nationalism has allowed a catastrophic extension of the war on terrorism from its original – and legitimate – targets of al Qaeda and the Taliban to embrace the Iraqi Ba'athist regime, anti-Israeli groups in Palestine and Lebanon and quite possibly other countries and forces in future. This reserve of embittered nationalism has also been tapped with regard to a wide range of international proposals which can be portrayed as hurting America or infringing on its national sovereignty, from the International Criminal Court to restrictions on greenhouse gas emissions.

A mixture of American energy interests and the addiction of most Americans to the automobile might well have killed the nation's adherence to the Kyoto Treaty in any case. The treaty's American opponents were however tremendously helped by that section of opinion whose political culture means that they see any international treaty involving sacrifices and commitments by the United States as a plot by hostile and deceitful aliens. Many Americans genuinely believe these ideas to be a matter of self-defense – of their economy, their way of life, their freedoms or their very nation.

This background helps explain tragicomic statistics such as the fact that the majority of Americans believe that their country spends more than 20 percent of its budget on foreign aid and that this figure should be reduced; the true figure is less than 1 percent and is the lowest in the developed world. Evidence like this allows international critics of American hegemony to portray the nation as a purely selfish imperial power, without generosity and without real vision. This pattern is strange, and very sad, when contrasted with the tremendous generosity of many Americans when it comes to domestic and private charity, and brings out the degree to which chauvinist nationalism can undermine even the noblest of impulses.

Under the administration of George W. Bush (Bush Jr.) the United States drove toward empire, but the domestic political fuel fed into the engine was that of a wounded and vengeful nationalism. After 9/11, this sentiment is entirely sincere as far as most Americans are concerned, and it is all the more dangerous for that. In fact, to judge by world history, there is probably no more dangerous element in the entire nationalist mix than a sense of righteous victimhood. In the past this sentiment helped wreck Germany, Serbia and numerous other countries, and it is now in the process of wrecking Israel.

THE TWO SOULS OF AMERICAN NATIONALISM

Like other nationalisms, American nationalism has many different faces. It concentrates on what I take to be the two most important elements in the historical culture of American nationalism and the complex relationship between them. Erik Erikson wrote that "every national character is constructed out of polarities." As I shall show, this is certainly true of the United States, which embodies among other things both the most modern and the most traditionalist society in the developed world.

The clash between those societies is contributing to the growing political polarization of American society. At the time of this writing, the American people more sharply and more evenly divided along party lines than at any time in modern history. This political division in turn reflects greater differences in social and cultural attitudes than at any time since the Vietnam War. White evangelical Protestants vote Republican rather than Democrat by a factor of almost two to one, with corresponding effects on the parties' stances on abortion and other moral issues. The gap is almost as great when it comes to nationalism, with 71 percent of Republicans in 2003 describing themselves as "very patriotic" compared to 48 percent of Democrats. This difference reflects in part racial political allegiances, with 65 percent of Whites describing themselves as "very patriotic" in that year to only 38 percent of Blacks. Gaps concerning attitudes to crime and faith in American business are even greater.

It is however not the opposition, but the combination of these different strands which determines the overall nature of the American national identity and largely shapes American attitudes and policies toward the outside world. This combination was demonstrated by the Bush administration, which [. . .] drew its rhetoric at least from both main strands of American nationalism simultaneously.

The first of these strands [. . .] stems from what has been called the "American Creed," an idea I also describe as the "American Thesis": the set of great democratic, legal and individualist beliefs principles on which the American state and constitution is founded. These principles form the foundation of American civic nationalism and also help bind the United States to the wider community of democratic states. They are shared with other democratic societies, but in America they have a special role in holding a disparate nation together. As the term "Creed" implies, they are held with an ideological and almost religious fervor.

The second element forms what I have called the American nationalist "antithesis" and stems above all from ethnoreligious roots. Aspects of this tradition have also been called "Jacksonian nationalism," after President Andrew Jackson (1767–1845). [. . .] Because the United States is so large and complex compared to other countries, and has changed so much over time, the nationalist tradition is correspondingly complex.

Rather than the simple, monolithic identity of a Polish or Thai ethnoreligious nationalism, this tradition in the United States forms a diffuse mass of identities and impulses, including nativist sentiments on the part of America's original White population, the particular culture of the White South and the beliefs and agendas ethnic lobbies. Nonetheless, these nationalist features can often be clearly distinguished from the principles of the American Creed and of American civic nationalism; and although many of their features are specifically American – notably, the role of fundamentalist Protestantism – they are also related to wider patterns of ethnoreligious nationalism in the world.

These strands in American nationalism are usually subordinate to American civic nationalism stemming from the Creed, which dominates official and public political culture. However, they have a natural tendency to rise to the surface in times of crisis and conflict. In the specific case of America's attachment to Israel, ethnoreligious factors have become dominant, with extremely dangerous consequences for the war on terrorism.

The reason why "civic nationalism," rather than "patriotism," is the appropriate name for the dominant strand in American political culture was well summed up in 1983 by one of the fathers of the neoconservative school in the United States, Irving Kristol:

> Patriotism springs from love of the nation's past; nationalism arises out of hope for the nation's future, distinctive greatness. . . . The goals of American foreign policy must go well beyond a narrow, too literal definition of "national security." It is the national interest of a world power, as this is defined by a sense of national destiny.

In drawing this distinction, Kristol echoed a classic distinction between patriotism and nationalism delineated by Kenneth Minogue, one of the great historians of nationalism. Minogue defined patriotism as essentially conservative, a desire to defend one's country as it actually is; whereas nationalism is a devotion to an ideal, abstract, unrealized notion of one's country, often coupled with a belief in some wider national mission to humanity. In other words, nationalism has always had a certain revolutionary edge to it. In American political culture at the start of the twenty-first century, there is certainly a very strong element of patriotism, of attachment to American institutions and to America in its present form; but as Kristol's words indicate, there is also a revolutionary element, a commitment to a messianic vision of the nation and its role in the world. [. . .]

As the American historian and social critic Richard Hofstadter (1917–1970) wrote, "The most prominent and pervasive failing [of American political culture] is a certain proneness to fits of moral crusading that would be fatal if they were not sooner or later tempered with a measure of apathy and common sense." This pattern has indeed repeated itself in our time, with the aftermath of the Iraq War leading to a new sobriety in American policies and the American public mood. In the meantime, however, the Bush administration's appeal to this crusading and messianic spirit had played a major part in getting the nation into Iraq in the first place.

If Minogue's and Kristol's distinction between patriotism and nationalism is valid, then it must be acknowledged that nationalism, rather than patriotism, is the correct word with which to describe the characteristic national feeling of Americans. And this feature also links the American nationalism of today to the unsatisfied, late-coming nationalisms of Germany, Italy and Russia, rather than to the satisfied and status-quo patriotism of the British. Thus this feature helps explain the strangely unsatisfied, Wilhelmine air of U.S. policy and attitudes at the start of the twenty-first century.

But if one strand of American nationalism is radical because it looks forward to "the nation's future, distinctive greatness," another is radical because it continuously looks backward, to a vanished and idealized national past. This "American antithesis" is a central feature of American radical conservatism: the world of the Republican Right and especially the Christian Right, with their rhetoric of "taking back" America and restoring an older, purer American society. [. . .] [T]his long-standing tendency in American culture and politics reflects the continuing conservative religiosity of many Americans; however, it also has always been an expression of social, economic, ethnic and above all racial anxieties.

In part, these anxieties stem from the progressive loss of control over society by the "original" White

Anglo-Saxon and Scots Irish populations, later joined by others. Connected to these concerns are class anxieties – in the past, the hostility of the small towns and countryside to the new immigrant-populated cities; today, the economic decline of the traditional White working classes. As a result of economic, cultural and demographic change, in America, the supremely victorious nation of the modern age, large numbers of Americans feel defeated. The domestic anxieties which this feeling of defeat generates spill over into attitudes to the outside world, with 64 percent of Americans in 2002 agreeing that "our way of life needs to be protected against foreign influence," compared to 51 percent of British and 53 percent of French. These figures lie between those for Western Europe and those for developing world countries such as India (76 percent) – which is piquant, because the "foreign influence" that Indian and other cultural nationalists in the developing world most fear is, of course, that of the United States.

These fears help give many American nationalists their curiously embittered, mean-spirited and defensive edge, so curiously at variance with America's image and self-image as a land of success, openness, wealth and generosity. Over the years, the hatred generated by this sense of defeat and alienation has been extended to both domestic and foreign enemies.

This too is a very old pattern in different nationalisms worldwide. Historically speaking, in Europe at least, radical conservatism and nationalism have tended to stem from classes and groups in actual or perceived decline as a result of socioeconomic change. One way of looking at American nationalism and the troubled relationship with the contemporary world which the nation dominates is indeed to understand that many Americans are in revolt against the world which America itself has made.

However, except for the extreme fringe among the various "militia" groups, the neo-Nazis and so on, these forces of the American antithesis are not in public revolt against the American Creed and American civic nationalism as such. Most radical nationalist and radical conservative movements elsewhere in the world in the past at least opposed democracy and demanded authoritarian rule. By contrast, Americans from this tradition generally believe strongly in the American democratic and liberal Creed. However, they also believe – consciously or unconsciously, openly or in private – that it is the product of a specific White Christian American civilization, and that both

are threatened by immigration, racial minorities and foreign influence. And I am not saying that they are necessarily wrong; a discussion of this point lies outside the scope of this book. I am only pointing out that people with this belief naturally feel embattled, embittered and defensive as a result of many contemporary trends.

American Protestant fundamentalist groups also do not reject the Creed as such. In terms of their attitude to culture and the intellect, however, their rejection of contemporary America is even deeper, for they reject key aspects of modernity itself. For them modern American mass culture is a form of daily assault on their passionately held values, and their reactionary religious ideology in turn reflects the sense of social, cultural and racial embattlement among their White middle-class constituency. Even as America is marketing the American Dream to the world, at home many Americans feel that they are living in an American nightmare.

America is the home of by far the most deep, widespread and conservative religious belief in the Western world, including a section possessed by wild millenarian hopes, fears and hatreds – and these two phenomena are intimately related. As a Pew Research Center Survey of 2002 demonstrates, at the start of the twenty-first century the United States as a whole is much closer to the developing world in terms of religious belief than to the industrialized countries (although a majority of believers in the United States are not fundamentalist Protestants but Catholics and "mainline," more liberal Protestants). The importance of religion in the contemporary United States continues a pattern evident since the early nineteenth century and remarked by Alexis de Tocqueville in the 1830s, when religious belief among the European populations had been shaken by several decades of the Enlightenment and the French Revolution but American religious belief was fervent and nearly universal.

As of 2002, with 59 percent of respondents declaring that "religion plays a very important role in their lives," the United States lies between Mexico (57 percent) and Turkey (65 percent) but is very far from Canada (30 percent), Italy (27 percent), or Japan (12 percent). In terms of sheer percentage points, it is indeed closer on this scale to Pakistan (91 percent) than to France (12 percent). As of 1990, 69 percent of Americans believed in the personal existence of the Devil, compared to less than half that number of

Britons. When a U.S. senator exclaimed (apocryphally) of the Europeans, "What common values? They don't even go to church!" he was expressing a truth, and this is as true of the U.S. political elites (but not of the cultural or economic ones) as of the population in general. Among the fundamentalist Protestant sections of the United States, there has been a strong historical inclination to a paranoid style, originally directed against Catholics, Freemasons and others, and perpetuated by the Cold War and the communist threat. In our own time, "the recent Evangelical engagement with public life reflects religious and cultural habits that Anglo-American Protestants, both liberal and Evangelical, learned when threatened by Americans of different religious and ethnic backgrounds."

The extreme tension between these fundamentalist religious values and the modern American mass culture which now surrounds them is an important cause of the mood of beleaguered hysteria on the American Right which so bewilders outside observers. Across large areas of America, these religious beliefs in turn form a central part of the identity of the original White American colonist population, above all in the Greater South, or what former First Lady Lady Bird Johnson described simply as "*us* – the simple American stock."

The religious beliefs of large sections of this core population are under constant, daily threat from modern secular culture, above all via the mass media. And perhaps of equal importance in the long term will be the relative decline in recent decades in the real incomes of the American middle classes, where these groups are situated socially. This decline and the wider economic changes which began with the oil shock of 1973 have had the side effect of helping force more and more women to go to work, thereby undermining traditional family structures even among those groups most devoted to them.

The relationship between this traditional White Protestant world on one hand and the forces of American economic, demographic, social and cultural change on the other may be compared to the genesis of a hurricane. A mass of warm, humid air rises from the constantly churning sea of American capitalism to meet a mass of cooler layers of air, and as it rises it sucks in yet more air from the sides, in the form of immigration. The cooler layers are made up of the White middle classes and their small-town and suburban worlds in much of the United States; the old White populations of the Greater South with their specific culture; and the especially frigid strata of old Anglo-Saxon and Scots Irish fundamentalist Protestantism.

The result of this collision is the release of great bolts and explosions of political and cultural electricity. Like a hurricane, the resulting storm system is essentially circular, continually chasing its own tail, and essentially self-supporting, generating its own energy – until, at some unforeseeable point in future, either the boiling seas of economic change cool down or the strata of religious belief and traditional culture dissolve. Among these bolts is hatred, including nationalist hatred.

Externally directed chauvinist hatred must therefore be seen as a byproduct of the same hatred displayed by the American Right at home, notably in their pathological loathing of President Bill Clinton. In Europe, Clinton was generally seen as a version of Tony Blair, a centrist who "modernized" his formerly center-Left party by stealing most of the clothes of the center-Right and adopting a largely right-wing economic agenda. To radical conservatives in America, this was irrelevant. They hated him not for what he *did*, but for what he is: the representative of a multiracial, pluralist and modernist culture and cultural elite which they both despise and fear, just as they hate the atheist, decadent, unmanly Western European nations not only for what they do, but for what they are.

In the U.S. context it is also crucial to remember that as in a hurricane or thunderstorm, rather than simply being opposing forces, the two elements which combine to produce this system work together. In a curious paradox, the unrestrained free market capitalism which is threatening the old conservative religious and cultural communities of Protestant America with dissolution is being urged on by the political representatives of those same communities.

This was not always so. In the 1890s and 1900s, this sector of America formed the backbone of the Populist protest against the excesses of American capitalism, and in the 1930s, it voted solidly for Roosevelt's New Deal. Today, however, the religious Right has allied itself solidly with extreme free market forces in the Republican Party – although it is precisely the workings of unrestrained American capitalism which are eroding the world the religious conservatives wish to defend.

The forces of radical capitalism in the United States may come to depend more and more on appeals to radical conservatism and nationalism to win votes and to defend their class interests. [. . .] The clash between cultural and social loyalties and the imperatives of capitalist change is an old dilemma for those social and cultural conservatives who at the same time are dedicated to the preservation of free market economics. As the distinguished U.S. political and ethical thinker Garry Wills has noted, "There is nothing less conservative than capitalism, so itchy for the new."

[. . .]

THE THREAT TO AMERICAN HEGEMONY

Because of a deep-rooted (and partly justified) belief in American exceptionalism and the decline of the study of history in American academia, Americans are not used to studying their own nationalism in a Western historical context – and it is vitally important that they begin to do so. For surely no sane person, looking at the history of nationalist Europe in the century prior to 1945, would suggest that the United States should voluntarily follow such a path. In particular, American nationalism is beginning to conflict very seriously with any enlightened, viable or even rational version of American imperialism; that is to say, with the interests of the United States as world hegemon and heir to the roles of ancient Rome and China within their respective regions.

Nationalism provides one clue to the difference between the strategy and philosophy of Clinton and those of George W. Bush and to the difference between an American approach which seeks legitimacy for American hegemony and one which makes a public cult of the unrestrained exercise of American will.

A number of highly distinguished American and other observers have, however, seen little basic difference between the international policies of Clinton and Bush. People on the Left view the policies of all U.S. administrations as reflecting above all the enduring dynamics and requirements of an imperial version of American capitalism: the domination of the world by capitalism and the primacy of the United States within the capitalist system. This analysis is indeed partly true, but in emphasizing common goals, left-wing analysts have a tendency to lose sight of certain other highly important factors: the means used to achieve these ends; the difference between intelligent and stupid means; and the extent to which the choice of means is influenced by irrational sentiments which are irrelevant or even contrary to the goals pursued. Of the irrational sentiments which have contributed to wrecking intelligent capitalist strategies – not only

today, but for most of modern history – the most important and dangerous is nationalism.

Walter Russell Mead, an American nationalist and no Marxist, sees Bush's globalization of the Monroe Doctrine as a process stretching back to World War II. Andrew Bacevich and Chalmers Johnson, basing their work in part on the analysis of the economic and institutional roots of American imperialism by William Appleman Williams, also see the administrations of Clinton and Bush as characterized by an essential continuity when it comes to the extension of American power.

For them, Bush's Iraq is just Clinton's Kosovo or Haiti on a much larger scale and with greatly increased risks. Clinton after all moved rather quickly to combat Russia's plans to retain a sphere of influence on the territory of the former Soviet Union and was not too scrupulous about the regimes he helped in the process. Clinton preserved the North Atlantic Treaty Organization (NATO) as what was then seen as the essential vehicle of U.S. strategic dominance in Europe and, as Basevich argues, fought the Kosovo war largely to justify NATO's continued existence as this vehicle.

Clinton, however, although dedicated to American hegemony, was not an American chauvinist. His vision of global order involved American hegemonic leadership rather than dictation and a desire to "place America at the center of every network" rather than simply to dictate in every situation. This at least was certainly the perception of his critics on the American Right, one of whose leaders accused Clinton of "moving us incrementally into a network of global organizations."

This desire to exercise American leadership through international institutions is an important strand in American international policy dating back to World War II. It stems in part from a conscious determination not to repeat the U.S. mistake of withdrawal from the world after 1919 and in part from the international needs and perceptions of American capitalism. Thus although partisans of the Bush administration repeatedly described its rhetoric of democratization and humanitarian intervention after 2001 as Wilsonian, such an attribution is quite wrong in historical terms, for President Woodrow Wilson also believed passionately in the creation of international institutions and in exerting U.S. power and influence through those institutions. Clinton, not Bush, therefore was the true Wilsonian of our time.

Moreover, Clinton's version of American hegemonic leadership, although often resented by the leaders of other states, was nonetheless far more acceptable to most than Bush's approach from 2001 to 2003. Clinton's strategy was detested by Russians and others who saw its content as a threat to their geopolitical interests and its democratizing language as arrogant, mendacious and hypocritical. Nonetheless, it was greatly preferred by most world governments to the Bush administration's approach in its first three years of power, since it paid some attention to their interests and, equally important, did not publicly humiliate them before their own populations by demanding ostentatiously servile displays of deference and obedience.

The dominant forces of the Bush administration in 2001–2003 were much more overt imperialists than their predecessors. Moreover, in response to their own sentiments but also to appeal to the American people, they made things worse by packaging imperialism as American nationalism, thereby adopting a number of gratuitously unilateralist measures and approaches. And this was no pose or piece of cynical manipulation of American nationalism. Bush, his leading officials and his intellectual and media supporters are genuinely motivated by nationalism, in a way that Clinton was not; and as nationalists, they are absolutely contemptuous of any global order involving any check whatsoever on American behavior and interests.

The harshly nationalist character of the Bush administration was evident from its coming to power at the start of 2001. A whole set of moves bitterly alienated much of the rest of the world and created a level of hostility to the administration in Europe which contributed greatly to the later rejection of the Iraq War by large majorities in most European countries. As antiterrorism coordinator Clarke remarked presciently in the summer of 2001, "If these guys in this administration are going to want an international coalition to invade Iraq next year, they are sure not making a lot of friends."

The rejection of vital international treaties on arms control seemed motivated by a blind nationalist desire for absolute American freedom of action and increased dangers to the United States from terrorism using weapons of mass destruction. The spirit behind these moves was described by John Bolton, later under secretary of state for arms control and international security, as "Americanism," but nationalism is a simpler description.

[. . .]

Most damaging of all to U.S. prestige in Europe was the outright rejection of the Kyoto Protocol on greenhouse gas emissions and the abandonment of early attempts by U.S. officials to find a substitute – a decision taken in a way which displayed utter contempt both for the international community and American allies in Europe, but also for moderate sections of Bush's own administration. This indifference to environmental threats will probably also be the strongest criticism leveled at the United States and its hegemony by future generations. The attitude to environmental policy displayed by the Bush administration therefore undermines the United States not only today, but in its role as the new Rome, a civilizational force transcending the current epoch.

For coupled with the growing craze for gas-guzzling sport utility vehicles (SUVs) in the U.S. middle classes, this more than anything else seemed to suggest that Americans are interested in using their power over the planet purely for their own most selfish and shortsighted interests and that talk of wider U.S. responsibilities was utter hypocrisy. The former Energy Secretary Paul O'Neill attributed the White House decision on Kyoto to a feeling of "the base likes this and who the hell knows anyway" – not a sentiment calculated to increase faith in American leadership and decision making elsewhere in the world.

In the vision set out in its new National Security Strategy of 2002 (NSS 2002), embodying the so-called Bush Doctrine, American sovereignty was to remain absolute and unqualified. The sovereignty of other countries, however, was to be heavily qualified by America, and no other country was to be allowed a sphere of influence, even in its own neighborhood. In this conception, "balance of power" – a phrase used repeatedly in the NSS – was a form of Orwellian doublespeak. The clear intention actually was to be so strong that other countries had no choice but to rally to the side of the United States, concentrating all real power and freedom of action in the hands of America.

This approach was basically an attempt to extend a tough, interventionist version of the Monroe Doctrine (the "Roosevelt Corollary" to the Doctrine, laid down by President Theodore Roosevelt) to the entire world. This plan is megalomaniac, completely impracticable (as the occupation of Iraq has shown) and totally unacceptable to most of the world. Because, however, this program was expressed in traditional American nationalist terms of self-defense and the messianic role of the United States in spreading freedom, many Americans found it entirely acceptable and indeed natural.

The accusation against the Bush administration then is that like the European elites before 1914, it has allowed its own national chauvinism and limitless ambition to compromise the security and stability of the world capitalist system of which America is the custodian and greatest beneficiary. In other words, members of the administration have been irresponsible and dangerous not in Marxist terms, but in their own. They have offended against the Capitalist Peace.

This difference is terribly important from the point of view of the stability of the world and of U.S. hegemony in the world. A relatively benign version of American hegemony is by no means unacceptable to many people around the world – both because they often have neighbors whom they fear more than America and because their elites are to an increasing extent integrated into a global capitalist elite whose values are largely defined by those of America. But American imperial power in the service of narrow American (and Israeli) nationalism is a very different matter and is an extremely unstable base for hegemony. It involves power over the world without accepting any responsibility for global problems and the effects of U.S. behavior on other countries – and power without responsibility was defined by Rudyard Kipling as "the prerogative of the harlot throughout the ages."

American nationalism has already played a key role in preventing America from taking advantage of the uniquely beneficent world-historical moment following the fall of communism. [. . .] [I]nstead of using this moment to create a "concert of powers" in support of regulated capitalist growth, world stability and the relief of poverty, preventable disease and other social ills, nationalism has helped direct America into a search for new enemies.

Such nationalism may encourage its adherents to cultivate not only specific national hatreds, but also hostility to all ideals, goals, movements, laws and institutions which aim to transcend the nation and speak for the general interests of humankind. This form of nationalism is therefore in direct opposition to the universalist ideals and ambitions of the American Creed – ideals upon which, in the end, rests America's role as a great civilizational empire and heir to Rome and China, and upon which is based America's claim to represent a positive example to the world. These ideals form the core of what Joseph Nye has called "soft power" in its specifically American form.

[. . .]

Nationalism therefore risks undermining precisely those American values which make the nation most admired in the world and which in the end provide both a pillar for its current global power and the assurance that future ages will look back on it as a benign and positive leader of humanity.

The historical evidence of the dangers of un-reflecting nationalist sentiments should be all too obvious and are all too relevant to U.S. policy today. Nationalism thrives on irrational hatreds and on the portrayal of other nations or ethnoreligious groups as congenitally, irredeemably wicked and hostile. Yesterday many American nationalists felt this way about Russia. Today, prejudices are likely to be directed against the Arab and Muslim worlds – and to a lesser extent any country that defies American wishes. Hence the astonishing explosion of chauvinism directed against France and Germany in the run-up to the war in Iraq. [. . .] Other nations are declared to be irrationally, incorrigibly and unchangingly hostile. This being so, it is obviously pointless to seek compromises with them or to accommodate their interests and views. And because they are irrational and barbarous, America is free to dictate to them or even to conquer them for their own good. This was precisely the discourse of nationalists in the leading European states toward each other and lesser breeds before 1914, which helped drag Europe into the great catastrophes of the twentieth century.

PART FOUR

The Geopolitics
of Global Dangers

INTRODUCTION TO PART FOUR

Simon Dalby

New threats stalk our world: the return of the plague, bioterrorism, environmental degradation, climate change, radioactive fallout, ozone layer depletion, violence over resources and conflict diamonds in particular. Some of them are obviously technological, others appear natural in some ways; perhaps the scariest of all are the diseases that have been sinisterly modified to use as weapons: genetically altered bacteria, anthrax weapons and dirty bombs are all part of current geopolitical fears. Most of them have imprecise geographies; they are only sometimes mapped, but are always understood as threats out there somewhere that threaten our well-being in the supposed safe domestic spaces of our lives and communities. This is the new agenda of global security in the age of geopolitical terror and "bio-anxiety" (Hartmann et al., 2005).

All these things are of pressing concern because we now apparently live in globalized times where information, people and shipping containers move round the world all the time. The possibility of carefully controlled frontiers where border guards check both people and packages seems to be slipping away no matter how many X-ray machines appear at airports and how huge the Department of Homeland Security is in the United States. Dangers from "over there" are now potentially "in here." Environmental disruptions are also setting people and plagues in motion. These fears are accelerated by instant television news, email and text message alerts right to your cell phone. The specific dangers are related to a changing understanding of the "global," a sense of living in a small interconnected planet, which the frequent use of the Apollo photograph of the whole earth and other global corporate logos have emphasized since the early 1990s.

ENVIRONMENTAL GEOPOLITICS

Ideas that "humanity" can collectively change the face of the planet and endanger vital biological systems have become widespread matters of concern only since World War II, and a serious matter for public geopolitical discussion in the last few decades. Recently additional environmental and scientific expertise in natural systems, climate change, ocean currents, stratospheric chemistry and many other fields has been added to the discussions of national and international security in Western states. In the process the important concepts of "global" problems and "global" security have become part of the geopolitical lexicon.

During the 1950s and 1960s episodes such as mercury poisoning at Minimata in Japan, fears about widespread use of pesticides in the United States, killer smogs in London and oil spills from a number of high profile tanker accidents introduced environmental themes onto the political agenda in many states (Sandbach, 1980). In the 1960s the issue of nuclear fallout from Cold War weapon "test" explosions which affected people worldwide connected the fate of all inhabitants to the consequences of geopolitical rivalry. The agreement to ban atmospheric tests by the United States, the Soviet Union and the United Kingdom in 1963 was an early international environmental agreement dealing with a problem that had global ramifications because radioactive fallout from weapons tests traveled round the world in the atmosphere. Environment and geopolitics were linked firmly then and have remained so since (Soroos, 1997).

Alarmist predictions of looming natural, resource, population and pollution "limits to growth" were published in the industrialized states in the late 1960s and early 1970s and drew considerable international attention (Ehrlich, 1968: Meadows et al., 1974). "Limits to growth" seemed to be immanent to many people in Western states following disruptions of oil supplies as a result of the oil embargo and price rises introduced by the Organization of Petroleum Exporting Countries (OPEC) and manipulations of oil prices by the transnational oil companies during and after the Yom Kippur/October war between Syria, Egypt and Israel in 1973. Increased attention was paid, by American foreign policy makers in particular, to the possibilities of using military intervention around the world to ensure that supplies of crucial resources, especially oil from the Middle East, would not be interrupted by either local political instability, or by Soviet political and military action. The geopolitical assumptions then, and subsequently codified in the Carter doctrine applied to the Persian Gulf at the end of the 1970s, were that the flow of oil from outside the West had to be maintained come what may and despite what people in the states that had the supplies might have to say about the matter (Yergin, 1991; Klare, 2004).

In the mid-1970s concerns were also raised in the United States about supplies of minerals from African states and elsewhere needed for military equipment production. These scenarios of resource "strangulation" fit well with the Cold War geopolitical understanding of the world as one of geopolitical rivalry between the Cold War blocs, with the "Third World" as the arena in which the contest for global domination was played out. But some prominent environmentalists in this period argued that the best method of ensuring resource security, at least in the sense of oil supplies, was to work hard at improving conservation measures and introducing such things as efficient building heating systems and automobile engines (Lovins, 1977). They argued that doing so would reduce the need for military interventions in OPEC states or elsewhere as part of the geopolitical rivalry of the Cold War, and simultaneously clean up pollution in the industrialized states while costing much less than military preparations for intervention.

Numerous things around the world can fairly easily be interpreted as threats requiring control and "management" by the dominant powers. Most alarming to those who think that the geopolitical priority is to maintain the political stability of modern states, is the potential for environmental and demographic changes to lead to destabilizing population movements and possibly military confrontations. Refugees are rapidly increasing in number, and while only some of the causes of their flight can be directly connected to environmental factors, there is widespread concern that environmental degradation may trigger many millions more environmental refugees. Worries about pollution and ozone depletion, diseases in the growing slums of Southern cities, and vague fears of terrorism are added into these concerns to suggest that we live in a world facing numerous complicated "global" threats.

But many of these are in important senses fabricated threats, ones that are a result of industrial production, pollution and accidents, nearly always in some way a consequence of the rapid technological and economic changes that are called globalization. The environment in which we live was for a long time assumed to be separate, or at least in important senses external to the lives of civilized people who live in cities, distant from nature (Latour, 2004). But both our technological acumen and the phenomenon of globalization now mean that we live in a world of complex dangers, a "risk society" in which our insecurities are in part as a result of our own actions (Beck, 1992). Ozone holes, radioactive fallout from Chernobyl, genetically modified organisms loose in many food chains, all raise various fears but also suggest a complex world of interconnections in a new context where the basic environmental context of our lives is increasingly artificial (McNeill, 2000).

But this focus on risk society and artificial environments does not mean that environmental problems are only political or technical; they very much "exist" in the real world. Forests are being cut down and people displaced. The potential for disruptions as a result of climate change needs to be taken seriously. Ozone holes are a real danger to ecosystems and both directly and indirectly to human health. But, as is clarified in this introduction, and in Part Four's readings, how these issues are described and who is designated as either the source of the problem, or provider of the potential solution to that problem, is an important matter in how environmental themes are argued about and in who gets to make decisions about what should be done by whom. If climate change is understood as being a problem caused mainly by car exhaust in industrial cities and their suburbs, or by international oil company policies in search of huge profits, or by peasants cutting down tropical rainforests to grow food, very different solutions are likely to be suggested. If energy conservation and environmental city planning are widely introduced and oil companies taxed heavily to provide an international "green tax" for environmental projects, results will be very different than if "Northern" states attempt to use economic sanctions to try to get "Southern" governments to stop tropical deforestation so that the trees will absorb carbon dioxide to slow climate change.

Once again we can see that geopolitics is about supposedly factual arguments and descriptions of the world that at first glance don't appear to be at all political. But careful analysis of the geographical assumptions in these arguments suggests that knowledge is not neutral, but frequently a political resource used in political arguments and in policy decisions.

RETHINKING SECURITY

Thinking about taken-for-granted geographies of security is important in analysing the discussions by academics and policy makers about rethinking the key concepts of national security and international security after the Cold War (Worldwatch Institute, 2005). Here are supposedly apparently obvious new threats to the political order requiring a geopolitical view of the global scene and management strategies in some cases backed by military preparedness. Concern has been widely expressed about the possibilities of international armed conflict over water resources which are being ever more heavily used by growing urbanized and increasingly industrialized populations using water directly and relying on irrigated crops for food. Climate change as a result of human activities in changing the global atmosphere may lead to weather changes that upset global agricultural productivity and induce political strife.

Some alarmist scenarios of the future draw from arguments about environmental degradation to suggest that resource shortages may lead to major conflict in places like China, where regional disparities and consequent political tensions are likely to be aggravated by pressures on the environment due to rapid industrialization and urbanization (see Boland, 2000). Many of these newly defined "threats" are interpreted as global phenomena, with population growth, ozone holes, biodiversity loss and climate change being only the most obvious matters of what is now often called "environmental security." But once one investigates specifically who is threatened by what changes where, the precise causes of change suggest that many "global" phenomena occur only in particular places.

Scenarios of declining resource bases leading to heightened awareness of communal identity and resulting group conflicts have been proposed by many writers since the early 1990s. Migrants in a number of places have come into conflict with host populations. Sorting out how environmental factors are influencing these processes is not easy as migrants usually move for complicated combinations of reasons. In addition doing detailed research in the middle of these conflicts is often very difficult. Researchers have undertaken a number of case studies in specific places since the early 1990s, but the precise role of the environment as a specific cause of conflict and refugee migration is not easy to figure out in general terms. Nonetheless Thomas Homer-Dixon's (1999) major research projects in the 1990s outlined some pathways from environmental disruptions to political conflict suggesting that environmental scarcity leads to conflict in some, but only some, circumstances.

An especially alarming article on these themes was written by Robert D. Kaplan and published by the *Atlantic Monthly* in February 1994 under the title "The Coming Anarchy". Widely read in Washington, this article crystallized concern about environmental causes of chaos and state breakdown. Kaplan is especially important as a journalist, travel writer and public intellectual, not least because his previous book, *Balkan Ghosts*, was influential in the Clinton White House policy discussions over Yugoslavia and the war in Bosnia. As the excerpts from his "Coming Anarchy" article reprinted here show (Reading 22), this is powerful prose that is compelling reading. But, as my critique of it suggests (Reading 23), it is inadequate as a rigorous analysis because of its many omissions and its failure to provide clear links between many of the things that it discusses and their apparent causes. It fails to look at the geography of environmental change in detail, preferring to use eye-witness accounts mixed in with references to scholarly work to create an impression of looming chaos. Nonetheless it has been an influential article in policy making circles, not least because it explicitly argues that the West faces new security threats in the form of crime, drugs, economic instabilities, diseases and "failed states." These themes dominated many discussions of global security in the 1990s. The concern about them was crystallized by Kaplan's lurid depictions of immanent disaster.

Many of the assumptions in the arguments about environment as a security threat or a cause of conflict are difficult to specify precisely and the overall political economy of globalization makes critics highly doubtful of the utility of this perspective (Peluso and Watts, 2001). General arguments about global environmental change are often so imprecise when applied to specific places as to be practically useless. Local economic situations, or the disruptions caused by development projects often generate poverty that is then blamed on environmental degradation. Specific environmental degradations are undoubtedly important in particular places, but generalizations as to how to respond are often not helpful. The potential for increased agricultural production in many parts of the world is still considerable, and statistics about arable land are often misleading because of large inaccuracies and generalizations.

But viewed from many places in "the South" the "discourses of danger" that structure the environmental security literature can be seen as little more than attempts to reassert Northern corporations' and political institutions' colonial domination of Southern societies, albeit now sometimes in the name of protecting the planet (Barnett, 2001). These specifications of the new "green" dimensions of geopolitics are not innocent constructions or "true" statements about how the world is organized. They are understandings of the world that relate to the traditional institutions of global politics but with new terms and language (Dalby, 2002). They are related to political power and enmeshed within the increasingly unequal global political economy.

There are also arguments that make the case that understanding all these things in terms of traditional Cold War themes of security are not helpful to either understanding the causes of contemporary problems or suggesting solutions. While most writers argue that any security crisis resulting from environmental degradation will need to be handled by cooperative measures rather than traditional "security" responses by the armed forces, nonetheless, as Daniel Deudney (1999) warned, if the military is seen as an essential institution in dealing with these problems, they are much more likely to be perpetuated than alleviated. This is the case in part because of the appalling records of environmental destruction by many militaries through the period of the cold war. But in addition security is often defined in terms of state security requiring a modern military armed with expensive industrial weapons. This leads to the perpetuation of industrial state policies as the "solution" to "security problems" when these are the very cause of much widespread environmental degradation.

RESOURCE WARS

In the latter part of the 1990s a number of investigations began to shift the focus of these concerns to look at the role of resources exported into the global economy played in violent conflict in the "South." Resources are traded globally, mines in one continent provide the raw materials for smelters elsewhere and manufacturing plants in distant states (Princen et al. 2002). As Philippe Le Billon (Reading 24) makes clear, the streams of resources exported from poor parts of the world often provide a very substantial source of wealth for those who can control the revenues generated by their exploitation and export. Not all resources have the same attributes. Diamonds are easy to smuggle, oil is less so. Minerals require more sophisticated mining technology than do trees, where a truck and a chainsaw are frequently the only equipment needed to extract timber either legally or illegally.

When the military is involved directly in the extraction of resources, or involved in a struggle with rebel forces to control such extraction, then violence and resources are directly interconnected (Bannon and Collier, 2003). These complicated connections between violence over local extraction and the international trade of these commodities are now often simply called "resource wars." The danger in using this terminology is that it obscures other complicated social phenomena and the failure of many states to provide even basic social services and opportunities for economic activity on the part of substantial parts of their populations (Boas 2004). Most of the time these wars are not between suppliers and consumers, not between forces in the North and those in the poor South, but rather struggles over the conditions of exploitation and export. States rarely go to war with each other over these issues, although this generalization does not necessarily seem to hold in the case of petroleum (Klare, 2001, 2004).

Economists frequently talk of these matters in terms of the resource curse because in many ways a rich local source of wealth and the disruptions to national economies as a result of economic distortions caused by heavy reliance on a single commodity with fluctuating prices tend to reduce innovation and the impetus to development in many places. Perhaps the classic example of how politics has been driven by squabbles over the wealth generated by oil revenues is the case of Nigeria where numerous political arrangements and considerable corruption and political violence are generated in arguments over the division of oil revenues (Watts, 2004). But it is important to note the global connections here between struggles to either resist or control the export of resources and the international corporations that profit from the transport and sale of these resources.

At the largest scale petroleum is a global commodity and one that has a geography that matters. The global economy relied on oil for much of the twentieth century as a portable and indispensable fuel (Reading 25). From ships to cars, electricity generating or home heating, oil has become ubiquitous in literally fueling globalization. Such immense wealth generates political struggles to control it, and the history of petroleum is also the history of war and struggle (Yergin, 1991). Nowhere is this more obviously the case than in the Middle East and Central Asia. But as Michael Renner (Reading 26) summarized in a short overview published prior to the American invasion of Iraq in 2003, keeping the tankers moving was part of the responsibilities of the US Navy in the Persian Gulf in the 1980s; it was a crucial factor in the decision by the first Bush presidency to use military force to remove Saddam Hussein's forces from Kuwait. American policy since the Carter doctrine of the late 1970s has been to declare the free passage of oil from the Middle East an essential national interest of the United States.

But this petroleum requires such "security measures" on the part of distant consumers only for so long as they are dependent on those supplies and they are potentially disrupted as a consequence of political and military actions in the region. And the United States, Europe, Japan and increasingly India and China which consume this petroleum are distant from the region. In so far as energy policy in the "developed" states relies on the uninterrupted supplies from the Middle East, and political conflict in the region is enmeshed with the control of these supplies, war and resources remain intertwined (Klare, 2004). The alternatives, so brilliantly articulated by Amory Lovins (1977) in terms of a strategy of a "soft energy path," were not taken seriously by the Reagan administration (with George Bush as Vice-President) in the 1980s. It abandoned many of the 1970s initiatives to conserve and reduce imports in the United States. A generation later, petroleum demand has grown and another George Bush in the White House in Washington is facing the ever larger need to find supplies to fuel the global economy and the American part of it in particular.

All this came to a head in the discussions in 2003 over why the United States invaded Iraq. Many protestors on the streets around the world objected to the invasion by carrying placards that posed the key geographical question of "How did our oil get under their sands?" Others demanded "No Blood for Oil." Political economy arguments suggested that this was a new form of imperialism (Harvey, 2003). Given the invasion of a country the other side of the world that apparently possessed a valuable resource, that's what it appeared to many to be. Others thought that a narrowed focus on the corporate links between the administration and Texas oil interests, including the Haliburton corporation, explained why the invasion was undertaken. The protestors were simply arguing that Americans, or indeed anyone else from outside the region, didn't have any right to these resources and hence there could be no justification for war.

Conservative scholar Andrew Bacevich's (2005) account of "The New American Militarism" suggests that in fact the Carter doctrine was the beginning of what he calls "World War IV" and that military control over the Middle East has been a gradually expanding part of American policy. Hence, he argues turning Iraq into a Western leaning democracy with a permanent American military presence on some of the largest oilfields in the world is a crucial part of a strategy to control the one natural resource that is absolutely essential to the functioning of the contemporary world economy. This is thus a "resource war" in the old-fashioned geopolitical sense of empire where Western states use military power to ensure control of remote sources of essential materials for their economy. Now that India and China's huge economies are growing rapidly and using more oil, this resource, and who controls it, becomes all the more important.

As Michael Klare (Reading 25) argues, in an article reflecting on the likely trajectory of all these matters in the second George W. Bush administration, dependence is likely to grow given the lack of attention to matters of consumption especially in the United States. Without some fundamental redirection of American technology in particular, these dangers of continued political violence remain. Despite increased rhetoric of energy independence and some initiatives on hybrid car engines most of the trends in the United States and in other "developed" economies are toward ever larger dependence on petroleum in the coming years. Confronting overconsumption in the wealthy part of the world is now an essential part of the politics of global environment (Princen et al. 2002; Clapp and Dauvergne, 2005). Once again the actions in one part of the biosphere enmesh those elsewhere in a global risk society, one now facing fears of new horrifying forms of threat in the form of bioterrorism (Reading 27).

"EMERGING DISEASES"

In the decade since Kaplan's dystopic vision of a "Coming Anarchy" was published, it became increasingly clear that HIV/AIDS is the disease that may have the worst consequences for Africans. As Gwyn Prins (Reading 28) outlines, the geopolitical literature on these matters is also concerned to address the issue in terms of security. In this case the complexity of how this might be done is illustrative of the politics of security and the importance of political language in getting attention at the United Nations. How something is specified as a threat of great enough importance to gain status as a security threat is a matter related to discussions of the extension of security issues after the Cold War to include environment and disease. Who can effectively "securitize" an issue is linked to the ability to specify the danger and hence the appropriate actions in response to the danger so specified. As was pointed out in the general introduction to this Reader, nearly three times as many people die each day from AIDS around the world than died in New York and Washington DC on September 11. Al-Qaeda is a threat requiring a huge military effort. In contrast a disease, two-thirds of whose victims are in sub-Saharan Africa, is clearly a concern but not a major security issue to decision makers in the capitals of the North.

But it is also very important to understand how war is an important factor in the spread of AIDS. Soldiers and military personnel are a major part of the problem. Ironically, as Gwyn Prins outlines, precisely those supposedly responsible for security are themselves one of the most vulnerable sectors of the population. But given that in many African states the military are the most organized social system, and now in many states the one losing the greatest number of competent people to disease, the tragedy of AIDS is compounded by the collapse of the ability of this institution to offer "security" in the sense of a functioning state. Linking this concern to matters of resource wars, and the increasing importance once again of external agencies in controlling the destiny of Africa, brings Prins' discussion back to the matter of resource wars and American intervention. The case for foreign intervention in Africa is linked to the increasing inability of local authorities, depleted of trained people by AIDS, to effectively deal with resource wars and related violence.

In Robert Kaplan's dark vision of Africa (Reading 22) he reflects on the importance of disease as a factor limiting the contact between Western and African peoples. His specification of the wild zones beyond the capabilities of civilization to tame suggests not only a region doomed to misery and violence, but also a zone from which diseases emerge to threaten populations elsewhere. The interactions of humans with many environments has raised new fears about diseases jumping species barriers and unleashing new plagues on human populations (Garrett, 1995). The very worrisome episode of anthrax in the US mail in late 2001, which killed five people and made others ill, means that this theme has now merged with post-9/11 fears of terrorist usage of weapons of mass destruction where biotechnologies might modify pathogens and use them as modes of killing large numbers of people.

Technological hazards now mean that risk society is interconnected with the "war on terror" and the political control mechanisms of homeland security in the United States. This other dimension of disease as a security issue is of constant concern in Washington. The more alarmist discussions of bioterrorism (see Garrett, 2001) paint a picture of terrorists loosing complicated modified disease organisms into American cities as part of a plot to damage the United States. Quite why this might be a useful strategy for terrorists is rarely discussed in much detail in the aftermath of September 11. The desire to protect the United States apparently requires all sorts of extreme measures regardless of asking questions of how and why what might be technically possible might actually be used (Hartmann et al. 2005). Diseases are hard to handle and quite as dangerous to those who try to use them as weapons as they are to the putative victims. To use them requires careful training and expert knowledge, which is not easy to come by on the part of terrorist networks. All of which probably explains why so very few people have actually died in bioterrorism episodes in the last few decades. While they may be horrifying, diseases don't obviously have the same political effects as bombs or hijackings, despite the popularity of these themes in horror movies, thrillers and such television series as the *X-Files*.

Jonathan B. Tucker suggests (Reading 27) that a science and technology based approach to this problem, which assumes that control can be maintained over microbes in perpetuity, and that secrecy concerning research into new disease agents is guaranteed, may in fact cause precisely the difficulties it sets out to defeat. After all the anthrax that killed people nearly at random in 2001 was apparently derived from American research stockpiles. But the geopolitical specification of the United States as a unique power, and one with the capabilities to dominate surveillance, but surrounded by potential enemies, suggests that the United States has no choice but to strive ahead rather than dealing with biological weapons as a matter of international cooperation and public health. Transparency and international arrangements to monitor and control response are an alternative that is not taken seriously by a technical definition of security as that which single states do alone.

Once again how a security danger is specified is crucial to understanding how geopolitics works. The security dilemma, where actions to provide specific groups with security end up reducing security, is rarely so well depicted as in this discussion of disease as a potential threat. The solution to these difficulties lies in transparency and widespread agreements to follow international guidelines, inspections and scientific openness. But such international cooperation is frequently anathema to the Bush administration. Once again global engagements run into the limitations of a very nationalist set of policy formulations (Smith, 2005). The perpetuation of technological security dilemmas in the early stages of the twenty-first century is tied into the geopolitical specification of external enemies and the need for eternal vigilance in the face of supposedly omnipresent dangers. Risk society and the global environmental disruptions caused by urban consumption societies are not within the geopolitical specifications of danger in the Bush doctrine; but if they were, very different policies would probably be in place.

WHOSE GLOBE? WHOSE SECURITY?

The importance of thinking carefully about all these themes should not be forgotten in thinking and studying international politics and economics. The shorthands and the apparent obviousness of the specifications of danger in discussions of "global security" combine with the importance of the issues to obscure quite as much as they reveal. Kaplan's text shows how the specification of places as dangerous is effective journalism without actually telling the reader much if anything about the causes of environmental change or violence. Fear and the representation of exotic places as dangerous is travel writing, not geopolitical analysis. But when such travel writing becomes the basis for policy it seems to perpetuate the stories that the travel writer tells much more effectively than it improves the lot of those suffering the kinds of violence that so horrifies readers in Northern cities.

When one compares how Kaplan suggests that violence in Africa is caused by environmental degradation with the careful analysis of the logic of resource wars the importance of getting the geography and the causations right becomes especially clear. Much of the violence in Africa is connected to the extraction of resources and who controls the revenues from these mines and forests, not to environmental degradation. This links the global economy directly to matters of who dies when diamonds and other resources are fought over (Le Billon, 2005). It raises questions of who supports and supplies weapons to which faction in the misnamed "civil wars" that continue in central Africa. When this is connected to how AIDS kills and how it is spread by soldiers on the move in these conflicts, as

Prins documents, the tragedies of Africa then appear in a very different light than that which Kaplan's account suggests. Instead of walls of disease separating peoples, the international links of arms trading and resource supplies tie them together in violent embrace.

At the larger scale these patterns are also to be seen in the discussion of petroleum and the military conflicts of the Middle East. Once again arms contractors do very nicely supplying regimes in the region with weapons, supplies and training. But the regimes there, sitting upon huge wealth, jealously guard their access to this wealth. At the biggest scale American power refuses to allow the flows of oil supplies to be disrupted; political threats to this flow, whether from the regime of Saddam Hussein, or the threat of internal collapse in Saudi Arabia, are a matter for military intervention with all the potential such efforts have for violence. Here the external world is a threat to the political order of carboniferous capitalism and as such must be disciplined and brought under control (Dalby, 2002). But to do so simply ignores the growing evidence of climate change and automobile pollution while perpetuating military adventures to control oilfields and pipelines in many places.

A similar technical response to the threats of bioterrorism is what Tucker so comprehensively criticizes. Contemporary American policy in particular is a series of scientific and military efforts which, due to their secrecy and their attempts to monopolize control, undermine the possibilities of cooperation and medical preparation against many potential health threats. The lesson from many diseases is that in a globalized world they can spread rapidly. But the related lesson is that diseases spread among poor and malnourished populations especially quickly. Here there is less access to preventative programs, poorer medical help once disease strikes, and compromised immune systems which increase susceptibility. Preventing disease spread in these circumstances improves the health of all in an interconnected world.

The lessons of risk society and globalization suggest that interconnectedness is our fate for the foreseeable future. Geopolitical solutions that try to separate causes and effects, that divide "our" secure space from threatening "external" spaces obscure the connections in a globalized world where consumption in the cities and suburbs of the North is so dependent on key economic supplies for all over the globe. Thinking about global security seriously is precisely a matter of thinking through these connections and understanding the consequences of actions in one place on people elsewhere. Thus global security has to be a common endeavor in many ways, one that requires policies of international cooperation in the place of simplistic geopolitical assumptions of safety here and danger there.

REFERENCES AND FURTHER READINGS

History, environment and politics

Diamond, J. (2005) *Collapse: How Societies Choose to Fail or Succeed.* New York: Viking.

Ehrlich, P.R. (1968) *The Population Bomb.* New York: Ballantine.

Lovins, A.B. (1977) *Soft Energy Paths: Toward a Durable Peace.* London: Penguin.

McNeill, J.R. (2000) *Something New Under the Sun: An Environmental History of the Twentieth Century.* New York: Norton.

Meadows, D.H., Meadows, D.L., Randers, J. and Behrens III, W.W. (1974) *The Limits to Growth.* London: Pan.

Sandbach, F. (1980) *Environment, Ideology and Policy.* Oxford: Basil Blackwell.

Soroos, M. (1997) *The Endangered Atmosphere: Preserving a Global Commons.* Columbia, SC: University of South Carolina Press.

Wapner, P. (1996) *Environmental Activism and World Politics*. Albany, NY: State University of New York Press.

World Commission on Environment and Development (1987) *Our Common Future*. New York: Oxford University Press.

Contemporary environmental politics

Clapp, J. and Dauvergne, P. (2005) *Paths to a Green World: The Political Economy of the Global Environment*. Cambridge, MA: MIT Press.

Conca, K. and Dabelko, G. (eds) (2002) *Environmental Peacemaking*. Baltimore, MD: Johns Hopkins University Press.

Conca, K. and Dabelko, G.D. (eds) (2004) *Green Planet Blues: Environmental Politics from Stockholm to Johannesburg*, 3rd edn. Boulder, CO: Westview.

Ehrlich, P. and Ehrlich, A. (2004) *One with Nineveh: Politics, Consumption and the Human Future*. Washington, DC: Island.

Elliott, L. (2004) *The Global Politics of the Environment*, 2nd edn. London: Palgrave Macmillan.

Latour, B. (2004) *Politics of Nature: How to Bring the Sciences into Democracy*. Cambridge, MA: Harvard University Press.

Lipschutz, R. (2004) *Global Environmental Politics: Power, Perspectives, and Practice*. Washington, DC: CQ Press.

Paterson, M. (2000) *Understanding Global Environmental Politics*. London: Palgrave Macmillan.

Princen, T., Maniates, M. and Conca, K. (eds) (2002) *Confronting Consumption*. Cambridge, MA: MIT Press.

Environmental security

Barnett, J. (2001) *The Meaning of Environmental Security: Ecological Politics and Policy in the New Security Era*. London: Zed.

Boland, A. (2000) "Feeding Fears: Competing Discourses of Interdependency, Sovereignty, and China's Food Security," *Political Geography*, 19(1): 55–76.

Dalby, S. (2002) *Environmental Security*. Minneapolis, MN: University of Minnesota Press.

Deudney, D. (1999) "Environmental Security: A Critique," in *Contested Grounds: Security and Conflict in the New Environmental Politics*, eds. D. Deudney and R. Matthew. Albany, NY: State University of New York Press.

Diehl, P.F. and Gleditsch, N.P. (eds) (2001) *Environmental Conflict*. Boulder, CO: Westview.

Homer-Dixon, T. (1999) *Environment, Scarcity and Violence*. Princeton, NJ: Princeton University Press.

Peluso, N. and Watts, M. (eds) (2001) *Violent Environments*. Ithaca, NY: Cornell University Press.

Pirages, D. and DeGeest, T.M. (2004) *Ecological Security: An Evolutionary Perspective on Globalization*. Lanham, MD: Rowman & Littlefield.

Suliman, M. (ed.) (1999) *Ecology, Politics and Violent Conflict*. London: Zed.

Worldwatch Institute (2005) *State of the World 2005: Redefining Global Security*. New York: Norton.

Resource wars

Bacevich, A.J. (2005) *The New American Militarism: How Americans are Seduced by War*. New York: Oxford University Press.

Bannon, I. and Collier, P. (eds) (2003) *Natural Resources and Violent Conflict: Options and Actions.* Washington, DC: The World Bank.

Berdal, M. and Malone, D.M. (eds) (2000) *Greed and Grievance: Economic Agendas in Civil Wars.* Boulder, CO: Lynne Rienner.

Boas, M. (2004) "Africa's Young Guerillas: Rebels with a Cause?" *Current History*, May: 211–214.

Dalby, S. (2004) "Ecological Politics, Violence, and the Theme of Empire," *Global Environmental Politics*, 4(2): 1–11.

Gedicks, A. (2001) *Resource Rebels: Native Challenges to Mining and Oil Corporations.* Boston, MA: South End Press.

Harvey, D. (2003) *The New Imperialism.* Oxford: Oxford University Press.

Jung, D. (ed.) (2003) *Shadow Globalization, Ethnic Conflicts and New Wars.* London: Routledge.

Klare, M.T. (2001) *Resource Wars: The New Landscape of Global Conflict.* New York: Metropolitan.

—— (2004) *Blood and Oil: The Dangers and Consequences of America's Growing Dependence on Imported Petroleum.* New York: Metropolitan.

Le Billon, P. (2005) *Fuelling War: Natural Resources and Armed Conflict.* Adelphi Paper 373. Oxford: Routledge.

Watts, M. (2004) "Antimonies of Community: Some Thoughts on Geography, Resources and Empire," *Transactions of the Institute of British Geographers*, NS 29: 195–216.

Yergin, D. (1991) *The Prize: The Epic Quest for Oil, Money and Power.* New York: Simon & Schuster.

Diseases, bioweapons, Africa and globalization

Beck, U. (1992) *Risk Society: Towards a New Modernity.* London: Sage.

Dunn, K. (2003) *Imagining the Congo: The International Relations of Identity.* New York: Palgrave Macmillan.

Garrett, L. (1995) *The Coming Plague: Newly Emerging Diseases in a World Out of Balance.* New York: Penguin.

—— (2001) "The Nightmare of Bioterrorism," *Foreign Affairs*, January/February: 76–89.

Guillemin, J. (2004) *Biological Weapons: From the Invention of State-sponsored Programs to Contemporary Bioterrorism.* New York: Columbia University Press.

Hartmann, B., Subramaniam, B. and Zerner, C. (eds) (2005) *Making Threats: Biofears and Environmental Anxieties.* Lanham, MD: Rowman & Littlefield.

Smith, N. (2005) *The Endgame of Globalization.* New York: Routledge.

Thomas, C. and Wilkin, P. (eds) (1999) *Globalization, Human Security and the African Experience.* Boulder, CO: Rienner.

Tucker, J.C. (ed.) (2000) *Toxic Terror: Assessing Terrorist Use of Chemical and Biological Weapons.* Cambridge, MA: MIT Press.

—— (2001) *Scourge: The Once and Future Threat of Smallpox.* New York: Grove Press.

Vale, P. (2003) *Security and Politics in South Africa.* Boulder, CO: Rienner.

22

The Coming Anarchy

Robert D. Kaplan

from *The Atlantic Monthly* (1994)

The Minister's eyes were like egg yolks, an after effect of some of the many illnesses, malaria especially, endemic in his country. There was also an irrefutable sadness in his eyes. He spoke in a slow and creaking voice, the voice of hope about to expire.

> In 45 years I have never seen things so bad. We did not manage ourselves well after the British departed. But what we have now is something worse – the revenge of the poor, of the social failures, of the people least able to bring up children in a modern society.

Then he referred to the recent coup in the West African country Sierra Leone. "The boys who took power in Sierra Leone come from houses like this." The Minister jabbed his finger at a corrugated metal shack teeming with children. "In three months these boys confiscated all the official Mercedes, Volvos and BMWs and wilfully wrecked them on the road." The Minister mentioned one of the coup's leaders, Solomon Anthony Joseph Musa, who shot the people who had paid for his schooling, "in order to erase the humiliation and mitigate the power his middle class sponsors held over him."
[. . .]
The cities of West Africa at night are some of the unsafest places in the world. Streets are unlit; the police often lack gasoline for their vehicles; armed burglars, carjackers and muggers proliferate. Direct flights between the United States and the Murtala Muhammed Airport, in neighboring Nigeria's largest city, Lagos, have been suspended by order of the US Secretary of Transportation because of ineffective security at the terminal and its environs. A State Department report cited the airport for "extortion by law enforcement and immigration officials." This is one of the few times the US government has embargoed a foreign airport for reasons that are linked purely to crime. In Abidjan, effectively the capital of the Cote d'Ivoire, or Ivory Coast, restaurants have stick and gun wielding guards who walk you the 15 feet or so between your car and the entrance, giving you an eerie taste of what American cities might be like in the future. An Italian ambassador was killed by gunfire when robbers invaded an Abidjan restaurant. The family of the Nigerian ambassador was tied up and robbed at gunpoint in the ambassador's residence.
[. . .]
"In the poor quarters of Arab North Africa," the minister continued,

> there is much less crime, because Islam provides a social anchor of education and indoctrination. Here in West Africa we have a lot of superficial Islam and superficial Christianity. Western religion is undermined by animist beliefs not suitable to a moral society, because they are based on irrational spirit power. Here spirits are used to wreak vengeance by one person against another, or one group against another.

Finally the minister mentioned polygamy. Designed for a pastoral way of life, polygamy continues to thrive in sub-Saharan Africa even though it is increasingly uncommon in Arab North Africa. Most youths I met on the road in West Africa told me that they were from "extended" families, with a mother in one place and a father in another. Translated to an urban environment, loose family structures are largely responsible for the world's highest birth rates and the explosion of the HIV virus on the continent. Like the communalism and animism, they provide a weak shield against the corrosive social effects of life in cities.

A PREMONITION OF THE FUTURE

West Africa is becoming the symbol of worldwide demographic, environmental and societal stress, in which criminal anarchy emerges as the real "strategic" danger. Disease, overpopulation, unprovoked crime, scarcity of resources, refugee migrations, the increasing erosion of nation states and international borders, and the empowerment of private armies, security firms and international drug cartels are now most tellingly demonstrated through a West African prism. West Africa provides an appropriate introduction to the issues, often extremely unpleasant to discuss, that will soon confront our civilization.

There is no other place on the planet where political maps are so deceptive – where, in fact, they tell such lies – as in West Africa. Start with Sierra Leone. According to the map, it is a nation state of defined borders, with a government in control of its territory. In truth the Sierra Leonian government, run by a 27-year-old army captain. Valentine Strasser, controls Freetown by day and by day also controls part of the rural interior. In the government's territory the national army is an unruly rabble threatening drivers and passengers at most checkpoints. In the other part of the country, units of two separate armies from the war in Liberia have taken up residence, as has an army of Sierra Leonian rebels. The government force fighting the rebels is full of renegade commanders who have aligned themselves with disaffected village chiefs. A premodern formlessness governs the battlefield, evoking the wars in medieval Europe prior to the 1648 Peace of Westphalia, which ushered in the era of organized nation states.

As a consequence, roughly 400,000 Sierra Leonians are internally displaced, 280,000 more have fled to neighboring Guinea, and another 100,000 have fled to Liberia, even as 400,000 Liberians have fled to Sierra Leone. The third largest city in Sierra Leone, Gondama, is a displaced-persons camp. With an additional 600,000 Liberians in Guinea and 250,000 in the Ivory Coast, the borders dividing these four countries have become largely meaningless. Even in quiet zones none of the governments except the Ivory Coast's maintains

the schools, bridges, roads and police forces in a manner necessary for functional sovereignty.

In Sierra Leone, as in Guinea, as in the Ivory Coast, as in Ghana, most of the primary rain forest and the secondary bush is being destroyed at an alarming rate. When Sierra Leone achieved its independence, in 1961, as much as 60 per cent of the country was primary rain forest. Now 6 per cent is. In the Ivory Coast the proportion has fallen from 38 per cent to 8 per cent. The deforestation has led to soil erosion, which has led to more flooding and more mosquitoes. Virtually everyone in the West African interior has some form of malaria.

Sierra Leone is a microcosm of what is occurring, albeit in a more tempered and gradual manner, throughout West Africa and much of the under-developed world: the withering away of central governments, the rise of tribal and regional domains, the unchecked spread of disease, and the growing pervasiveness of war. West Africa is reverting to the Africa of the Victorian atlas. It consists now of a series of coastal trading posts, such as Freetown and Conakry, and an interior that, owing to violence, volatility, and disease, is again becoming, as Graham Greene once observed, "blank" and "unexplored." However, whereas Greene's vision implies a certain romance, as in the somnolent and charmingly seedy Freetown of his celebrated novel *The Heart of the Matter*, it is Thomas Malthus, the philosopher of demographic doomsday, who is now the prophet of West Africa's future. And West Africa's future, eventually, will also be that of most of the rest of the world.

Consider "Chicago." I refer not to Chicago, Illinois, but to a slum district of Abidjan, which the young toughs in the area have named after the American city. ("Washington" is another poor section of Abidjan.) Chicago, like more and more of Abidjan, is a slum in the bush: a checkerwork of corrugated zinc roofs and walls made of cardboard and black plastic wrap. It is located in a gully teeming with coconut palms and oil palms, and is ravaged by flooding. Few residents have easy access to electricity, a sewage system or a clean water supply. [. . .]

Fifty-five per cent of the Ivory Coast's population is urban, and the proportion is expected to reach 62 per cent by 2000. The yearly net population growth is 3.6 percent. This means that the Ivory Coast's 13.5 million people will become 39 million by 2025, when much of the population will consist of urbanized peasants like those of Chicago. But don't count on the Ivory Coast's still existing then. Chicago, which is more indicative of Africa's and the Third World's demographic present – and even more in the future – than any idyllic junglescape of women balancing earthen jugs on their heads, illustrates why the Ivory Coast, once a model of Third World success, is becoming a case study in Third World catastrophe. [. . .]

Because the military is small and the non-Ivorian population large, there is neither an obvious force to maintain order nor a sense of nationhood that would lessen the need for such enforcement. The economy has been shrinking since the mid-1980s. Though the French are working assiduously to preserve stability, the Ivory Coast faces a possibility worse than a coup: an anarchic implosion of criminal violence – an urbanized version of what has already happened in Somalia. [. . .]

As many internal African borders begin to crumble, a more impenetrable boundary is being erected that threatens to isolate the continent as a whole: the wall of disease. [. . .] Africa may today be more dangerous in this regard than it was in 1862, before antibiotics. [. . .] Of the approximately 12 million people worldwide whose blood is HIV positive, 8 million are in Africa. In the capital of the Ivory Coast, whose modern road system only helps to spread the disease, 10 per cent of the population is HIV positive. And war and refugee movements help the virus break through to more remote areas of Africa. It is malaria that is most respon-sible for the disease wall that threatens to separate Africa and other parts of the Third World from more developed regions of the planet in the twenty-first century. Carried by mosquitoes, malaria, unlike AIDS, is easy to catch. Most people in sub-Saharan Africa have recurring bouts of the disease throughout their entire lives, and it is mutating into increasingly deadly forms.

Africa may be as relevant to the future character of world politics as the Balkans were a hundred years ago, prior to the two Balkan wars and the First World War. Then the threat was the collapse of empires and the birth of nations based solely on tribe. Now the threat is more elemental: *nature unchecked*. Africa's immediate future could be very bad. The coming upheaval, in which foreign embassies are shut down, states collapse, and contact with the outside world takes place through dangerous, disease ridden coastal trading posts, looms large in the century we are entering. Precisely because much of Africa is set to go over the edge at a time when the Cold War has ended, when environmental and demographic stress in other parts of the globe is becoming critical, and when the post-First World War system of nation states – not just in the Balkans but perhaps also in the Middle East – is about to be toppled, Africa suggests what war, borders, and ethnic politics will be like a few decades hence. [. . .]

THE ENVIRONMENT AS A HOSTILE POWER

For a while the media will continue to ascribe riots and other violent upheavals abroad mainly to ethnic and religious conflict. But as these conflicts multiply, it will become apparent that something else is afoot, making more and more places like Nigeria, India and Brazil ungovernable. [. . .]

It is time to understand "the environment" for what it is: the national security issue of the early twenty-first century. The political and strategic impact of surging populations, spreading disease, deforestation and soil erosion, water depletion, air pollution and, possibly, rising sea levels in critical, overcrowded regions such as the Nile Delta and Bangladesh – developments that will prompt mass migrations and, in turn, incite group conflicts – will be the core foreign policy challenge from which most others will ultimately emanate, arousing the public and uniting assorted interests left over from the Cold War. In the twenty-first century, water will be in dangerously short supply in such diverse locales as Saudi Arabia, Central Asia and the south-western United States. A war could erupt between Egypt and Ethiopia over Nile River water. Even in Europe tensions have arisen between Hungary and Slovakia over the damming of the Danube, a classic case of how environmental disputes fuse with ethnic and historical ones.

Our Cold War foreign policy truly began with George F. Kennan's famous article, signed "X," pub-lished in *Foreign Affairs* in July of 1947, in which Kennan argued for a "firm and vigilant containment" of a Soviet Union that was imperially, rather than ideologically, motivated. It may be that our post-Cold War foreign policy will one day be seen to have had its begin-nings in an even bolder and more detailed piece of written analysis: one that appeared in the journal *International Security*. The article, published in the fall of 1991 by Thomas Eraser Homer-Dixon, who is the head of the Peace and Conflict Studies Program at the University of Toronto, was titled "On the Threshold:

Environmental Changes as Causes of Acute Conflict." Homer-Dixon has, more successfully than other analysts, integrated two hitherto separate fields – military conflict studies and the study of the physical environment.

In Homer-Dixon's view, future wars and civil violence will often arise from scarcities of resources such as water, cropland, forests and fish. Just as there will be environmentally driven wars and refugee flows, there will be environmentally induced praetorian regimes – or, as he puts it, "hard regimes." Countries with the highest probability of acquiring hard regimes, according to Homer-Dixon, are those that are threatened by a declining resource base yet also have "a history of state (read 'military') strength." Candidates include Indonesia, Brazil and, of course, Nigeria. Though each of these nations has exhibited democratizing tendencies of late, Homer-Dixon argues that such tendencies are likely to be superficial "epiphenomena" having nothing to do with long term processes that include soaring populations and shrinking raw materials. Democracy is problematic; scarcity is more certain.

Indeed, the Saddam Husseins of the future will have more, not fewer, opportunities. In addition to engendering tribal strife, scarcer resources will place a great strain on many peoples who never had much of a democratic or institutional tradition to begin with. Over the next 50 years the Earth's population will soar from 5.5 billion to more than 9 billion. Though optimists have hopes for new resource technologies and free market development in the global village, they fail to note that, as the National Academy of Sciences has pointed out, 95 per cent of the population increase will be in the poorest regions of the world, where governments now – just look at Africa – show little ability to function, let alone to implement even marginal improvements. Homer-Dixon writes ominously, "neo-Malthusians may underestimate human adaptability in today's environmental social system, but as time passes their analysis may become ever more compelling."

While a minority of the human population will be, as Francis Fukuyama would put it, sufficiently sheltered so as to enter a "post-historical" realm, living in cities and suburbs in which the environment has been mastered and ethnic animosities have been quelled by bourgeois prosperity, an increasingly large number of people will be stuck in history, living in shantytowns where attempts to rise above poverty, cultural dysfunction

and ethnic strife will be doomed by a lack of water to drink, soil to till and space to survive in. In the developing world, environmental stress will present people with a choice that is increasingly among totalitarianism (as in Iraq), fascist tending mini states (as in Serb-held Bosnia) and road warrior cultures (as in Somalia). Homer-Dixon concludes that "as environmental degradation proceeds, the size of the potential social disruption will increase."

Quoting Daniel Deudney, another pioneering expert on the security aspects of the environment, Homer-Dixon says that for too long we've been prisoners of "social-social" theory, which assumes there are only social causes for social and political changes, rather than natural causes, too. This social-social mentality emerged with the Industrial Revolution, which separated us from nature. But nature is coming back with a vengeance, tied to population growth. It will have incredible security implications.

Think of a stretch limo in the potholed streets of New York City, where homeless beggars live. Inside the limo are the air conditioned post-industrial regions of North America, Europe, the emerging Pacific Rim and a few other isolated places, with their trade summitry and computer information highways. Outside is the rest of mankind, going in a completely different direction.

SKINHEAD COSSACKS, JUJU WARRIORS

In the summer 1993 issue of *Foreign Affairs*, Samuel P. Huntington, of Harvard's Olin Institute for Strategic Studies, published a thought-provoking article called "The Clash of Civilizations?" The world, he argues, has been moving during the course of this century from nation state conflict to ideological conflict to, finally, cultural conflict. I would add that as refugee flows increase and as peasants continue migrating to cities around the world – turning them into sprawling villages – national borders are the most tangible and intractable ones: those of culture and tribe. Huntingdon writes, "First, differences among civilizations are not only real; they are basic, " involving, among other things, history, language, and religion. "Second, . . . interactions between peoples of different civilizations are increasing; these interactions intensify civilization consciousness." Economic modernization is not necessarily a panacea, since it fuels individual and group ambitions while weakening traditional loyalties to the state. It is worth noting, for example, that it is precisely the wealthiest and fastest developing city in India, Bombay, that has seen the worst intercommunal violence between Hindus and Muslims. Consider that Indian cities, like African and Chinese ones, are ecological timebombs – Delhi and Calcutta, and also Beijing, suffer the worst air quality of any cities in the world – and it is apparent how surging populations, environmental degradation and ethnic conflict are deeply related.

Huntington points to interlocking conflicts among Hindu, Muslim, Slavic Orthodox, Western, Japanese, Confucian, Latin American and possibly African civilizations: for instance, Hindus clashing with Muslims in India, Turkic Muslims clashing with Slavic Orthodox Russians in Central Asian cities, the West clashing with Asia. (Even in the United States, African-Americans find themselves besieged by an influx of competing Latinos.) Whatever the laws, refugees find a way to crash official borders, bringing their passions with them, meaning that Europe and the United States will be weakened by cultural disputes. . . .

Most people believe that the political Earth since 1989 has undergone immense change. But it is minor compared with what is yet to come. The breaking apart and remaking of the atlas is only now beginning. The crack up of the Soviet empire and the coming end of Arab-Israeli military confrontation are merely prologues to the really big changes that lie ahead. Michael Vlahos, a long range thinker for the US Navy, warns, "We are not in charge of the environment, and the world is not following us. It is going in many directions. Do not assume that democratic capitalism is the last word in human social evolution." [. . .]

THE PAST IS DEAD

Built on steep, muddy hills, the shantytowns of Ankara, the Turkish capital, exude visual drama. Altindag, or "Golden Mountain," is a pyramid of dreams, fashioned from cinder blocks and corrugated iron, rising as though each shack were built on top of another, all reaching awkwardly and painfully toward heaven – the heaven of wealthier Turks who live elsewhere in the city. For reasons that I will explain, the Turkish shack town is a psychological universe away from the African one.

Slum quarters in the Ivory Coast's Abidjan terrify and repel the outsider. In Turkey it is the opposite.

Golden Mountain was a real neighborhood. The inside of one house told the story: The architectural bedlam of cinder block and sheet metal and cardboard walls was deceiving. Inside was a *home* – order, that is, bespeaking dignity. I saw a working refrigerator, a television, a wall cabinet with a few books and lots of family pictures, a few plants by a window, and a stove. Though the streets become rivers of mud when it rains, the floors inside this house were spotless.

My point in bringing up a rather wholesome, crime free slum is this: Its existence demonstrates how formidable is the fabric of which Turkish Muslim culture is made. A culture this strong has the potential to dominate the Middle East once again. Slums are litmus tests for innate cultural strengths and weaknesses. Those peoples whose cultures can harbor extensive slum life without decomposing will be, relatively speaking, the future's winners. Those whose cultures cannot will be the future's victims. [. . .]

In Turkey, [. . .] Islam is painfully and awkwardly forging a consensus with modernization a trend that is less apparent in the Arab and Persian worlds (and virtually invisible in Africa). In Iran the oil boom – because it put development and urbanization on a fast track, making the culture shock more intense – fuelled the 1978 Islamic revolution. But Turkey, unlike Iran and the Arab world, has little oil. Therefore, its development and urbanization have been more gradual. Islamists have been integrated into the parliamentary system for decades.

Resource distribution is strengthening Turks in another way vis-à-vis Arabs and Persians. Turks may have little oil, but their Anatolian heartland has lots of water – the most important fluid of the twenty-first century. Turkey's Southeast Anatolia Project, involving 22 major dams and irrigation systems, is impounding the waters of the Tigris and Euphrates rivers. Much of the water that Arabs and perhaps Israelis will need to drink in the future is controlled by Turks. The project's centerpiece is the mile wide, 16 story Ataturk Dam, upon which are emblazoned the words of modern Turkey's founder: "Ne Mutlu Turkum Diyene" ("Lucky is the one who is a Turk"). [. . .] Power is certainly moving north in the Middle East, from the oil fields of Dhahran, on the Persian Gulf, to the water plain of Harran, in southern Anatolia – near the site of the Ataturk Dam. But will the nation state of Turkey, as presently constituted, be the inheritor of this wealth? I very much doubt it.

THE LIES OF MAPMAKERS

According to the map, the great hydropower complex emblemized by the Ataturk Dam is situated in Turkey. Forget the map. This southeastern region of Turkey is populated almost completely by Kurds. About half of the world's 20 million Kurds live in "Turkey." The Kurds are predominant in an ellipse of territory that over-laps not only with Turkey but also with Iraq, Iran, Syria and the former Soviet Union. The Western enforced Kurdish enclave in northern Iraq, a consequence of the 1991 Persian Gulf War, has already exposed the fictitious nature of that supposed nation state.

On a recent visit to the Turkish–Iranian border, it occurred to me what a risky idea the nation state is. Here I was on the legal fault line between two clashing civilizations, Turkic and Iranian. Yet the reality was more subtle: As in West Africa, the border was porous and smuggling abounded, but here the people doing the smuggling, on both sides of the border, were Kurds. In such a moonscape, over which peoples have migrated and settled in patterns that obliterate borders, the end of the Cold War will bring on a cruel process of natural selection among existing states. No longer will these states be so firmly propped up by the West or the Soviet Union. Because the Kurds overlap with nearly everybody in the Middle East, on account of their being cheated out of a state in the post-First World War peace treaties, they are emerging, in effect, as the natural selector – the ultimate reality check. They have destabilized Iraq and may continue to disrupt states that do not offer them adequate breathing space, while strengthening states that do.

Because the Turks, owing to their water resources, their growing economy and the social cohesion evinced by the most crime free slums I have encoun-tered, are on the verge of big power status, and because the 10 million Kurds within Turkey threaten that status, the outcome of the Turkish–Kurdish dispute will be more critical to the future of the Middle East than the eventual outcome of the recent Israeli–Palestinian agreement.

A NEW KIND OF WAR

To appreciate fully the political and cartographic implications of postmodernism – an epoch of theme-less juxtapositions, in which the classificatory grid of nation states is going to be replaced by a jagged glass pattern of city states, shanty states, nebulous and anarchic regionalisms – it is necessary to consider, finally, the whole question of war.

The intense savagery of the fighting in such diverse cultural settings as Liberia, Bosnia, the Caucasus and Sri Lanka – to say nothing of what obtains in American inner cities – indicates something very troubling that those of us concerned with issues such as middle-class entitlements and the future of interactive cable television lack the stomach to contemplate. It is this: A large number of people on this planet, to whom the comfort and stability of a middle-class life are utterly unknown, find war and a barracks existence a step up rather than a step down.

"Just as it makes no sense to ask 'why people eat' or 'what they sleep for,'" writes Martin van Creveld, a military historian at the Hebrew University in Jerusalem, in "The Transformation of War,"

so fighting in many ways is not a means but an end. Throughout history, for every person who has expressed his horror of war there is another who found in it the most marvellous of all the experiences that are vouchsafed to man, even to the point that he later spent a lifetime boring his descendants by recounting his exploits.

[. . .]

Van Creveld's book begins by demolishing the notion that men don't like to fight. "By compelling the senses to focus themselves on the here and now," van Creveld writes, war "can cause a man to take his leave of them." As anybody who has had experience with Chetniks in Serbia, "technicals" in Somalia, Tontons Macoutes in Haiti or soldiers in Sierra Leone can tell you, in places where the Western Enlightenment has not penetrated and where there has always been mass poverty, people find liberation in violence. Physical aggression is a part of being human. Only when people attain a certain economic, educational and cultural standard is this trait tranquillized. In light of the fact that 95 per cent of the Earth's population growth will be in the poorest areas of the globe, the question is not whether there will be war (there will be a lot of it) but what kind of war. And who will fight whom?

Debunking the great military strategist Carl von Clausewitz, van Creveld, who may be the most original thinker on war since that early nineteenth-century Prussian, writes, "Clausewitz's ideas . . . were wholly rooted in the fact that, ever since 1648, war had been waged overwhelmingly by states." But, as van Creveld explains, the period of nation states and, therefore, of state conflict is now ending, and with it the clear "threefold division into government, army and people" which state directed wars enforce. Thus, to see the future, the first step is to look back to the past immediately prior to the birth of modernism – the wars in medieval Europe that began during the Reformation and reached their culmination in the Thirty Years' War.

Van Creveld writes:

> In all these struggles political, social, economic and religious motives were hopelessly entangled. Since this was an age when armies consisted of mercenaries, all were also attended by swarms of military entrepreneurs. . . . Many of them paid little but lip service to the organizations for whom they had contracted to fight. Instead, they robbed the countryside on their own behalf. . . . Given such conditions, any fine distinctions . . . between armies on the one hand and peoples on the other were bound to break down. Engulfed by war, civilians suffered terrible atrocities.

Back then, in other words, there was no "politics" as we have come to understand the term, just as there is less and less "politics" today in Liberia, Sierra Leone, Somalia, Sri Lanka, the Balkans and the Caucasus, among other places.

Because, as van Creveld notes, the radius of trust within tribal societies is narrowed to one's immediate family and guerrilla comrades, truces arranged with one Bosnian commander, say, may be broken immediately by another Bosnian commander. The plethora of short lived cease fires in the Balkans and the Caucasus constitute proof that we are no longer in a world where the old rules of state warfare apply. [. . .]

Also, war making entities will no longer be restricted to a specific territory. Loose and shadowy organisms such as Islamic terrorist organizations suggest why borders will mean increasingly little and sedimentary layers of tribalistic identity and control will mean more. "From the vantage point of the present, there appears every prospect that religious . . . fanaticisms will play a larger role in the motivation of armed conflict" in the West than at any time "for the last 300 years," van Creveld writes. [. . .]

Future wars will be those of communal survival, aggravated or, in many cases, caused by environmental scarcity. These wars will be sub-national, meaning that it will be hard for states and local governments to protect their own citizens physically. This is how many states will ultimately die. [. . .]

THE LAST MAP

In "Geography and the Human Spirit," Anne Buttimer, a professor at University College, Dublin, recalls the work of an early nineteenth-century German geographer, Carl Ritter, whose work implied "a divine plan for humanity" based on regionalism and a constant, living flow of forms. The map of the future, to the extent that a map is even possible, will represent a perverse twisting of Ritter's vision. Imagine cartography in three dimensions, as if in a hologram. In this hologram would be the overlapping sediments of group and other identities atop the merely two dimensional color markings of city states and the remaining nations, themselves confused in places by shadowy tentacles, hovering overhead, indicating the power of drug cartels, mafias and private security agencies. Instead of borders, there would be moving "centers" of power, as in the Middle Ages. Many of these layers would be in motion. Replacing fixed and abrupt lines on a flat space would be a shifting pattern of buffer entities, like the Kurdish and Azeri buffer entities between Turkey and Iran, the Turkic Uighur buffer entity between Central Asia and Inner China (itself distinct from coastal China), and the Latino buffer entity replacing a precise US Mexican border. To this protean cartographic hologram one must add other factors, such as migrations of populations, explosions of birth rates, vectors of disease. Henceforward the map of the world will never be static. This future map – in a sense, the "Last Map" – will be an ever mutating representation of chaos.

Indeed, it is not clear that the United States will survive the next century in exactly its present form. Because America is a multiethnic society, the nation state has always been more fragile here than it is in more homogeneous societies such as Germany and Japan. James Kurth, in an article published in *The National Interest* in 1992, explains that whereas nation state societies tend to be built around a mass conscription army and a standardized public school system, "multicultural regimes" feature a high tech, all volunteer army (and, I would add, private schools that teach competing values), operating in a culture in which the international media and entertainment industry have more influence than the "national political class." In other words, a nation state is a place where everyone has been educated along similar lines, where people take their cue from national leaders, and where everyone (every male, at least) has gone through the crucible of military service, making patriotism a simpler issue. Writing about his immigrant family in turn of the century Chicago, Saul Bellow states, "The country took us over. It was a country then, not a collection of 'cultures.'" During the Second World War and the decade following it, the United States reached its apogee as a classic nation state. During the 1960s, as is now clear, America began a slow but unmistakable process of transformation. The signs hardly need belaboring: racial polarity, educational dysfunction, social fragmentation of many and various kinds. "Patriotism" will become increasingly regional as people in Alberta and Montana discover that they have far more in common with each other than they do with Ottawa or Washington, and Spanish speakers in the Southwest discover a greater commonality with Mexico City. As Washington's influence wanes, and with it the traditional symbols of American patriotism, North Americans will take psychological refuge in their insulated communities and cultures.

Returning from West Africa last fall was an illuminating ordeal. After leaving Abidjan, my Air Afrique flight landed in Dakar, Senegal, where all passengers had to disembark in order to go through another security check, this one demanded by US authorities before they would permit the flight to set out for New York. Once we were in New York, despite the midnight hour, immigration officials at Kennedy Airport held up disembarkation by conducting quick interrogations of the aircraft's passengers – this was in addition to all the normal immigration and customs procedures. It was apparent that drug smuggling, disease and other factors had contributed to the toughest security procedures I have ever encountered when returning from overseas.

Then, for the first time in over a month, I spotted businesspeople with attaché cases and laptop computers. When I had left New York for Abidjan, all the business people were boarding planes for Seoul and Tokyo, which departed from gates near Air Afrique's. The only non-Africans off to West Africa had been relief workers in T-shirts and khakis. Although the borders within West Africa are increasingly unreal, those separating West Africa from the outside world are in various ways becoming more impenetrable.

But Afrocentrists are right in one respect: We ignore this dying region at our own risk. When the Berlin Wall was falling, in November of 1989, I happened to be in Kosovo, covering a riot between Serbs and Albanians. The future was in Kosovo, I told myself that night, not in Berlin. The same day that Yitzhak Rabin and Yasser Arafat clasped hands on the White House lawn, my Air Afrique plane was approaching Bamako, Mali, revealing corrugated zinc shacks at the edge of an expanding desert. The real news wasn't at the White House, I realized. It was right below.

Cartoon 8 Trouble in the Global Village
Numerous issues in global politics apparently require urgent attention, but the stories of dangers requiring action are frequently more about political and moral imperatives than they are careful analyses of situations in particular places.

Source: M. Wuerker

23

Reading Robert Kaplan's "Coming Anarchy"

Simon Dalby

from *Ecumene* (1996)

ROBERT KAPLAN'S GEOPOLITICAL IMAGINATION

The world is not quite so conveniently simple as Kaplan's popularization of environmental degradation as the key national security issue for the future suggests. His article for all its dramatic prose and empirical observation is vulnerable to numerous critiques. Read as a cultural production of considerable political importance it is fairly easy to see how the logic of the analysis, premised on "eye witness" empirical observation, and drawing on an eclectic mixture of intellectual sources, leaves so much of significance unsaid. But the impression, as has traditionally been the case in geopolitical writing, generated from the juxtaposition of expert sources and empirical observation is that this is an "objective" detached geopolitical treatise. The focus in what follows is on the political implications of the widely shared geopolitical assumptions that structure this text and ultimately render the environment as a threat.

The most important geopolitical premise in the argument posits a "bifurcated world", one in which the rich in the prosperous "post-historical" cities and suburbs have mastered nature through the use of technology, while the rest of the population is stuck in poverty and ethnic strife in the shanty towns of the underdeveloped world. The presentation of the article in the magazine supports this basic formulation of the world into the rich, who read magazines like *Atlantic*, and the rest who don't. The closing image in the text of New York airport with its business people flying to Asia, but not to Africa, is very strongly reinforced through the article by the juxtaposition of the advertisements in the original magazine version of the article with the violent imagery of the photographs, and the themes in the text. The affluence of New York airport contrasts sharply with the poverty and dangers elsewhere.

But these phenomena are treated as completely separate in terms of economics. Poverty and affluence are only connected where poverty is seen as a threat to the affluence of the *Atlantic's* North American readers. In all of Kaplan's article matters of international trade are barely mentioned. The wall of disease he writes about may bar many foreigners from all except some coastal "trading posts" of Africa in the future, but the significance of what is being traded and with what implications for the local environment is not investigated. "Hot cash", presumably laundered drug moneys from African states, apparently does flow to Europe we are told, but this has significance only because of the criminal dimension of the activity, not as part of a larger pattern of political economy. While the lack of business people flying to Africa is noted, comments about the high rate of logging are never connected to the export markets for such goods, or to the economic circumstances of indebted African states that distort local economies to pay international loans and meet the requirements for structural adjustment programs. Logging continues apace, but it is apparently driven only by some indigenous local desire to strip the environment of trees, not by any exogenous cause. A focus on the larger political economy driving forest destruction would lead the analysis in a very different direction, but it is a direction that is not taken by the focus on West Africa as a quasi-autonomous geopolitical entity driven by internal developments.

The political violence and environmental degradation are not related to larger economic processes anywhere in this text. This is not to suggest that the legacy of colonialism, or the subsequent neo-colonial economic arrangements, are solely to "blame" for current crises, although the history cannot be ignored as Kaplan is wont to do. It is to argue that these sections of Kaplan's text show a very limited geopolitical imagination, one that focuses solely on local phenomena in a determinist fashion that ignores the larger trans-boundary flows and the related social and economic causes of resource depletion. Kaplan ignores the legacy of the international food economy which has long played a large role in shaping the agricultural infrastructures, and the nutritional levels of many populations of different parts of the world in specific ways. He also ignores the impact of the economic crisis of the 1980s and the often deleterious impact of the debt crisis and structural adjustment policies. He completely misses their important impact on social patterns and the impact on rural women upon whom many of the worst impacts fell (Mackenzie, 1993). Ironically, given his repeated comments about the inadequacies of cartographic designations of state boundaries in revealing cross-border ethnic and criminal flows, Kaplan effectively establishes economic boundaries precisely by not investigating economic phenomena that supposedly ought to be crucial to his specification of various regions in Malthusian terms. While Kaplan emphasizes the inadequacies of maps for understanding ethnic and cultural clashes, he never investigates their similar inadequacies for understanding economic interconnections as an important part of either the international relations or the foreign policies of these states. The crucial failure to do this allows for the attribution of the "failure" of societies to purely internal factors. Once again the local environment can be constructed as the cause of disaster without any reference to the historical patterns of development that may be partly responsible for the social processes of degradation (Crush, 1995; Slater, 1993).

Given the focus of most Malthusians on the shortage of "subsistence" and resources in general, there is remarkably little investigation of how the burgeoning populations of various parts of the world actually are provided for either in terms of food production or other daily necessities. Despite accounts of trips across Africa by "bush-taxi", agricultural production remains invisible to Kaplan's "eye witness". While cities are dismissed as "dysfunctional" the very fact that they continue to grow despite all their difficulties suggests that they do "function" in many ways. Informal arrangements and various patterns of "civil society" are ignored. People move to the cities, but quite why is never discussed in this article. There is no analysis here of traditional patterns of subsistence production and how they and access to land may be changing in the rural areas, particularly under the continuing influence of modernization. While it is made clear that traditional rural social patterns fray when people move to the very different circumstances of the city, the reasons for migration are assumed but never investigated. In Homer-Dixon's language, absolute scarcity is assumed and the possibilities of relative scarcity, with the negative consequences for poor populations due to unequal distribution or the marginalization of subsistence farmers as a result of expanded commercial farming, is never investigated. Why Malthus, in particular, should be the prophet of West Africa, given the complete failure to investigate the changing patterns of these rural economies, is far from clear. Disease and crowding there may be in the shanty towns of many cities, a phenomenon that is not exactly new, but not all the new urban population are dispossessed forest dwellers or refugees from criminal activities.

The focus on environment as the key factor in triggering violent changes is not entirely consistent with Kaplan's arguments elsewhere about the cohesive force of Islam, identified ironically in a few places, given the usual orientalizations in practice when discussing Islam, as a Western religion. His discussion of Turkey suggests that while urbanization is occurring rapidly, social cohesion and resistance to crime are being maintained by Islam, even as new geopolitical identities are being forged in the slums. While he suggests that these identities may transcend the force of Islam in the ongoing conflict between Turks and Kurds, his emphasis on non-environmental factors of social cohesion suggests that his argument is perhaps more concerned with traditional matters of ethnic identity and "civilizational clashes", than with environmental degradation.

Here resurgent cultural fears of "the Other" and assumptions about the persistence of cultural patterns of animosity and social cleavage are substituted for analysis of resources and rural political ecology. Precisely where the crucial connections between environmental change, migration and conflict should be investigated, the analysis turns away to look at ethnic rivalries and the collapse of social order. The connections are asserted, not demonstrated, and in so far as this is done the opportunity for detailed analysis is missed and the powerful rhetoric of the argument retraces familiar political territory instead of looking in detail at the environment as a factor in social change. In this failure to document the crucial causal connections in his cases Kaplan ironically follows Malthus who relied on his unproven key assumption that subsistence increases only at an arithmetic rate in contrast to geometric population growth.

Political angst about the collapse of order is substituted for an investigation of the specific reasons for rapid urbanization, a process that is by default rendered as a "natural" product of demographic pressures. This unstated "naturalization" then operates to support the Malthusian fear of poverty stricken mobs, or in Kaplan's terms, young homeless and rootless men forming criminal gangs, as a threat to political order. Economics becomes nature, nature in the form of political chaos becomes a threat, the provision of security from such threats thus becomes a policy priority. In this way "nature unchecked" can thus be read directly as a security threat to the political order of post-modernity.

GEOPOLITICS, MALTHUS AND KAPLAN

Kaplan explicitly links the Malthusian theme in his discussion of Africa to matters of national security, where a clear "external" threatening dimension of crime and terrorism is linked to the policy practices of security and strategic thinking. The logic of a simple Malthusian formulation is complicated by the geographical assumptions built into Kaplan's argument, while he has simultaneously avoided any explicit attempt to deal at all with the political economy of rural subsistence or contemporary population growth. Thus, in his formulation, the debate is shifted from matters of humanitarian concern, starvation, famine relief and aid projects and refocused on matters of military threat and concern for political order within Northern states.

What ultimately seems to matter in this new designation is whether political disorder and crime will spill over into the affluent North. The affluent world of the *Atlantic* advertisements with their high-technology consumer items (Saabs, Mazdas and Bose stereos etc.) is implicitly threatened by the spreading of "anarchy". The article implies that it has done so already in so far as American inner cities are plagued with violent crime. The reformulation once again posits a specific geopolitical framework for security thinking.

Kaplan himself suggests that by his own logic the US may become more fragmented. What cannot be found in this article is any suggestion that the affluence of those in the limousine might in some way be part of the same political economy that produces the conditions of those outside. This connection is simply not present in the text of the article because of the spatial distinctions Kaplan makes between "here" and "there". He notes the dangers of the criminals from "there" compromising the safety of "here" but never countenances the possibility that the economic affluence of "here" is related to the poverty of "there". The spatial construction of his discourse precludes such consideration, only some factors violate the integrity of cartographic boundaries.

Although Kaplan is particularly short on policy prescription in his *Atlantic* article, some of the implications of his reworked Malthusianism do have clear policy implications. Instead of repression and the use of political methods to maintain inequalities in the face of demands for reform, Kaplan's implicit geopolitics suggest abandoning Africa to its fate. If more Northern states withdraw diplomatic and aid connections and, as he notes, stop direct flights to airports such as Lagos, the potential to isolate this troubled region may be considerable. If contact is restricted to coastal trading posts then the "wall of disease" will become a wall of separation keeping non-Africans out and restricting the possibilities for Africans to migrate. Once again security is understood in the geopolitical term of containment and exclusion.

In a subsequent article in the *Washington Post* (17 April 1994) Kaplan explicitly argues against US military interventions in Africa. He suggests that intervention in Bosnia would do some good, because the developed nature of the societies in conflict there allows some optimism that a political settlement is workable. The chances of intervention having much effect in Africa are dismissed because of the illiterate poverty stricken populations there. However, the pessimism of the *Atlantic* article is muted here by a contradictory suggestion that all available foreign policy money for Africa be devoted to population control, resource management and women's literacy. These programs will, Kaplan hopes, in the very long term resolve some of the worst problems allowing development to occur and "democracy" eventually to emerge.

The ethnocentrism of the suggestion that Africa's problems are solvable in terms of modernization, is coupled to the implication that West Africa is of no great importance to the larger global scheme of power and economy, and therefore can be ignored, at least as long as the cultural affinities between Africans and African Americans don't cause political spillovers into the United States. In this geopolitical argument Kaplan parallels Saul Cohen's geopolitical designation of Sub-Saharan Africa as part of a "quartersphere of marginality" consigning it to irrelevance in the post-Cold War order (Cohen, 1994). Precisely this marginalization is of concern to many African leaders and academics. But in stark contrast to Kaplan, many Africans emphasize the need to stop the export of wealth from the Continent, and the need to draw on indigenous traditions to rebuild shattered societies and economies (Adadeji, 1993; Amin, 1990; Taylor and Mackenzie, 1992).

Spatial strategies of containment are a long standing component of security thinking. Cutting anarchy ridden regions loose in the hopes that their political turmoil will remain internal makes sense in an argument that constructs these places as clearly external to the political arrangements that one wishes to render secure from threats. Given the specification of the political turmoil as caused internally within these areas, this argument makes logical sense. Also given the startling failure in this analysis to consider matters of international economics as a possible cause for some of the phenomena that are involved in the dissolution of political order, no sense of external responsibility applies. Kaplan deals with deterritorialized phenomena when they suit his argument, but conveniently ignores trans-boundary flows when they don't fit his cartographic scheme. They suit it here because they emphasize political violence and threats across frontiers that are in some cases disappearing.

Large scale geopolitical isolation as a *cordon sanitaire* might work as a Western security strategy in these circumstances; it seems less likely to help Africans, but that point is not high on Kaplan's scheme of priorities. But to advocate these "solutions" is once again to specify complex political phenomena in territorial terms, a strategy that is, as John Agnew argues, falling into the familiar "territorial trap" in international relations thinking where boundaries are confused with

barriers and flows and linkages are obscured by the widespread assumption of autonomous states as the only actors of real importance in considering global politics (Agnew, 1994).

There is an ironic twist in Kaplan's geopolitical specifications of "wild zones". He argues that they are threats to political stability and in the case of Africa probably worth cutting loose from conventional political involvement. In the subsequent *Washington Post* article he argues against military interventions in Africa on the basis of their uselessness in the political situation of gangs, crime and the absence of centralized political authority. His suggestions imply that interventions are only considered in terms of political attempts to resolve conflicts and provide humanitarian aid. In this assumption Kaplan is at odds with Cold War geopolitical thinking. While ignoring the political economy of underdevelopment as a factor in the African situation, he also ignores the traditional justifications for US political and military involvement in Africa and much of the Third World. Through the Cold War these focused on questions of ensuring Western access to strategic minerals in the continent. This theme continues to appear in many other discussions of post-Cold War foreign policy and in US strategic planning. But Kaplan ignores both these economic interconnections and their strategic implications, preferring an oversimplified geopolitical specification of Malthusian-induced social collapse as the sole focus of concern.

But the specification of danger as an external "natural" phenomena works in an analogous way to the traditional political use of Neo-Malthusian logic. Once again threats are outside human regulation, inevitable and natural in some senses – if not anarchic in the neo-realist sense of state system structure then natural in a more fundamental sense of "nature unchecked". By the specific spatial assumptions built into his reasoning Kaplan accomplishes geopolitically what Malthusian thinking did earlier in economic terms. Coupled to prevalent American political concerns with security as "internal" vulnerability to violent crime, and "external" fears of various foreign military, terrorist, economic, racial, and immigration "threats", Kaplan re-articulates his modified Malthusianism in the powerful discursive currency of geopolitics. His themes fit neatly with media coverage of Rwanda and Somalia where his diagnosis of the future appeared in many media accounts to be occurring nearly immediately.

Understood as problems of "tribal" warfare such formulations reproduce the earlier tropes of "primitive savagery". As other commentators on contemporary conflict have noted, detailed historical analysis suggests that the formation of "tribes", and many of the "tribal wars" that European colonists deplored, were often caused by the sociological disruptions triggered by earlier European intrusions. Denial or failure to understand the causal interconnections of this process allowed for the attribution of "savagery" to "Others" inaccurately specified as geographically separate. Kaplan notes that the disintegration of order is not a matter of a "primitive" situation, but following van Creveld, a matter of "reprimitivized" circumstances in which high-technology tools are used for gang and "tribal" rivalries. But the economic connections that allow such "tools" to become available are not mentioned. Thus re-primitivization is specified as the indirect result of environmental degradation, a process that is asserted frequently but not argued, demonstrated or investigated in any detail. Once again geopolitical shorthand is substituted for detailed geographical analysis. In Ó Tuathail and Agnew's (1992) terms, the irony of the policy discourse of geopolitics, as the antithesis to detailed geographical understanding, is in play once again in this text, although this time with environment as a reified concept.

BEYOND MALTHUS AND MACKINDER?

The continued possibilities of using Malthusian themes as ideological weapons by the powerful in justifying repression, or at the least, justifying inaction in the face of gross inequities, now have to be complimented by a recognition that these themes can be mobilized in foreign policy discourse to suggest the appropriateness of military solutions to demographic and "environmental" problems. At least in the earlier version of his famous essay, Malthus argued that population growth is inevitable, natural, and largely beyond human regulation. Politics is thus rendered as just a reaction to the consequences of the unchangeable patterns of fecundity. If the political consequences of population growth are disruptive to the Northern geopolitical order that is judged to be the only acceptable one, then Neo-Malthusianism acts as a powerful intellectual weapon in formulating policies to repress and politically control reformist demands for greater

equality or economic redistribution. It can do so on the grounds that such policies only aggravate adverse demographic trends. When coupled to Kaplan's assertions that population growth is related to environmental degradation, the argument is strengthened.

If the more alarmist versions of some of Kaplan's arguments gain credence in Washington, or if the formulation of politics in terms of the Rest and the West becomes prominent, then the dangers of a new Cold War against the poor are considerable. The discussions of illegal immigration in the US in the early 1990s, and suggestions that the solution is increased border guards, denial of services to immigrants incapable of proving legal residence, and deportations, suggest that the geopolitical imagination of spatial exclusion is dominating the policy discourse once again. In particular this may be because of the propensity among American politicians to formulate American identity in antithesis to external perceived dangers. Through the history of the last two centuries this has been a powerful theme in the formulation of American foreign policy which has drawn on the related discourses of American exceptionalism (Agnew, 1983).

This geopolitical imagination has been frequently coupled to assertions of cultural superiority and ideological rectitude in the form of various articulations of moral certainty. The dangers of ethnocentrism, when coupled to geopolitical reasoning, are greatest precisely where they assert strategic certainty in ways that prevent analysis of the complex social, political and economic interactions that might lead to assessments that in at least some ways "the problem is us" (Hentsch, 1992). Through the course of the Cold War and subsequently in the 1991 Gulf War, these formulations have fueled arms races, the global politics of deterrence and "security" understood in terms of violent containment and military superiority (Campbell, 1992, 1993; Dalby, 1990). This is done by privileging territorial sovereignty over other modes of human organization.

But it is the focus on the failures of these strategies in many places that makes Kaplan's vision so troubling to conventional analysis. In Shapiro's (1991) terms he focuses on some flows or "exchanges" that transgress the frontiers of sovereignty unsettling the possibilities of political order constrained in the spatial imaginations of modern sovereignty. While the fear that traditional military protection of borders is no longer efficacious, and that social disorder will spread despite the spatial demarcations of boundaries, induces fear, it can also ironically draw on the traditional thinking to suggest that if current efforts are inadequate then what is needed is redoubled actions in the military sphere to reassert control. Such a policy of militarization suggests escalating violence rather than attempts to tackle large scale problems in more cooperative ways.

Kaplan's posing of these problems in terms of national security suggests such a strategy. Once again the sovereignty problematic can lead to specifications of dangers and violent solutions, rather than to any consideration of an ethics of post-sovereignty (Shapiro, 1994; Walker, 1993). The construction of the threat as "nature unchecked" simply adds to the specification of danger as beyond the possibilities of simple inter-ventions and amelioration, hence a long lasting security threat that is particularly intractable. Kaplan's analysis doesn't escape classical geopolitical thinking. While his analysis of the collapse of geopolitical boundaries suggests a new departure in understanding politics, one that looks at the necessity of rethinking warfare and that gets beyond themes of geopolitical boun-daries, his focus on organic communities and on Malthusian environmental causes of turmoil, phrased as security threats, leads the analysis back to the need to keep the feared threats at bay by strategies of spatial exclusion. A form of geographical determinism is once again linked to threats of geopolitical violence.

REFERENCES

Adadeji, A. (ed.) (1993) *Africa Within the World: Beyond Dispossession and Dependence*, London: Zed.

Agnew, J. (1983) "An Excess of 'National Exceptionalism': Towards a New Political Geography of American Foreign Policy", *Political Geography Quarterly*, 2(2). 151–66.

Agnew, J. (1994) "The Territorial Trap: The Geographical Assumptions of International Relations Theory", *Review of International Political Economy*, 1(1). 53–80.

Amin, S. (1990) *Maldevelopment: Anatomy of a Global Failure*, London: Zed.

Bennet, O. (ed.) (1991) *Greenwar: Environment and Conflict*, London: Panos.

Campbell, D. (1992) *Writing Security: American Foreign Policy and the Politics of Identity*, Minneapolis: University of Minnesota Press.

Campbell, D. (1993) *Politics Without Principle: Sovereignty, Ethics, and the Narratives of the Gulf War*, Boulder, Colorado: Lynne Rienner.

Cohen, S. (1994) "Geopolitics in the New World Era: A New Perspective to an Old Discipline" in G.J. Demko and W.B. Wood (eds) *Reordering the World: Geopolitical Perspectives on the Twenty-first Century*, Boulder, Colorado: Westview, pp. 15–48.

Crush, J. (ed.) (1995) *Power of Development*, London: Routledge.

Dalby, S. (1990) *Creating the Second Cold War: The Discourse of Politics*, London: Pinter and New York: Guilford.

Hentsch, T. (1992) *Imagining the Middle East*, Trans. F.A. Reed, Montreal: Black Rose.

Mackenzie, F. (1993) "Exploring the Connections: Structural Adjustment, Gender and the Environment", *Geoforum*, 24(1). 71–87.

Ó Tuathail, G. and J. Agnew (1992) "Geopolitics and Discourse: Practical Geopolitical Reasoning in American Foreign Policy", *Political Geography*, 11(2). 190–204.

Shapiro, M.J. (1991) "Sovereignty and Exchange in the Orders of Modernity", *Alternatives*, 16(4). 447–477.

Shapiro, M.J. (1994) "Moral Geographies and the Ethics of Post-Sovereignty", *Public Culture*, 6. 479–502.

Slater, D. (1993) "The Geopolitical Imagination and the Enframing of Development Theory", *Transactions of the Institute of British Geographers*, New Series 18. 419–437.

Taylor, D.R.F. and F. Mackenzie (eds) (1992) *Development From Within: Survival in Rural Africa*, London: Routledge.

Walker, R.B.J. (1993) *Inside/Outside: International Relations as Political Theory*, Cambridge: Cambridge University Press.

The Geopolitical Economy of 'Resource Wars'

Philippe Le Billon
from *Geopolitics* (2004)

Resources have provided some of the means and motive of global European power expansion, while also being the focus of inter-state rivalry and strategic denial of access. Western geopolitical thinking about resources has been dominated by the equation of trade, war, and power, at the core of which were overseas resources and maritime navigation. During the mercantilist period of 15th century, trade and war became intimately linked to protect or interdict the accumulation of the 'world riches', mostly in the form of bullion, enabled by progresses in maritime transport and upon which much of the balance of power was perceived to be determined.[1] For example, the decision to pursue 'commerce warfare', in effect piracy, by French military engineer Vauban aimed, but failed, at precipitating the downfall of English and Dutch power by targeting their maritime trading.[2]

Writing on the wake of the three consecutive wars that opposed England to the Dutch in 17th century, John Evelyn commented that, 'Whoever commands the ocean commands the trade of the world, and whoever commands the trade of the world commands the riches of the world, and whoever is master of that commands the world itself'.[3] Since sea power itself rested on access to timber and naval timber supply became a major preoccupation for major European powers from the 17th century onward. Besides motivating overseas alliances, trade, or even imperialist rule, England in particular pursued a policy of open sea 'at all costs' that led to several armed interventions in the Baltic; a situation that would bear similarities with the case of oil in the 20th century.[4] With growing industrialisation and increasing dependence on imported materials during the 19th century, western powers intensified their control over raw materials, leading along with many other factors such as political ideologies to an imperialist 'scramble' over much of the rest of the world.[5]

Late imperial initiatives also influenced the Prussian strategy of consolidating their economic self-sufficiency through a resource access provided by a 'vital space', or *Lebensraum*, while the potential role of railways to create a land-based transcontinental control of resources raised threat to maritime-based power, giving way to the idea of 'Heartland' developed by Halford Mackinder. The significance of imported resources, and in particular oil, during the First World War reinforced the idea of resource vulnerability, which was again confirmed during the Second World War.[6]

Strategic thinking about resources during the Cold War continued to focus on the vulnerability of rising resource supply dependence, and to consider the potential for international conflicts resulting from competition over access to key resources.[7] In their search for resource security and strategic advantage, industrialised countries continued to take a diversity of initiatives, including military deployment near exploitation sites and along shipping lanes, stockpiling of strategic resources, diplomatic support, 'gunboat' policies, proxy wars or coup d'etat to maintain allied regimes in producing countries, as well as support to transnational corporations and favourable international trade agreements.

Geopolitical discourses and practices of resource competition were not only defined at an international scale but as well as at a sub-national one, especially in reference to the territorial legacy surviving the decolonisation process and its implications in terms of resource control. By the 1970s, concerns also came to encompass the potential threat of political instability resulting from population growth, environmental degradation, and social inequalities in poor countries, leading to a redefinition of national security.[8] The ensuing concept of 'environmental security' came about to reflect ideas of global interdependence, illustrated through the debates on global warming, environmental 'limits to growth', or the political instability caused by environmental scarcity in the South.[9] Traditional western strategic thinking remained, however, mostly concerned about supply vulnerability within the framework of the two blocks, notably about Soviet threats over the Western control of oil in the Persian Gulf or 'strategic minerals' in Southern and Central Africa.[10]

The decolonisation process, the 1956 Suez crisis, the 1973 Arab oil embargo, and the 1979 Iranian revolution also clearly focused western strategic concerns on the part of Western governments as well as resource businesses, over domestic and regional political stability and alliances.[11] The end of the Cold War and disintegration of the Soviet empire, and the Iraqi invasion of Kuwait further reinforced this view. If the security of supply continues to inform governmental and corporate decisions in the management of several minerals, in particular with regard to high-tech and radioactive materials, oil stands largely alone in terms of global strategic importance.[12]

As more attention was again devoted to the internal mechanisms of wars in the early 1990s, a view emerged that a new and violent scramble for resources amongst local warlords, regional hegemons, and international powers was becoming 'the most distinctive feature of the global security environment'.[13] Noting the growth of mass consumerism and the 'economisation' of international affairs in the 1990s, political scientist Michael Klare associates 'resource wars' with a combination of population and economic growth leading to a relentless expansion in the demand for raw materials, expected resource shortages, and contested resource ownership.[14] Asia's growing mass consumerism and energy demand, for example, are of specific concern for the militarised control of the South China Sea and Spratly islands. The control of the oil and pipelines in the Caspian region is another. If market forces and technological progress can mitigate some of these problems, Klare remains essentially pessimist given the

readiness of countries claiming resources or importing them, especially the US, to secure their access to resources through military force, as well as the political instability of many producing regions. Indeed, the strategic military posturing of the US in the Arabian Peninsula, the maritime deployment of the US-led Multinational Interception Force enforcing UN sanctions on Iraq, as well as the US military occupation of Iraq and the deployment in Central Asia give to the geopolitics of oil in this region a strong military tone. [. . .]

In the aftermath of the Second World War and decolonisation, much hope was placed in the promise that extractive sectors would assist poor countries in developing economically and politically.[15] The successful development path of countries benefiting from rich natural endowment, such as Australia, Canada and the US frequently served to justified some views; even though development largely preceded and enabled the relatively positive role of mineral resources, for example, and most poor countries have been facing vastly different domestic and international contexts to see resources contributing to their development.[16] Since the oil shocks of the 1970s, resource wealth appears to leave large numbers of people in developing countries worse off than otherwise. Resource-dependent countries tend to have lower social indicators and their states tend to be more corrupt, ineffective, prioritise military expenditures, and more authoritarian.[17] They also appear to be amongst the most conflict-ridden countries.[18] Although some argue that these problems characterise all poor countries and that resource dependence is simply a symptom of economic underdevelopment, others believe that a rich resource endowment is more a curse than a blessing.[19]

Well-managed resources can prove a valuable development asset, but resources can also prove a source of vulnerabilities and excesses negatively influencing the domestic politics and economy of exporting countries, as well as foreign relations.[20] Becoming of 'strategic' importance to domestic or foreign economic and political concerns, resource access and exploitation can become highly contested issues. Because of their territorialisation, resources generate more than many other economic sectors territorial stakes, centred on the definition of political boundaries and local representation or alliances with foreign powers. Exportable on the international market, resources give rise to stakes over access and control of *filières* or commodity networks, trading routes, and markets. Generating large financial rents, the control of resources often provide a crucial link between the economy and politics, in particular through relations of co-optation or patronage that often come to replace the taxation/representation nexus, while the impact of resources on development is itself highly sensitive to the institutional context in which they are exploited. [. . .]

Respectively, people and informal business groups lightly or not taxed by a government relying on the resource rents would be less concerned by a government's lack of accountability and legitimacy than heavily taxed ones; thereby being less motivated to promote political changes. Rulers can play on this by ignoring corruption and leaving most of the economy to become informal. Mobutu did precisely that when urging citizens to 'fend for yourselves' and to 'steal a little in a nice way', without aiming to become rich overnight or to transfer funds overseas; a 'policy' that became popularly known as 'Article 15' of the Zairian constitution and served as a justification for all forms of trafficking.[21] These policies reflected as much Mobutu's pragmatism in the face of an economic meltdown, as the instrumentalisation of disorder by local political and economic actors.[22] Smuggling and the unofficial economy did provide the marginalised population in general and the political opposition with an alternative political economy that delayed political polarisation, but they also further weakened the fiscal base of the state apparatus, and promoted corruption or demobilisation among officials.[23] [. . .]

In a worst case scenario, resource revenues monopolised by a corrupt elite or squandered by mismanagement justifiably feed grievances amongst marginalised groups, while resentment may also easily grow out of other resource related issues, such as pollution, labour conditions, or the social muddle frequently accompanying resource exploitation.[24] Competing businesses convinced of their own powerlessness assert their neutrality and continue to serve as intermediaries between local tragedies and global consumers; leaving a wide gap of accountability that an economically disempowered population cannot easily fill. Importing countries too often accommodate or even support predatory states, as long as access to cheap or strategic resources is secure. As resources become depleted, prices collapse, or corruption-weary businesses leave, the legitimacy and capacity of local rulers are further eroded. Disavowed

by their population, rulers face the challenge of political change and the temptation of their own radicalisation. At this juncture, violence and exclusionary identity politics become seductive means of empowerment and survival for most parties.

As natural resources gain in importance for belligerents, so the focus of military activities becomes centred on areas of economic significance. This has a critical effect on the location of military deployment, type of conflict, and intensity of confrontations.[25] Complementing guerrilla strategies of high mobility, concentration of forces, and location along international borders, rebel groups seek to establish permanent strongholds or areas of 'insecurity' wherever resources and transport routes are located. Government troops generally attempt to prevent this by extending counter-insurgency to these areas, occasionally displacing and 'villagising' populations. In many cases, however, government troops join in the plunder. Distinctions between soldiers and rebels then often become blurred, as both groups entertain the same economic agendas, occasionally cooperating to keep trading routes open and to maximise gains while minimise their costs. As demonstrated by the coalition formed by many elements of the Sierra Leone Army and the rebel Revolutionary United Front in 1997, both groups can also have similar social backgrounds, similar grievances towards the traditional ruling elite and a shared goal of empowerment through force. Beyond politico-military entrepreneurs turning into warlords building their power in part out of the (violent) control of valuable resources, many ordinary people may also use violence as a deliberate means of accessing resources, thereby increasing the spatial and social diffusion of a conflict.[26]

History as well as political culture, institutions, the individual personality of leaders and the availability of weapons intervene at least as much than the political economy of natural resources *per se* in these conflicts and their violent escalation, but the exploitation of nature represents a source of power and conflicts that should not be ignored. [. . .]

The geopolitics of natural resources has long been a strategic concern for both exporting and importing states. Western powers' concerns over 'resource wars' have been largely put at ease with the end of the Cold War and greater flexibility of international trade, even if their continued supply dependence, rising demand for raw materials, and recent armed confrontations and instability in key areas such as the Persian Gulf continue to place this item on their geopolitical agenda. But this apparent progress has not resolved and may even have aggravated several other strategic issues about resources, this time mostly of concern to exporting countries.

The first issue relates to the political economy and governance of resource-dependent countries, many of which face a similar pattern of growth collapse, corruption, and delegitimated state authority. Given the importance of natural resources in the economy or the economic potential of most developing countries, the issue of translating resource exploitation into political stability and economic development will remain central in the years to come, often for entire regions.

The second issue relates to the scale and number of economic, environmental or socio-cultural conflicts related to resource exploitation that increasingly oppose local populations, business interests, the state, and global environmental and human rights networks.[27] While most conflicts are either peacefully negotiated or limited to social protest movements and small-scale skirmishes, in other cases customs of violence and a radicalisation of ideologies turned them into full-scale civil wars. Organised opposition to processes of globalisation unaccountable to local interests and growing demand for raw materials should only increase such adversarial politics and the need for more effective dialogue.

The third, and often related strategic issue is that natural resource revenues have become the economic mainstay of most wars in the post Cold War context. Accessible and internationally marketable resources such as diamonds and timber, not to mention drugs, have played a significant role in conflicts in at least twenty countries during the 1990s. This is not to argue that those wars are only financed or motivated by the control of resources, but that resources figure prominently in their agendas, at least economically. Given the concentration of wars in poor countries with few foreign-earning sources, resources are likely to remain the economic focus of most belligerents in years to come. Even if 'conflict resources' come under greater regulatory pressure, there is a likelihood that criminal networks and unscrupulous businesses will pursue trading, especially those already involved in arms trafficking.

There is no simple and comprehensive measure that can reduce the prevalence of conflicts in resource-dependent countries, but several factors can assist in this regard. The specificities of resources and licit character of their trade demand a new type of engagement and set of regulations on the part of businesses and policy makers to tackle their contribution to war economies. The Security Council, governments, business associations, and advocacy NGOs have been developing an array of rules, investigation, sanctions, and implementation measures targeted at specific commodities over the past few years, and these initiatives need support and encouragement.[28] Most noticeably, diamonds have been the targets of unprecedented regulatory measures that, however, in the absence of sustained monitoring efforts will most probably remain plagued by difficulties inherent to the physical and market specificities of this commodity. In other cases, vested commercial and geopolitical interests, as well as by the potential humanitarian impact on the targeted 'conflict resource' have continued to refrain the use of sanctions, with a mixed effect.

Beyond targeting the access of belligerents to resource revenues, three areas are particularly important: fair and more stable prices for resources; tighter domestic and international regulation of resource-derived revenues focusing on transparency; and a change in the culture of impunity in international resource trade. In the first case, producing countries and the international community should consider how revitalised commodity agreements and complementary mechanisms such as insurances might improve revenue flows to producing countries and contribute to positive economic and political improvements. Accordingly, revisiting pricing mechanisms should take place in tandem with an international framework for the regulation of resource revenues, which would seek not only greater stability in revenues, but also greater transparency, and increased accountability to local populations. Finally, international instruments used to prevent or terminate conflicts financed by natural resource exploitation would move from 'shaming' international actors to formalising punishments and sanctions, against individuals as well as corporations. These measures will take time to develop. In the interim, confronted with the likelihood of continued resource-fuelled wars, the international community should seek to develop and apply frameworks through which the 'economic demobilisation' of combatants could break the current pernicious relationships between natural resources, underdevelopment, and armed conflict.

NOTES

1   Ian O. Lesser, *Resources and Strategy* (Basingstoke: Macmillan 1989) p. 9.

2   Raymond Aron, *Peace and War* (London: Weidenfeld and Nicolson 1966) pp. 244–245.

3   John Evelyn, *Navigation and Commerce* (1674); cited in Aron *Peace and War* p. 245.

4   Lesser *Resources and Strategy* pp. 11–12.

5   On the case of Africa, see Thomas Pakenham, *The Scramble for Africa: White Man's Conquest of the Dark Continent from 1876–1912* (London: Weidenfeld and Nicholson 1991).

6   Hitler's party programme demanded as early as 1920 "land and territory for the sustenance of our people, and the colonisation of our surplus population." Cited in http://www.yale.edu/lawweb/avalon/imt/proc/judnazi.htm.

7   Hanns W. Maull, *Raw Materials, Energy and Western Security* (Basingstoke: Macmillan 1984); Arthur H. Westing (ed.), *Global Resources and International Conflict: Environmental Factors in Strategy Policy and Action* (Oxford: Oxford University Press 1986).

8   Lester R. Brown, *Redefining National Security* (Washington DC: Worldwatch Institute 1977); Richard H. Ullman, 'Redefining Security', *International Security* 8/1 (Summer 1983) pp. 129–153; Jessica T. Mathews, 'Redefining Security', *Foreign Affairs* 68/2 (Spring 1989) pp. 162–177.

9   World Commission on Environment and Development, *Our Common Future* (Oxford: Oxford University Press 1987); Norman Myers, *Ultimate Security: The Environmental Basis of Political Stability* (New York: W.W. Norton 1993) pp. 17–30; for a comprehensive critique of this concept, see Simon Dalby, *Environmental Security* (Minneapolis: University of Minnesota Press 2002).

10   James E. Sinclair and Robert Parker, *The Strategic Metals Wars* (New York: Arlington House 1983); Oye Ogunbadejo, *The International Politics of Africa's Strategic Minerals* (London: Frances Pinter 1985); Ruth W. Arad (ed.), *Sharing Global Resources* (New York: McGraw-Hill 1979).

11   Bruce Russett, 'Security and the Resources Scramble: Will 1984 be Like 1914?' *International Affairs* 58/1 (Winter 1981–82) pp. 42–58.

12   Ewan W. Anderson and Liam D. Anderson, *Strategic Minerals: Resource Geopolitics and Global Geoeconomics* (Chichester: Wiley, 1998).

13   Kofi Annan, *The Causes of Conflict and the Promotion of Durable Peace and Sustainable Development in Africa* (New York: United Nations 1998) para. 14; William Reno, *Warlord Politics and African States* (Boulder CO: Lynne Rienner 1998), Klare, (note 3) p. 213.

14   Michael Klare, *Resource Wars: The New Landscape of Global Conflict* (New York: Metropolitian 2001), pp. 10, 23, 25.

15   Canada and the US provided examples of success and led to the staple theory of growth, see Harold A. Innis, *Essays in Canadian Economic History* (Toronto: University of Toronto Press 1956) and Douglas C. North, 'Location Theory and Regional Economic Growth', *Journal of Political Economy* 63 (April 1955).

16   For a comparative historical study of the contribution of mining sectors to development, see Thomas M. Power, *Digging to Development? A Historical Look at Mining and Economic Development*, Oxfam America Report (New York: Oxfam 2002).

17   On the manipulation of resource sectors, see, Robert H. Bates *Markets and States in Tropical Africa: the Political Basis of Agricultural Policies* (Berkeley, CA: University of California Press 1981); Raymond L. Bryant and Michael J. P. Parnwell (eds), *Environmental Change in South-East Asia: People, Politics and Sustainable Development* (London: Routledge 1996). On the characteristics of oil and mineral dependent states, see, Michael L. Ross, 'Does Oil Hinder Democracy?', *World Politics* 53 (April 2001) pp. 325–341; and Michael L. Ross, *Extractive Sectors and the Poor*, Oxfam America Report (New York: Oxfam 2001).

18   Paul Collier, 'Economic Causes of Civil Conflict and Their Implications for Policy', in Chester A. Crocker, Fen Osler Hampson, and Pamela Aall (eds), *Turbulent Peace: The Challenges of Managing International Conflict* (Washington DC: United States Institute for Peace Press 2001).

19   Alan H. Gelb (ed.), *Oil Windfalls: Blessing or Curse* (Oxford: Oxford University Press 1989); Richard M. Auty, *Sustaining Development in Mineral Economies: the Resource Curse Thesis* (London: Routledge 1993); Michael L. Ross, 'The Political Economy of the Resource Curse', *World Politics*, 51/2 (January 1999) pp. 297–322.

20   For a discussion of the 'excesses' of authority and connections with the West on the underdevelopment of Africa, see James D. Sidaway, 'Sovereign Excesses? Portraying Postcolonial Sovereigntyscapes', *Political Geography*, 22 (2003) pp. 157–178.

21   Cited in Kisangani N. F. Emizet, 'Zaire After Mobutu: A Case of a Humanitarian Emergency', Research for Action no. 32 (Helsinki: WIDER 1997) p. 35.

22   Patrick Chabal and Jean-Pierre Daloz, *Africa Works:*

Disorder as Political Instrument (Oxford: James Currey 1999).

23 Crawford Young and Thomas Turner, *The Rise and Decline of the Zairian State* (Madison WI: University of Wisconsin Press 1985); Emizet "Zaire after Mobuto" p. 39.

24 On the cases of Liberia and Sierra Leone, see Stephen Ellis, *The Mask of Anarchy. The Destruction of Liberia and the Religious Dimension of an African Civil War* (London: Hurst 1999); William Reno, *Corruption and State Politics in Sierra Leone* (Cambridge: Cambridge University Press 1995); Paul Richards, *Fighting for the Rain Forest: War Youth and Resources in Sierra Leone* (Oxford: James Currey 1996).

25 Philippe Le Billon, 'The Political Ecology of War: Natural Resources and Armed Conflicts', *Political Geography* 20/5 (2001) pp. 561–584.

26 David Keen, *The Economic Functions of Violence in Civil Wars*, Adelphi Paper no. 320 (Oxford: Oxford University Press for the IISS 1998), p. 45.

27 Al Gedicks, *Resource Rebels: Native Challenges to Mining and Oil Corporations* (Cambridge MA: South End Press 2001).

28 Philippe Le Billon, Jake Sherman, Marcia Hartwell, 'Controlling Illicit Resource Flows to Civil Wars: A Review and Analysis of Current Policies and Legal Instruments' (New York: International Peace Academy 2002).

25

No Escape from Dependency

Looming Energy Crisis Overshadows Bush's Second Term

Michael T. Klare

from *TomDispatch.com* (2004)

When George W. Bush entered the White House in early 2001, the nation was suffering from a severe "energy crisis" brought on by high gasoline prices, regional shortages of natural gas, and rolling blackouts in California. Most notable was the artificial scarcity of natural gas orchestrated by the Enron Corporation in its rapacious drive for mammoth profits. In response, the President promised to make energy modernization one of his top concerns. However, aside from proposing the initiation of oil drilling in Alaska's Arctic National Wildlife Refuge, he did little to ameliorate the country's energy woes during his first four years in office. Luckily for him, the energy situation improved slightly as a national economic slowdown depressed demand, leading to a temporary decline in gasoline prices. But now, as Bush approaches his second term in office, another energy crisis looms on the horizon – one not likely to dissipate of its own accord. The onset of this new energy crisis was first signaled in January 2004, when Royal Dutch/Shell – one of the world's leading energy firms – revealed that it had overstated its oil and natural gas reserves by about 20%, the net equivalent of 3.9 billion barrels of oil or the total annual consumption of China and Japan combined. Another indication of crisis came only one month later, when the *New York Times* revealed that prominent American energy analysts now believe Saudi Arabia, the world's largest oil producer, had exaggerated its future oil production capacity and could soon be facing the wholesale exhaustion of some of its most prolific older fields. Although officials at the U.S. Department of Energy (DoE) insisted that these developments did not foreshadow a near-term contraction in the global supply of energy, warnings increased from energy experts of the imminent arrival of "peak" oil – the point at which the world's known petroleum fields will attain their highest sustainable yield and commence a long, irreversible decline.

How imminent that peak oil moment may in fact be has generated considerable debate and disagreement within the specialist community, and the topic has begun to seep into public consciousness. A number of books on peak oil – *Out of Gas* by David Goodstein, *The End of Oil* by Paul Roberts, and *The Party's Over* by Richard Heinberg, among others – have appeared in recent months, and a related documentary film, *The End of Suburbia*, has gained a broad underground audience. As if to acknowledge the seriousness of this debate, the *Wall Street Journal* reported in September that evidence of a global slowdown in petroleum output can no longer be ignored. While no one can say with certainty that recent developments portend the imminent arrival of peak oil output, there can be no question that global supply shortages will prove increasingly common in the future.

Nor is the evidence of a slowdown in oil output the only sign of an unfolding energy crisis. Of no less significance is the dramatic increase in energy demand from newly industrialized nations – especially China. As recently as 1990, the older industrialized countries (including the former Soviet Union) accounted for approximately three-quarters of total worldwide oil consumption. But the consumption of petroleum in developing nations is growing so rapidly – at three times the rate for developed countries – that it is soon expected to draw even.

To meet the needs of their older customers and satisfy the rising demand from the developing world, the major oil producers will have to boost production at breakneck speed. According to the DoE, total world petroleum output will have to grow by approximately 44 million barrels per day between now and 2025 – an increase of 57% – to satisfy anticipated world demand. This increase represents a prodigious amount of oil, the equivalent to total world consumption in 1970, and it is very difficult to imagine where it will all come from (especially given indications of a global slowdown in daily output). If, as appears likely, the world's energy firms prove incapable of satisfying higher levels of international demand, the competition among major consumers for access to the remaining supplies will grow increasingly more severe and stressful.

To further complicate matters, many of the countries the Bush administration considers potential suppliers of additional petroleum, including Angola, Azerbaijan, Colombia, Equatorial Guinea, Iran, Iraq, Kazakhstan, Nigeria, Saudi Arabia, and Venezuela, are torn by ethnic and religious conflict or are buffeted by powerful anti-American currents. Even if these countries possess sufficient untapped reserves to sustain an increase in output, as long as they remain chronically unstable, the desired increases are unlikely to appear. After all, any significant increase in day-to-day energy output requires substantial investment in new infrastructure – investment that is not likely to materialize in countries suffering from perpetual disorder. At best, production in such countries will remain flat or rise sluggishly; at worst, as in Iraq today, it may even threaten to fall. Indeed, the persistence of political turmoil in countries like Angola, Colombia, Iraq, Nigeria, and Venezuela has largely been responsible for the higher gasoline prices still evident, despite recent modest decreases, at the neighborhood pump.

If anything, the potential for conflict in such countries is likely to grow as demand for their petroleum rises. The reason is simple. Increased petroleum output in otherwise impoverished nations tends to widen the gap between haves and have-nots – a divide that often falls along ethnic and religious lines – and to sharpen internal political struggles over the distribution of oil revenues. Because the wealth generated by oil production is so vast, and because few incumbent leaders are willing to abandon their positions of privilege, internal struggles of this sort are prone to trigger violent clashes between competing claimants to national power.

In many cases, these clashes may take the form of attacks on the oil infrastructure itself, further jeopardizing the global availability of energy. As shown in Colombia and Iraq, where raids on oil pipelines and pumping stations have become a near-daily occurrence, such infrastructure – stretched out over miles and miles of jungle or desert – represents an unusually vulnerable and inviting target for terrorism. Not only do such attacks deprive the prevailing regime of vital revenues, but they also constitute an assault on the United States and the large multinational corporations that are deemed responsible for so many of the developing world's afflictions.

With oil demand regularly outpacing supply and disorder spreading in major producing areas, global shortages and resulting high prices are likely to become the norm, not the exception. Ideally, the United States could compensate for any shortfalls in the global availability of petroleum by increasing its reliance on other sources of energy. When producing electricity, for example, it is often possible to switch from coal to natural gas and back again. But most of our petroleum supplies are used in transportation – mainly to power cars, trucks, buses, and planes – and, for this purpose, oil has no readily available substitutes. Indeed, we have so organized our economy and society around the availability of cheap and abundant petroleum that we are severely ill-equipped to deal with the sort of shortages and supply disruptions that are likely to become the norm in the years ahead.

It is here that the performance of the Bush administration should come in for close scrutiny. In response to the earlier energy crisis of 2001, the President appointed a National Energy Policy Development Group (NEPDG), headed by Vice President Dick Cheney, to analyze America's energy predicament and devise appropriate solutions. The NEPDG issued its final report, the National Energy Policy (also known as the Cheney Report), in May, 2001. How the group arrived at its final assessment is a matter of some speculation, as the administration has refused to make its deliberations public, but its conclusions are incontrovertible: rather than stress conservation and the rapid development of renewable energy sources, the report called for increased U.S. reliance on petroleum. And because domestic oil production is in an irreversible decline, any rise in American oil usage necessarily entails an increased reliance on imported petroleum.

In a crude attempt to mislead the public about the nature of our oil dependency, the Cheney Report called for increasing U.S. energy "independence" by exploit-ing the untapped oil reserves of Alaska's Arctic National Wildlife Refuge (ANWR) and other protected wilderness areas. But ANWR only possesses sufficient petroleum to provide this country with (at most) 1 million barrels per day for an estimated 15–20 years, a tiny fraction of the 20 million barrels of additional oil that will be needed to supplement domestic output in 2025. What this suggests is that the overwhelming bulk of this additional energy will have to be acquired from foreign sources. To obtain all this imported energy, the Cheney Report calls on the President and his chief associates to place a high priority on acquiring additional petroleum from producers in the Persian Gulf, the Caspian Sea basin, Africa, and Latin America – that is, from regions especially susceptible to instability and anti-Americanism.

As a result, we are more dependent on foreign oil in 2004 than we were in 2001, and all the indicators suggest that this dependency will only become more pronounced during Bush's second term. Yes, the administration has proposed modest investment in the development of hydrogen-powered fuel cells and other new energy systems; but, at current rates of development, these new technologies will not prove capable of substituting for oil on a significant scale during the next few decades. This means that we will face our looming energy crisis with no viable fallback measures in sight. We remain trapped in our dependence on imported oil. In the long run, the only conceivable result of this will be sustained crisis and deprivation.

When, and in just what form, the United States enters the coming energy crisis cannot be foreseen. Perhaps it will be provoked by a coup d'état in Nigeria, a civil war in Venezuela, or a feud among senior princes in the Saudi royal family (possibly brought on by the impending death of King Fahd). Or it could be thanks to a major act of terrorism or a catastrophic climate event. Whatever the case, our existing energy system, already stretched to its limits, will not be able to absorb a major blow like this without considerable readjust-ment and pain – or worse. While President Bush is likely to respond to a new energy crisis, as he has in the past, with renewed calls for drilling in ANWR and the further relaxation of U.S. environmental standards, nothing he has proposed to date even suggests a viable exit strategy from perpetual crisis.

Cartoon 9 SUV on Planet

Gasoline consumption continues to rise around the world despite clear indications that supplies of petroleum are limited. The passion for SUVs in contemporary car culture only speeds up the depletion of resources and threatens to aggravate potential conflicts over control of the remaining supplies.

Source: M. Wuerker

26

Oil and Blood

The Way to Take over the World

Michael Renner

from *World Watch Magazine* (2003)

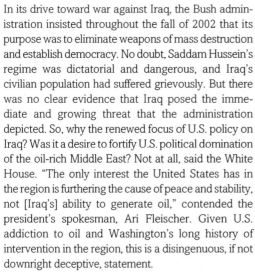

In its drive toward war against Iraq, the Bush administration insisted throughout the fall of 2002 that its purpose was to eliminate weapons of mass destruction and establish democracy. No doubt, Saddam Hussein's regime was dictatorial and dangerous, and Iraq's civilian population had suffered grievously. But there was no clear evidence that Iraq posed the immediate and growing threat that the administration depicted. So, why the renewed focus of U.S. policy on Iraq? Was it a desire to fortify U.S. political domination of the oil-rich Middle East? Not at all, said the White House. "The only interest the United States has in the region is furthering the cause of peace and stability, not [Iraq's] ability to generate oil," contended the president's spokesman, Ari Fleischer. Given U.S. addiction to oil and Washington's long history of intervention in the region, this is a disingenuous, if not downright deceptive, statement.

The Middle East – and specifically the Persian Gulf region – accounts for about 30 percent of global oil production. But it has about 65 percent of the planet's known reserves, and is therefore the only region able to satisfy any substantial rise in world oil demand – an increase that the administration's energy policy documents say is inevitable. Saudi Arabia, with 262 billion barrels, has a quarter of the world's total reserves and is the single largest producer. But Iraq, despite its pariah status for the past 12 years, remains a key prize. At 112 billion barrels, its known reserves are second only to Saudi Arabia's. And, given that substantial portions of Iraqi territory have never been fully explored, there is a good chance that actual reserves are far larger.

For half a century, the United States has made steadily increasing investments in keeping the Gulf region in its geopolitical orbit. The investments have included the overthrow of "hostile" governments and support of client regimes, massive arms transfers to allies, acquisition of military bases, and direct and indirect forms of intervention – many of these activities involving shifting alliances and repeated large-scale violence. In Washington's calculus, securing oil supplies has consistently trumped the pursuit of human rights and democracy. This is still the case today, as the Bush administration prepares for a more openly imperial role in the region.

Saudi Arabia has had a close relationship with the United States since the 1940s. But it has long been vulnerable to pressures from the far more populous Iraq and Iran. Iran was brought firmly into the Western orbit by a 1953 CIA-engineered coup against the Mossadegh government, which had nationalized Iran's oil. The coup re-installed the Shah on the Persian throne. Armed with modern weaponry by the United States and its allies, the Shah became the West's regional policeman once the military forces of Britain – the former colonial power – were withdrawn from the Gulf area in 1971. Iraq, on the other hand, was a pro-Western country until 1958, when its British-installed monarchy was overthrown. Fearing that Iraq might turn communist under the new military regime, the United States dabbled in a temporary alliance of convenience with the Ba'ath (Renaissance) Party in its efforts to grab power. CIA agents provided critical logistical information to the coup plotters and supplied lists with the names of hundreds of suspected Communists to be eliminated.

Even so, in 1972 the Ba'ath regime signed a treaty of friendship and cooperation with the Soviet Union. Baghdad turned to Moscow both for weapons and for help in deterring any U.S. reprisals against Iraq for nationalizing the Iraq Petroleum Company, which had been owned by Royal Dutch-Shell, BP, Exxon, Mobil, and the French firm CFP. Iraq was the first Gulf country to successfully nationalize its oil industry. Saddam Hussein, a strongman of the Ba'ath regime who formally took over as President in 1979, was instrumental in orchestrating the pro-Moscow policy. But as it became apparent that the Soviet Union could not deliver the technologies and goods (both civilian and military) needed to modernize Iraq he shifted to a more pro-Western policy. Western governments and companies were eager to soak up the rising volume of petrodollars, and to lure Iraq out of the Soviet orbit. During the 1980s, this eagerness extended to supplying Baghdad with the ingredients needed to make biological, chemical, and nuclear weapons.

The year 1979 turned out to be a watershed, as the Shah of Iran was swept aside by an Islamic revolution that brought Ayatollah Khomeini to power. One of Washington's main geopolitical pillars had crumbled, and the new regime in Teheran was seen as a mortal threat by the conservative Persian Gulf states. The Carter administration responded by pumping rising quantities of weapons into Saudi Arabia, and began a quest for new military bases in the region (Bahrain eventually became the permanent home base for the U.S. Fifth Fleet). But there was no escaping the fact that neither Saudi Arabia nor any of the smaller Gulf states was strong enough to replace Iran as a proxy. Instead, Iraq became a surrogate of sorts. Iran and Iraq had long been at loggerheads. Seeing a rival in revolutionary disarray, and sensing an opportunity for an easy victory that would propel him to leadership of the Arab world, Saddam Hussein invaded Iran in September 1980. Eager to see Teheran's revolutionary regime reined in, the United States turned a blind eye to the aggression, opposing UN Security Council action on the matter.

But instead of speeding the Iranian regime toward collapse, the attack consolidated Khomeini's power. And marshalling revolutionary fervor, Iran was soon turning the tide of battle. With the specter of an Iraqi defeat looming, the United States went much farther in its support of Saddam:

▦ To facilitate closer cooperation, the Reagan administration removed Iraq from a list of nations that it regarded as supporters of terrorism. Donald Rumsfeld, now secretary of defense, met with Saddam in Baghdad in December 1983. His visit paved the way to the restoration of formal diplomatic relations the following year, after a 17-year hiatus.

▦ The United States made available several billion dollars' worth of commodity credits to Iraq to buy U.S. food, alleviating severe financial pressures that had threatened Baghdad with bankruptcy. The food purchases were a critical element in the regime's attempts to shield the population as much as possible from the war's repercussions – and hence limiting the likelihood of any challenges to its rule. The U.S. government also provided loan guarantees for an oil export pipeline through Jordan (replacing other export routes that had been blocked because of the war).

▦ Though not selling weapons directly to Iraq, Reagan administration officials allowed private U.S. arms dealers to sell Soviet-made weapons purchased in Eastern Europe to Iraq. U.S. leaders permitted Saudi Arabia, Kuwait, and Jordan to transfer U.S.-made weapons to Baghdad. And they abandoned earlier opposition to the delivery of French fighter jets and Exocet missiles (which were subsequently used against tankers transporting Iranian oil).

▦ From the spring of 1982 on, the Reagan administration secretly transmitted highly classified military intelligence – battlefront satellite images, intercepts of Iranian military communications, information on Iranian troop deployments – to Saddam Hussein's regime, staving off its defeat.

▦ As the war went on, the United States took an increasingly active military role. It tilted toward Iraq in the "war on tankers" by protecting oil tankers in the southern Gulf against Iranian attacks, but did not provide security from Iraqi attacks for ships docking at Iran's Kharg Island oil terminal. Later, the United States even launched attacks on Iran's navy and Iranian offshore oil rigs.

Washington's immediate objective was to prevent an Iranian victory. In a larger sense, though, U.S. policymakers were intent on keeping both Iraq and Iran bogged down in war, no matter how horrendous the human cost on both sides – hundreds of thousands were killed. (The Reagan administration secretly allowed Israel to ship several billion dollars' worth of U.S. arms and spare parts to Iran.) Preoccupied with fighting one another, Baghdad and Teheran would be unable to challenge U.S. domination of the Gulf region. Reflecting administration sentiments, Henry Kissinger said in 1984 that "the ultimate American interest in the war [is] that both should lose."

Oil and geopolitical interests translated into U.S. support for Saddam Hussein when he was at his most dangerous and murderous — not only committing an act of international aggression by invading Iran, but also by using chemical weapons against both Iranian soldiers and Iraqi Kurds. U.S. assistance to Baghdad was provided although top officials in Washington knew at the time that Iraq was using poison gas.

Undoubtedly, U.S. support emboldened Saddam Hussein to invade Kuwait in 1990. But the United States would never consent to a single, potentially hostile, power gaining sway over the Gulf region's massive oil resources. When its regional strongman crossed that line, U.S. policy shifted to direct military intervention.

Following the Gulf War, the United States supplied Saudi Arabia and other allies among the Gulf states with massive amounts of highly sophisticated armaments. Washington and other suppliers delivered more than $100 billion worth of arms from 1990 to 2001. In the late 1980s, Saudi Arabia had imported 17 percent, by dollar value, of worldwide weapons sales to developing countries. In the 1990s, the Saudi share rose to 38 percent.

But rather than becoming independent military powers, Riyadh and the other Gulf states are at best beefed-up staging grounds for the U.S. military: Washington has for many years been "pre-positioning" military equipment and supplies and expanding logistics capabilities to facilitate any future intervention. And although political sensitivities rule out a visible, largescale U.S. troop presence, more than 5,000 U.S. troops have been continuously deployed in Saudi Arabia, and more than 20,000 in the Gulf region as a whole.

Despite insinuations by the Bush administration, there is no evidence that Saddam Hussein's regime is in any way linked to the events of September 11, 2001. However, the terrorist attacks facilitated a far more belligerent, unilateralist mood in Washington and set the stage for the Bush administration doctrine of preemptive war.

Installing a U.S. client regime in Baghdad would give American and British companies (ExxonMobil, Chevron-Texaco, Shell, and BP) a good shot at direct access to Iraqi oil for the first time in 30 years — a windfall worth hundreds of billions of dollars. And if a new regime rolls out the red carpet for the oil multinationals to return, it is possible that a broader wave of de-nationalization could sweep through the world's oil industry, reversing the historic changes of the early 1970s.

Rival oil interests were a crucial behind-the-scenes factor as the permanent members of the UN Security Council jockeyed over the wording of a new resolution intended to set the parameters for any action against Iraq. The French oil company TotalFinaElf has cultivated a special relationship with Iraq since the early 1970s. And along with Lukoil of Russia and China's National Petroleum Corp., it has for years positioned itself to develop additional oil fields once UN sanctions are lifted. But there have been thinly veiled threats that these firms will be excluded from any future oil concessions unless Paris, Moscow, and Beijing support the Bush policy of regime change. Intent on constraining U.S. belligerence, France, Russia, and China nonetheless are eager to keep their options open in the event that a pro-U.S. regime is installed in Baghdad — and accordingly voted in favor of the U.S.-drafted resolution in November. But the stakes in all this maneuvering involve much more than just the future of Iraq. The Bush energy policy is predicated on growing consumption of oil, preferably cheap oil. Given rising depletion of U.S. oil fields, most of that oil will have to come from abroad, and indeed primarily from the Gulf region. Controlling Iraqi oil would allow the United States to reduce Saudi influence over oil policy and give Washington enormous leverage over the world oil market.

Both in the Middle East and in other regions, securing access to oil goes hand in hand with a fast-expanding U.S. military presence. From Pakistan to Central Asia to the Caucasus, and from the eastern Mediterranean to the Horn of Africa, a dense network of U.S. military facilities has emerged — with many bases established in the name of the "war on terror."

Only in the most direct sense is the Bush administration's Iraq policy directed against Saddam Hussein. In a broader sense, it aims to reinforce the world economy's reliance on oil – undermining efforts to develop renewable energy sources, boost energy efficiency, and control greenhouse gas emissions. The same administration that decided to slash annual spending for energy efficiency and renewables R&D has no problem with preparing for a war that could cost as much as $200 billion.

By rejecting U.S. participation in the Kyoto Protocol early in his tenure, George W. Bush sought to throw a wrench into the international machinery set up to address the threat of climate change. By securing the massive flow of cheap oil, he may hope to kill Kyoto. In a perverse sense, a war on Iraq reinforces the assault against the Earth's climate.

27

Biological Threat Assessment

Is the Cure Worse than the Disease?

Jonathan B. Tucker

from *Arms Control Today* (2004)

In the three years since the September 11 terrorist attacks and the subsequent mailings of anthrax bacterial spores, federal spending to protect the U.S. civilian population against biological terrorism has soared more than 18-fold. For the 2005 fiscal year, the Bush administration has requested about $7.6 billion for civilian biodefense, up from $414 million at the time of the 2001 attacks.[1] Several federal agencies are involved in biodefense research and development (R&D),[2] and the huge increase in funding from the National Institutes of Health for work on "select agents," or pathogens and toxins of bioterrorism concern, has attracted thousands of academic scientists.[3]

Of growing concern to U.S. biodefense officials is the possibility that rapid advances in genetic engineering and the study of pathogenesis (the molecular mechanisms by which microbes cause disease) could enable hostile states or terrorists to create "improved" biowarfare agents with greater lethality, environmental stability, difficulty of detection, and resistance to existing drugs and vaccines.[4] (See *ACT*, July/August 2004.) It is known, for example, that the Soviet biological weapons program did extensive exploratory work on genetically engineered pathogens.[5] The Bush administration's response to this concern has been to place a greater emphasis on "science-based threat assessment," which involves the laboratory development and study of offensive biological weapons agents in order to guide the development of countermeasures. This approach is highly problematic, however, because it could undermine the ban on offensive development enshrined in the Biological Weapons Convention (BWC) and end up worsening the very dangers that the U.S. government seeks to reduce.

BIOLOGICAL THREAT ASSESSMENT – WEIGHING THE RISKS

The Bush administration contends that science-based threat assessment is needed to shorten the time between the discovery of new bioterrorist threats, such as pathogens engineered to be resistant to multiple antibiotics, and the development of medical countermeasures, such as vaccines and therapeutic drugs. This rationale is flawed, however, for three reasons.

First, the administration's biodefense research agenda credits terrorists with having cutting-edge technological capabilities that they do not currently possess nor are likely to acquire anytime soon. Information in the public domain suggests that although some al Qaeda terrorists are pursuing biological weapons, these efforts are technically rudimentary and limited to standard agents such as the anthrax bacterium and ricin, a widely available plant toxin. Assistance from a country with an advanced biological weapons program may be theoretically possible, but no state has ever transferred weaponized agents to terrorists, and the risks of retaliation and loss of control make this scenario unlikely. Although more sophisticated bioterrorist threats may emerge someday from the application of modern biotechnology, they are unlikely to materialize for several years.

Second, prospective threat-assessment studies involving the creation of hypothetical pathogens are of limited value because of the difficulty of correctly predicting technological innovations by states or terrorist organizations. Distortions such as "mirror-imaging" – the belief that an adversary would approach a technical problem in the same way as the person doing the analysis – make such efforts a deeply flawed basis for the development of effective countermeasures.

Third, by blurring the already hazy line between offensive and defensive biological R&D, science-based threat assessment raises suspicions about U.S. compliance with the BWC and fosters a "biological security dilemma" that could lead to a new biological arms race. At the same time, the novel pathogens and related know-how generated by threat-assessment work could be stolen or diverted for malicious purposes, exacerbating the threat of bioterrorism.

CURRENT THREAT ASSESSMENT ACTIVITIES

Although biological threat-assessment studies have been under way for several years, they have received a major boost under the Bush administration. On April 21, after a 10-month policy review of national biodefense programs, President George W. Bush signed Homeland Security Presidential Directive 10 (HSPD-10). In addition to allocating roles and responsibilities among various federal agencies, this directive requires the Department of Homeland Security (DHS) to conduct a national risk assessment of new biological threats every two years and a "net assessment" of biodefense effectiveness and vulnerabilities every four years. Under HSPD-10, significant resources will be devoted to projecting future threats, not just addressing current ones. According to an unclassified summary of the directive, the U.S. government is "continuing to develop more forward-looking analyses, to include Red Teaming efforts, to understand new scientific trends that may be exploited by our adversaries to develop biological weapons and to help position intelligence collectors ahead of the problem."[6]

The expression "Red Teaming" dates back to the Cold War, when "red" symbolized the Soviet Union and its Warsaw Pact allies; the term now refers to any simulation involving the actions of a hostile country or subnational group. In the biodefense context, Red Teaming covers a variety of activities including scenario writing and paper studies, computer modeling of hypothetical biological attacks, and the development and testing of novel pathogens and weaponization techniques in the laboratory in order to guide the preparation of defenses. To expand U.S. government capabilities in the field of biological threat assessment,

DHS recently established a new multi-agency organization called the National Biodefense Analysis and Countermeasures Center (NBACC), headquartered at Fort Detrick, Maryland.[7] NBACC comprises four specialized centers, including a Biothreat Characterization Center whose mission is to "conduct science-based comprehensive risk assessments to anticipate, prevent, and respond to and recover from an attack."[8] The biothreat characterization program at NBACC will explore how bioterrorists might use genetic engineering and other advanced technologies to make viruses or bacteria more deadly or contagious.[9] During a White House online discussion forum on April 28, 2004, DHS Assistant Secretary for Science and Technology Penrose "Parney" Albright stated, "We are very concerned about genetically modified pathogens that might be, for example, vaccine-resistant or an attempt to elude our detection abilities. We have efforts underway within [DHS, the Department of Health and Human Services (HHS)], and the Department of Defense [to] think through carefully the kinds of genetic modifications and genetic engineering that might be done so we can get ahead of the emerging threat."[10]

In another published interview, Colonel Gerald W. Parker, director of the Science-Based Threat Analysis and Response Program Office at DHS, explained that the laboratory component of threat characterization "will be focused on addressing high-priority information gaps in either understanding the threat or our vulnerabilities." When asked if NBACC would conduct exploratory research on genetically engineered pathogens, Parker replied,

> We will not be intentionally enhancing pathogenicity of organisms to do "what-if" type studies. . . . [But] if there is information either in the classified or open literature, and it is validated information, that indicates that somebody may have [enhanced pathogenicity], and that we believe indicates that we might have a vulnerability in our defensive posture, we may have to, in fact, evaluate the technical feasibility and the vulnerability of our countermeasures.[11]

THE BIOLOGICAL SECURITY DILEMMA

Even if, as Parker asserts, threat-assessment studies at NBACC involving the creation of genetically modified pathogens will be carried out only in response to "validated" intelligence that a state or terrorist organization has already done so, other countries may perceive such efforts as a cover for illicit, offensively oriented activities. The reason is that the distinction between defensive and offensive biological R&D is largely a matter of intent, giving rise to a "security dilemma" in which efforts by some states to enhance their biological security inadvertently undermine the security of others.[12] Because intent is so hard to judge reliably, states tend to err on the side of caution by reacting to the capabilities, rather than the stated intentions, of potential adversaries. As a result, threat-assessment activities that a country pursues for defensive purposes may be perceived as offensive, particularly if those studies involve the genetic modification of pathogens to enhance their harmful properties.

Although the Bush administration has expressed concern about alleged biological weapons development activities in North Korea, Syria, Iran, and Cuba, it appears to have a blind spot with regard to how its own biological threat-assessment efforts are perceived abroad. Rival nations, fearing that the U.S. exploration of emerging biological weapons threats could generate scientific breakthroughs that would put them at a strategic disadvantage, may decide to pursue or expand similar activities. Even if these programs are initially defensive in orientation, they could acquire a momentum of their own that eventually pushes them over the line into the offensive realm.

The biological security dilemma has been inadvertently deepened by policies that the United States adopted after the September 11 attacks to tighten physical security and access controls at laboratories that possess, store, or transfer select agents.[13] Although these new regulations aim to prevent the theft or diversion of dangerous pathogens and toxins for malicious purposes, they have had the undesirable side effect of reducing the transparency of biodefense R&D at a time when greater openness is needed to reassure outsiders of the benign intent behind such activities.[14] Moreover, since the mid-1990s, the U.S. government has conducted an unknown number of classified threat-assessment studies, three of which were reported by *The New York Times* in September 2001. [. . .] The stated rationale for classification is to prevent terrorists from learning about and exploiting U.S. vulnerabilities to biological attack, but secrecy has the pernicious effect of increasing suspicions about U.S. intentions and worsening the security dilemma.

The most serious risk associated with science-based threat assessment is that the novel pathogens and information it generates could leak out to rogue states and terrorists. To prevent such proliferation, the United States will have to impose even more stringent security measures. Yet history suggests that the greatest risk of leakage does not come from terrorists breaking into a secure laboratory from the outside, but rather from trusted insiders within the biodefense community who decide, for various motives, to divert sensitive materials or information for sale or malicious use.

The expanded pool of researchers currently engaged in biological threat-assessment studies could well include a few spies, terrorist sympathizers, or sociopaths. Moreover, because a pathogen culture can be smuggled out of a laboratory in a small, easily concealable plastic vial, the odds of getting caught are fairly low. Security background checks on scientists working with select agents can reduce the threat of diversion but not eliminate it, as suggested by the cases of CIA or FBI insiders who became spies, such as Aldridge Ames and Robert Hanssen. Indeed, although the perpetrator of the mailings of anthrax bacterial spores in the fall of 2001 remains unknown, the technical expertise needed to prepare the highly refined material points to someone with experience inside the biodefense research complex.

Thus, rather than enhancing U.S. national security, science-based threat-assessment projects involving the development of novel pathogens are likely to create a vicious circle that ends up worsening the problems of biological warfare and bioterrorism. Prospective threat assessment entails two simultaneous risks: (1) developing dangerous new technologies that will leak out to proliferators and terrorists and create a self-fulfilling prophecy, and (2) undermining the norms in the BWC and provoking a biological arms race at the state level, even if the countries involved merely seek to anticipate and counter offensive developments by potential adversaries.

BREAKING OUT OF THE VICIOUS CIRCLE

In order to break out of the vicious circle created by the biological security dilemma, the United States should reduce its current emphasis on science-based threat assessment and pursue a number of strategies to build confidence in the strictly peaceful nature of its biodefense program.

Enhanced Transparency. The U.S. government should promote greater international transparency in biodefense R&D by including in its annual confidence-building measure (CBM) declarations under the BWC a comprehensive list of all of its biodefense activities, including classified projects, while omitting sensitive technical details that could assist proliferators or terrorists. (The fact that the United States had not declared the three secret threat-assessment studies uncovered by *The New York Times* suggested to some that it wished to avoid international scrutiny of legally dubious biodefense work.) In those rare cases where the risk of proliferation warrants classification, U.S. officials should explain why the experiments were done and provide a clear rationale for the limits on transparency. As a rule, however, openness should be considered the default condition, and any U.S. government agency seeking to classify specific biodefense projects or activities should be required to justify the need for secrecy.

International Collaboration. A second approach to building confidence would be for the United States to conduct biological threat-assessment studies jointly with other countries. NATO allies such as Canada, France, Germany, and the United Kingdom (as well as non-NATO countries such as Sweden) have advanced biodefense programs. Although the U.S. government conducts some joint R&D with allies, these efforts are currently pursued on an ad hoc basis. Integrating Canada, the European Union, and the United States into a formal system of collaborative biodefense R&D that includes effective oversight would give the international community greater confidence that Washington is not pursuing a unilateral path in this highly sensitive area and that its biodefense R&D program is fully compliant with the BWC.

Russia is also a potential U.S. partner in the biodefense field because of the large number of former bioweapons scientists and facilities remaining from the Soviet biowarfare program and the existence of several areas in which the two countries have complementary expertise and pathogen strain collections. To date, however, U.S.–Russian biodefense collaboration has been undermined by Moscow's refusal to share a genetically modified strain of the anthrax bacterium and the fact that biodefense facilities under the control of the Russian Ministry of Defense remain off-limits to Western scientists.[15] These issues will have to be resolved before joint U.S.–Russian R&D can become a source of greater international confidence in the BWC compliance of both countries.

Domestic Oversight. A third approach to breaking out of the vicious circle is to improve the domestic oversight of biological threat assessment. In October 2003, the National Research Council of the U.S. National Academy of Sciences released the report of an expert panel chaired by Dr. Gerald R. Fink on preventing the malicious application of "dual-use" research in the life sciences.[16] This report identified seven types of experiments that could result in information with a potential for misuse, including the genetic modification of pathogens to explore the mechanisms by which microbes cause disease. The Fink committee recommended the creation of a voluntary system for reviewing the security implications of federally funded biological research at the proposal stage. Such oversight would be performed at the local level by Institutional Biosafety Committees and at the national level by a new oversight board made up of scientists and security experts.

On March 4, 2004, the Bush administration responded to the Fink committee report by announcing the planned establishment of a National Science Advisory Board for Biosecurity (NSABB) under the auspices of the National Institute of Health. This new entity will establish guidelines for the security review of sensitive biological research projects in academia and, on a voluntary basis, in private industry.[17] Although the administration announced this initiative with much fanfare, the creation of the NSABB has proceeded at a snail's pace, and its first meeting has not yet been scheduled. Moreover, the advisory board will have no binding regulatory authority, and its mandate explicitly excludes the review of classified biodefense research initiated by the U.S. government and conducted at federal facilities with federal money. According to the NSABB web site, "Government-sponsored research that is classified at its inception . . . will be outside of the purview of the NSABB. This research is subject to other institutional and federal oversight, and is not the target of this biosecurity initiative."[18]

In fact, "other" federal oversight of classified biodefense R&D is extremely limited. Each U.S. government agency involved in such research is responsible for policing its own compliance with the BWC. The Defense Department, for example, has a Compliance Review Group that subjects the department's biodefense programs to internal legal review for consistency with the treaty.[19] Yet this committee is not accountable to the National Security Council or to other federal agencies such as the Department of State, which has the lead on the negotiation and legal interpretation of arms control treaties.

As a matter of principle, U.S. departments and agencies should not be responsible for reviewing the BWC compliance of their own biodefense programs because of the clear potential for conflict of interest. For example, lawyers employed by agencies with an institutional and budgetary stake in biological threat assessment may come under pressure to find loopholes so that legally questionable projects can go forward. For this reason, an interagency review process is needed to create internal checks and balances and build international confidence in the U.S. biodefense program.

Given the tenacity with which federal agencies defend their autonomy and turf, presidential leadership will be required to ensure adequate oversight and accountability for biological threat-assessment studies. Improved oversight mechanisms should be introduced by the executive and the legislative branches. For example, the Homeland Security Advisory Council in the White House might establish an interagency oversight board for biodefense consisting of representatives of the Defense and State Departments, CIA, DHS, HHS, and the intelligence community. This board would review the treaty compliance of all federal threat-assessment programs, including special-access ("black") projects whose existence is not acknowledged publicly. Congress should also pass legislation requiring all federal agencies involved in biodefense work to submit detailed reports on any classified threat-assessment activities to the House and Senate Select Committees on Intelligence, whose members and staff hold high-level clearances. These committees might also conduct closed hearings to review "black" biodefense projects on an annual basis.

Internal government oversight is not a panacea, however, because it can be corrupted by interagency collusion or a lack of good faith on the part of senior administration officials, particularly in an atmosphere of extreme secrecy. Prior to the Abu Ghraib prison-abuse scandal in Iraq, for example, the Department of Justice's Office of Legal Counsel prepared a memorandum arguing that the president's authority as commander in chief enabled him to disregard domestic laws and international treaties banning torture during interrogations of enemy combatants, thereby nullifying the existing checks and balances.[20]

Unilateral Restraint. Perhaps the most effective way for the United States to build international confidence in the peaceful nature of its biodefense program would be for the president to make a public statement renouncing the prospective development of genetically modified microorganisms with increased pathogenicity for threat-assessment purposes and urging all other countries to follow suit. As noted above, because the utility of prospective studies of genetically modified pathogens is severely limited by mirror-imaging and other sources of error, abandoning such studies would entail little risk to U.S. national security. On rare occasions, it may be necessary to test the efficacy of standard drugs or vaccines against genetically engineered pathogens that have already been developed by other countries. In these cases, the study should require a special authorization from the president following a careful interagency review to ensure that the proposed work complies with the letter as well as the spirit of the BWC.

To enforce the proposed unilateral ban on the prospective development of new pathogens with increased pathogenicity, the president should encourage scientists within the biodefense community to "blow the whistle" if they become aware of unauthorized studies that violate this policy, regardless of whether the work is being conducted in an academic setting or in a top-secret government laboratory. Confidential reporting channels and legal protections should also be established to shield scientists who expose illicit activities. To bolster the norm of professional responsibility further, scientists working in federal biodefense programs should be required to sign a code of conduct, similar to the Hippocratic oath, that precludes them from deliberately developing agents with enhanced pathogenicity or other harmful properties and requires them to report any deviations from this norm.

At the same time that the United States renounces the prospective development of pathogens for threat-assessment purposes, it should pursue less provocative ways of getting a jump on defending against new bioterrorist threats, such as bacteria that have been genetically modified to make them resistant to multiple antibiotics. One approach would be for researchers to focus on developing broad-spectrum therapeutic and preventive measures that are not agent-specific. A second strategy would be for the U.S. government to invest in building an R&D and industrial infrastructure that can assess novel biological threat agents as soon as they are detected and then develop, test, and manufacture safe and effective countermeasures. With these systems in place, it should become possible to move "from bug to drug" in a matter of weeks or months, rather than years.[21] (The recent identification of the SARS virus and the rapid development of a diagnostic test and a candidate vaccine is a case in point.) Because the two alternative strategies would be unequivocally defensive, they would build confidence that the U.S. biodefense program is fully consistent with the BWC.

These practical steps are needed to prevent the Bush administration's growing emphasis on science-based threat assessment from increasing biological weapons proliferation risks, exacerbating the security dilemma, weakening the BWC, and drawing the United States into a dangerous biological arms race. It is time to break the vicious circle before it starts.

NOTES

1   Ari Schuler, "Billions for Biodefense: Federal Agency Biodefense Funding, FY2001-FY2005," *Biosecurity and Bioterrorism*, 2004, pp. 86–96.

2   Federal agencies active in biodefense research include the Department of Homeland Security; the National Institutes of Health and Centers for Disease Control and Prevention under the Department of Health and Human Services; the National Laboratories operated by the Department of Energy; the U.S. Army Medical Research Institute of Infectious Diseases, Dugway Proving Ground, and the Naval Medical Research Center under the Department of Defense; the U.S. Department of Agriculture; and the U.S. intelligence community.

3   As of June 2004, more than 300 facilities and 11,000 individuals nationwide had been granted permission to work with "select agents." See Scott Shane, "Bioterror Fight May Spawn New Risks," *Baltimore Sun*, June 27, 2004.

4   Mark Wheelis, "Will the 'New Biology' Lead to New Weapons?" *Arms Control Today*, July/August 2004, pp. 6–13; Jennifer Couzin, "Active Poliovirus Baked From Scratch," *Science*, 2002, pp. 174–175.

5   See Ken Alibek with Stephen Handelman, *Biohazard* (New York: Random House, 1999).

6   Homeland Security Council, The White House, *Biodefense for the 21st Century*, April 2004, p. 4.

7   NBACC is currently being housed in temporary quarters until new buildings are constructed on a seven-acre plot

within a new National Interagency BioDefense Campus at Fort Detrick. When completed, the facility will include high- and maximum-containment laboratories (Biosafety Levels 3 and 4) capable of handling the most dangerous and incurable pathogens.

8   DHS Science and Technology Directorate, "Notice of Intent to Prepare and Environmental Impact Statement for the National Biosecurity Analysis and Counter-measures Center Facility at Fort Detrick, Maryland," *Federal Register* 69, June 7, 2004, pp. 31830–31831.

9   In a February 2004 briefing titled "The Leading Edge of Biodefense," Lieutenant Colonel George W. Korch Jr., NBACC deputy director, stated that the Biothreat Characterization Center will conduct studies in 16 subject areas, including genetic engineering of pathogens, susceptibility to current therapeutics, environmental stability, aerosol dynamics, novel delivery systems, and bioregulators and immunoregulators. The center will also study biothreat-agent use and the effectiveness of countermeasures "across the spectrum of potential attack scenarios" through "high-fidelity modeling and simulation." Milton Leitenberg, James Leonard, and Richard Spertzel, "Biodefense Crossing the Line," *Politics and the Life Sciences* 22, 2004.

10 Dr. Penrose "Parney" Albright, "Ask the White House," April 28, 2004, available at http://www.whitehouse.gov/ask/20040428.html (online interactive forum).

11 Laurie Goodman, "Biodefense Cost and Consequence," *Journal of Clinical Investigation* 114, July 1, 2004, pp. 2–3.

12 The security dilemma with regard to biodefense R&D is discussed in Gregory Koblentz, "Pathogens as Weapons: The International Security Implications of Biological Warfare," *International Security* 28, Winter 2003/04, p. 118. The concept of the security dilemma is presented in Robert Jervis, *Perception and Misperception in International Politics* (Princeton, NJ: Princeton University Press, 1976), pp. 169–170.

13 These regulations were mandated by the Public Health Security and Bioterrorism Preparedness and Response Act of 2002. For background, see Jonathan B. Tucker, "Preventing the Misuse of Pathogens: The Need for Global Biosecurity Standards," *Arms Control Today*, June 2003, pp. 3–10.

14 Gerald L. Epstein, "Controlling Biological Warfare Threats: Resolving Potential Tensions Among the Research Community, Industry, and the National Security Community," *Critical Reviews in Microbiology* 27, 2001, p. 346.

15 Kenneth N. Luongo et al., "Building a Forward Line of

Defense: Securing Former Soviet Biological Weapons," *Arms Control Today*, July/August 2004, pp. 18–23.

16 National Research Council, Committee on Research Standards and Practices to Prevent the Destructive Application of Biotechnology, *Biotechnology Research in an Age of Terrorism* (Washington, DC: National Academies Press, 2004).

17 HHS Press Office, "HHS Will Lead Government-Wide Effort to Enhance Biosecurity in 'Dual Use' Research," *HHS News*, March 4, 2004.

18 National Science Advisory Board for Biosecurity, "FAQ," March 4, 2004, available at http://www.biosecurityboard.gov/faq.htm.

19 Department of Defense, "Directive No. 2060.1," January 9, 2001 (Implementation of, and Compliance With, Arms Control Agreements).

20 Adam Liptak, "Legal Scholars Criticize Memos on Torture," *The New York Times*, June 25, 2004, p. A14.

21 Representative Jim Turner, *Beyond Anthrax: Confronting the Future Biological Weapons Threat* (May 2004), pp. 10–14.

FOUR

28

AIDS and Global Security

Gwyn Prins

from *International Affairs* (2004)

During the last twenty years, there has often been an uneasy relationship between the claim that an issue is important and that it is a security issue. That is because the political benefits of so doing are high, but so too are the costs. If an issue can be "securitised", it is the equivalent of playing a trump at cards, for at once it leap-frogs other issues in priority. But the unavoidable cost of this is first, that to obtain that priority, people must be persuaded to be afraid of the threat, and to see it as a "real and present danger;" secondly, it throws the solution into the hands of state – or state-derived and mediated – structures, for they alone command the resources to satisfy the scale and the urgency of the "securitised" threat once accepted as such.[1]

The most frustrating aspect of the most serious global security threats is that they are often insidious: slow of onset and demanding acts of extrapolation from fragmentary data. This is especially the case with global climate change. There, confronted with an unwillingness to confer political priority, there was an open attempt made in the early 1990s to ramp up the issue as a "real and present danger" security threat. The issue was that of "water wars"; and when they did not happen, the effect of over-selling to try and win the "securitisation" bonus, was to discredit the scientific underpinning of the entire subject in the eyes of political sceptics, and to deflect analysis into barren areas, putting back the arrival of effective action by at least a decade.[2]

At an early stage the HIV/AIDS issue also encountered securitisation, but inversely to the experience of the climate. Part of the inclusion of the disease within a discourse of social exclusion was the complaint, especially when the African epidemic became apparent, that whereas it was a security issue, it was being deliberately denied that status by the Rich World. In the most extreme form of the accusation, it was suggested that the rich might not be unhappy to see so many black deaths arising through inaction.

The positive engagement of AIDS with the security claim has proceeded in two stages. First, AIDS was represented by analogy: "AIDS is as destabilising as any war". Attempts to move beyond this were resisted on much the same grounds that the traditional "hard" security community and the "peace" activists had both earlier resisted the embrace of environmental threats. The former feared a dilution of the meaning of security to the point that everything in the world could claim this title; the latter, notably Daniel Deudney, feared that the environment would become "militarised" by such associations.[3] In the AIDS case, the claim was seen as a threat to the division of labour within the UN, and became swept into generalised suspicion of American power, given that the Clinton Administration pushed the issue into the Security Council so forcefully.

The move to the second stage was the moment when the case began to be made in high political circles that AIDS was, indeed, a direct security threat. It was made in both a broad and a narrow version. It can be dated precisely. In December 1999, the Clinton Administration's ambassador to the UN, Richard Holbrooke, travelled to southern Africa with Senator Russ Feingold. After he had seen AIDS orphans in Lusaka at a day centre, who were then turned out for the night, to forage in the streets, Holbrooke was galvanised to act.

Back on the plane, Holbrooke picked up the telephone. "It was at this point," Feingold says, "that he started doing what Dick Holbrooke does.

I watched him call up the Secretary-General (Kofi Annan) and tell him that we had to have a Security Council meeting on AIDS. The Secretary-General said, 'We can't do that. AIDS isn't a security issue.'[4]

That view was initially held by the other Permanent Members. Recollecting the 4087th meeting of the Security Council a year later, the British Permanent Representative, Sir Jeremy Greenstock, admitted as much.[5] France and China relented, and eventually Russia agreed, but only because everyone else wanted it; not on positive grounds.

The formal case to accept that AIDS should be visible to the Security Council as a threat to international peace and security was made by Vice President Gore when presiding over the first meeting of the 21st century on 10 January 2000. Gore elaborated a three point anchorage for the proposition, which has wide ramification:[6]

1 "The heart of the security agenda is protecting lives"
2 "when a single disease threatens everything from economic strength to peacekeeping, we clearly face a security threat of the greatest magnitude"
3 "It is a security crisis because it threatens not just individual citizens, but the very institutions that define and defend the character of a society"

Point (1) inscribes the tendency which Kofi Annan himself has done much to promote in recent years, to redefine the operational as well as the conceptual understanding of sovereignty. Taking as his text the Preface to the UN Charter, where the rights of men and women are given precedence over those of states, Annan argued that

> States are now widely understood to be instruments at the service of their peoples and not vice-versa. At the same time, individual sovereignty – by which I mean the fundamental freedom of each individual, enshrined in the Charter of the UN and subsequent international treaties – has been enhanced by a renewed and spreading consciousness of individual rights.[7]

Point (2) pre-figures the narrow issue area that was to be employed a little later that year, to produce the first Security Council resolution on AIDS. This proved to be a very touchy subject, for the Council confronted

the fact that with militaries often preferentially infected with HIV, the possibility is real that UN Peace-Keeping troops recruited from areas of high infection may actually be vectors to the populations whom they are sent to protect.[8] Thus Resolution 1308 of 17 July 2000 expressly included the encouragement of member states to test their soldiers within "the Council's primary responsibility for the maintenance of international peace and security."[9] Several African nations were strongly hostile to this advice. That pressure-point is a sign-post to one of the under-explored new avenues in broadening the front upon which AIDS is to be attacked.[10]

Resolution 1308 also reiterates in its Recital the essence of point (3) and frames the broad construction of the AIDS/security nexus: "recognizing that the HIV/AIDS pandemic is also exacerbated by conditions of violence and instability" and the converse: "stressing that the HIV/AIDS pandemic, if unchecked, may pose a risk to stability and security." The prevention of AIDS and of political violence are represented dramatically by the International Crisis Group as ". . .the two blades of the scissors required to cut the strangler's cord choking Africa."[11]

The Resolution, and more particularly the touchiness about it, point beyond the paradoxical impact of UN Peace-Keeping Forces to two structural facts which give the AIDS/security nexus such growing importance. The first is a leading feature of the next wave of AIDS and is a bitter paradox. AIDS has been spread by war, but also by the ending of war. The blood-to-blood contacts consequent upon the use of edge weapons to wound as well as to kill in the Rwanda genocide and the way in which the war in Guinea-Bissau served to accelerate the spread of HIV-2 variant dolefully illustrated the first point.

The ending of the Ethiopian/Eritrean war has released soldier and prostitute vectors from the confines of the battle-lines to return home, which served to accelerate the spread of the major variant in that region.[12] "Now our conscience is assailed by the pitiful and profitless phenomenon of war between 'have-nots' and 'have-nots'. War began," Sir John Keegan dryly observed,

> as an attack by the poor on the rich, persisted as an exercise between the rich and the rich, resolved itself with a counter attack by the rich against the poor. Now war is exclusively an occupation of the poor alone.[13]

World Bank statistics underscore this. Civil wars are much more common than they were in 1960. Indeed, Paul Collier and colleagues find that the strongest predictor of proneness to more civil war is previous civil war. Once fallen into the "conflict trap" it is as if new rules apply: a ghastly social version of a Black Hole in space.[14]

This is reflected in the stark exception that Africa (along with the Middle East, for different reasons) presents against the global trend of world-wide decline in all forms of armed violence since 1991.[15] The other structural fact is tied to a widespread characteristic which is also a deep flaw in the construction of the post colonial state apparatus in most of sub-Saharan Africa, which makes the region so vulnerable to AIDS.

Use of the military coup as a way of giving a bureaucratic skeleton to systems of patronage politics was not only common but also, in its own terms, rather logical.[16] Disapproved externally, military rule was politically a non-pathological way of using the natural hierarchy and command chains of military organisation to give form and function to inheritor governments. That, naturally, makes the link between AIDS and the military central; and it frames part of the broad construction of the nexus.

Zimbabwe is a case in point. Covert testing by a Swedish haematologist of blood samples from the Army revealed 80% positive rates.[17] If one extrapolates even considerably lower rates for the Zimbabwean political elite at large than the blood study revealed for the Army, we are nonetheless confronted with a circumstance where a decisive part of an embattled elite probably knows that it is condemned to early death, or that substantial part of any given individual's extended family is. What does such knowledge do to the life choices of such individuals, and most specifically, to their political behaviour, when there is nothing to lose? These are sensitive but essential dimensions of the effects of pandemic that have been, of course, matters for consideration within the closed walls of the intelligence and government analysis services, but which should properly be placed in the open debate also. They engage actively the advancing agenda of the responsibility to protect human rights. Without question, they raise tricky ethical and legal dimensions of the management of state failure.

Reliance upon the administrative capacities of military structure increases as AIDS stokes the fires of social unrest and saps the less resilient sinews of the state. Yet the military (and the Police) are preferentially

infected and therefore the ability to be able to perform may stand in inverse relationship to increasing need.[18] It is also suggested that by hollowing-out the officer corps, eroding discipline and making soldiers lethargic and irresponsible, AIDS may delay the resolution of conflicts.[19] To date, the most closely studied case of this trap has been that of the South African armed forces. Rising rates of infection, especially among young blacks, coincide with a policy to promote "fast track" young blacks to replace middle ranking whites, while at the same time there is freezing of the employment profile of the forces by granting security of tenure. All this, Dr Heinecken suggests, may well exacerbate already tense race relations and will pretty inexorably decommission the SANDF as a competent military force within a few years. Seven out of ten deaths in the SANDF are from AIDS, and largely because of AIDS, it can at present field only one operational brigade.

HIV/AIDS in itself is not a security problem. It is the collective impact of the disease on the social structure of society and on state strength that creates the problem, and this is where the armed forces play a crucial role in the catastrophe.[20]

This comment directs us into unexplored territory. To what extent will the cultural reproduction of societies be impaired or prevented? One simply lists the cumulative and pervasive effects of so many deaths, of foreshortened life expectancies – all the improvements of the previous forty years reversed in the last ten in southern Africa – of the presence of orphans (Whiteside calculates a rise from zero to 2 million, 1995–2010, for South Africa).[21] The only comparative cases are speculative and from distant history.[22] But amazing coping strategies in great adversity are the story of 20th century southern Africa;[23] and so this is a taxing and important area for future research.

In sum, in the story of how AIDS has come to be linked to both the narrow and broad constructions of security, by late 2001 we had reached a point where, for the first time – and due to the efforts of Richard Holbrooke more than to anyone else – a formal case for the securitisation of AIDS had been made and activated in international fora.

Furthermore, the three anchorage points in Gore's speeches located mandates for action under a securitised analysis at three scales: The first is that of *human security interpreted principally as an extension of the responsibility to protect human rights.* That was the subject of the last and best of the recent major UN studies.[24]

Secondly, the authority to breach state sovereignty under the ICISS "Just Cause" criterion means that *the internal consequences of chaos become, for the first time since the end of the European colonial empires in the mid 20th century, judiciable* in the wider community. Plainly this will be one of the issues to be grasped in the successor to the ICISS, the Annan High Level Panel established in December 2003.

Thirdly, the UN process of 2000–01 served to set down markers for the arena of the largest scale, where AIDS may be a driver for *threats to international peace and security in the direct and traditional economic and political senses.*

Arguably this formalisation and the uses which, starting with the 2004 High Level Panel, may be made of it, may be as great a mile-stone in the campaign against this pandemic as the increases in the provision of funds to UNAIDS and the Global Fund and from direct national sources. Indeed, set against the remorseless demographic dynamics of the pandemic, an underlying presumption is that without efficient and timely capitalisation upon this double achievement, we may find ourselves having lost a window of opportunity that will not return. There is one further aspect of the recent developments in the securitisation of AIDS which must be recorded; for it is of fundamental importance in the world of power politics to which we have returned, and is the announcement by the White House, coincident with the UNSC Resolution, that AIDS posed a threat to US national security.

That announcement was a logical development from the major shifts in the methodologies of security analysis which have been under way within the US intelligence community since the mid 1990s. Following five years or so after similar changes had occurred in the British intelligence community, the American move to multi-variate scenario development was published in the December 2000 study of global trends, 2015.[25] It identified seven global drivers (population; natural resources and the environment; science and technology; the global economy and globalisation; national and international governance; the nature of conflict; the role of the USA). It examined their potential interactions and registered them in four "scenarios": inclusive globalisation; pernicious globalisation; regional competition and post-polar (i.e. US isolationist) world. In all scenarios, it expected a net decline of American influence; in most it saw the triggering role

of pandemic and/or climate stress leading to state failure. It signalled some possible major discontinuities, including prominently the effect of accelerated and/or new pandemic (as well as several of the recent developments in Middle Eastern and Euro-American affairs).

"AIDS, and other diseases and health problems will hurt prospects for transition to democratic regimes as they undermine civil society, hamper the evolution of sound political and economic institutions, and intensify the struggle for power and resources."[26] The point to grasp is that both the statement and the changes in intelligence methodologies which underpin it contributed to the change in US approach, as stated in the UN process of 2000–01.

Speaking at a conference in May 2001, David Gordon, the officer in the NIC with responsibility at that time for global and economic issues, provided the most concrete summary of the position on AIDS as a US national security issue then pertaining within the intelligence community. He listed them under five headings.

1 There is a direct threat to the health of US citizens from pandemic viruses (AIDS, SARS etc)
2 There is a threat of biological terrorism or warfare employing such agents. The West Nile virus scare in New York was mentioned illustratively
3 There is a threat to US forces operating in both developing and post-communist countries on peace-keeping and humanitarian missions
4 The exacerbation of military conflicts by the presence of AIDS, which may draw on US resources
5 The friction with key US trading partners that may arise from restrictive immigration controls and disputes over intellectual property rights for drugs.

To these he then added seven indirect but potent linkages between AIDS and conflict in Africa, as he and his colleagues saw them. Deepening immiseration leads to a breakdown of social bonds: above 10% prevalence, Mr Gordon was doubtful that the structures of the extended family could hold. Thirdly, he emphasised the fate and role of AIDS orphans, especially if/when recruited as child soldiers. The withdrawal of children from education in turn exacerbates the general undermining of civil society. But that was especially the consequence of the preferential infection of the elites: ". . . that weakening of civil society leads to a context in which the maintenance and sustainability of effective governance declines dramatically." That in turn erodes economic growth which leads to increased conflicts over power, resources and patronage.[27] What might also be added is that in a world returned to power politics, the ability of the USA to deploy vast resources may in itself be a factor in turning victim countries and regions towards its sphere and interests. This has, indeed, been the experience of AIDS since the introduction of the President's Emergency Plan for AIDS Relief (PEPFAR).[28]

NOTES

1   See G. Prins, *The Heart of War: On Power, Conflict and Obligation in the 21st Century*, Routledge, 2002, p. 24.

2   Prins, *Heart of War*, pp. 21–22; 41–44, esp Figure at p. 44; M. Harrison, "Knowledge, Environment and Security" Final Report of the Wilton Park Conference, Hadley Centre, Met. Office, March 2003.

3   Daniel Deudney, "The Case Against Linking Environmental Degradation and National Security," *Millennium*, Vol. 19, No. 3 (Winter 1990), pp. 461–476.

4   S. Sternberg, "Former diplomat Holbrooke takes on global AIDS", *USA Today*, 10 June 2002.

5   "The impact of HIV/AIDS on Peace and Security," Statement by Sir Jeremy Greenstock in the Security Council, 19 January 2001, UK Mission to the UN.

6   Quotations are taken from both the Vice President's speeches (the Opening Statement and the Statement), delivered within the session.

7   K. Annan, "Two concepts of sovereignty", *The Economist*, 18 September 1999, on this see also Prins, *Heart of War*, Chapters 4 & 5.

8   The risks can also run the other way. All South African troops sent on peace-keeping duties to highly infected Burundi were tested HIV negative: many were tested positive upon return. In reporting this, Dr Heinecken also added the results of questioning of SA troops: 95% were aware of the risks of HIV and 80% engaged in casual sex. Lecture by Dr L. Heinecken (SA Military Academy) at Royal Military College of Science conference "Post-Modern Military: Rethinking the Future," 24 April 2003.

9   UNSC 4172nd meeting, 17 July 2001, Resolution 1308 (2000) on the responsibility of the Security Council in the maintenance of international peace and security: HIV/AIDS and International Peace-keeping operations, clause 4.

10 M Schoofs, "The Security Council declares AIDS in Africa a Threat to World Stability," *The Body: AIDS/HIV Information Service*, 12–18 January 2000; P.W. Singer, "AIDS and international security," *Survival*, vol. 44 (1) Spring 2002, pp. 145–158

11 International Crisis Group, "HIV/AIDS as a security issue in Africa: lessons from Uganda," *ICG Issues Report* No. 3, 16 April 2004, p. i.

12 I am indebted to Dr Awash Tecklehaimanot of the Columbia Earth Institute for first highlighting this point to me.

13 J. Keegan, "A brief history of warfare – past, present and future," in (eds) H. Tromp and G. Prins, *The Future of War*, Kluwer, 2000, p. 179.

14 "The global menace of local strife," *The Economist*, 367 (8325) 24 May 2003, pp. 25–7; P. Collier *et al.*, *Breaking the Conflict Trap: Civil War and Development Policy*, World Bank and Oxford University Press, 2003. Lecture and subsequent discussion with Paul Collier, SSRC symposium "The economic analysis of conflict" Washington DC, 19 April 2004.

15 Håvard Strand, Lars Wilhelmsen & Nils Petter Gleditsch, *Armed Conflict Dataset, Codebook*, International Peace Research Institute, Oslo, 2002; A. Mack (ed) First annual *Human Security Report*, Human Security Centre, University of British Columbia, Oxford University Press, 2004.

16 R. First, *The Barrel of a Gun: Political Power in Africa and the Coup d'état*, Allan Lane, 1970, reviews the early sequence, especially of West African coups and was one of the first studies to note the in-built cascade effect within them. For the logic of patrimonial patronage politics, P. Chabal and J-P. Daloz, *Africa Works: Disorder as Political Instrument*, African Issues, James Currey, 1999.

17 G. Mills, "AIDS and the South African military: Timeworn cliché or Time-bomb?" Konrad Adenauer Stiftung Occasional Papers, "HIV/AIDS: A Threat to the African Renaissance?" available at www.kas.org.za/publica tions/occasionalpapers/Aids/mills.pdf.

18 This bitter paradox is forcefully stated in Michael Moodie and William Taylor, *Contagion and Conflict: Health as a Global Security Challenge*, Report of the Chemical & Biological Arms Control Institute and CSIS International Security Program, Washington, January 2000, p. 15 and p. 22 fn 30. The wider governance question begins to be explored in A. de Waal, "Modeling the governance implications of the HIV/AIDS pandemic in Africa: first thoughts," AIDS and Governance Discussion Paper No. 2, Justice Africa, June 2002 and further in R. Pharoah and M. Schönteich, "AIDS, security and governance in

Southern Africa: exploring the impact," Occasional Paper No. 65, Institute for Security Studies, Pretoria, January 2003 p. 9.

19 ICG, "Lessons from Uganda," p. 17.

20 L. Heinecken, "Facing a merciless enemy: HIV/AIDS and the South African Armed Forces," *Armed Forces & Society*, 29 (2), Winter 2003, pp. 281–300. Quotation at p. 296.

21 A. Whiteside, "The causes and consequences of the AIDS crisis in southern Africa," lecture at the Pugwash Conferences workshop on "Threats without Enemies," Betty's Bay, Cape Province, South Africa, February 2004, available at www.heard.org.za.

22 N. Wachtel, *La Vision des Vaincus: Les Indiens du Pérou devant la conquête espagnol, 1530–1570*, Editions Gallimard, Paris, 1971.

23 As, for example, illuminated with humane brilliance in C. van Onselen, *The Seed in Mine: The Life of Kas Maine, a South African Share-cropper, 1894–1985*, James Currey, 1996.

24 Chairmen, G. Evans & M. Sahnoun, *The Responsibility to Protect: Report of the International Commission on Intervention and State Sovereignty*, January 2002.

25 National Intelligence Council, "Global Trends 2015: A dialogue about the future with non-government experts," NIC 2000–02 December 2000.

26 *ibid* p. 16.

27 D. Gordon speaking at "Plague upon plague: AIDS and violent conflict in Africa," United States Institute of Peace, 8 May 2001.

28 G. Prins, "AIDS, power, culture and multilateralism – a case study," in (eds) E. Newman & J. Tirman, *Multilateralism under Challenge? Power, International Order and Structural Change*, UNU Press, 2005 (forthcoming).

FOUR

THE NEW SCARLET LETTER

M. WUERKER

Cartoon 10 The New Scarlet Letter
In the early years of White settlement in what later became the United States, women charged with "crimes" of adultery were often banished and required to wear a scarlet letter A. Now this cartoon implies victims of the disease in many parts of the world are treated no better.

Source: M. Wuerker

PART FIVE

Anti-geopolitics

INTRODUCTION TO PART FIVE

Paul Routledge

Throughout this volume we have argued that geopolitical knowledge tends to be constructed from positions and locations of political, economic and cultural power. Hence the histories of geopolitics have tended to focus upon the actions of states and their elites, overemphasizing statesmanship and understating rebellion. However, the geopolitical policies enacted by states and the discourses articulated by their policy makers have rarely gone without some form of contestation by those who have faced various forms of domination, exploitation and/or subjection which result from such practices. As Foucault has noted: 'there are no relations of power without resistances . . . like power, resistance is multiple and can be integrated in global strategies' (1980: 142).

Indeed, myriad alternative stories can be recounted which frame history from the perspective of those who have engaged in resistance to the state and/or the practices of geopolitics. These histories keep alive the memory of people's resistances, and in so doing, suggest new definitions of power that are not predicated upon military strength, wealth, command of official ideology and cultural control (Zinn, 1980). These histories of resistance can be characterized as a "geopolitics from below" emanating from subaltern (i.e. dominated) positions within society that challenge the military, political, economic and cultural hegemony of the state and its elites. These challenges are counter-hegemonic struggles in that they articulate resistance to the coercive force of the state – in domestic and/or foreign policy – as well as withdrawing popular consent to be ruled "from above." They are expressions of what we term "anti-geopolitics."

Anti-geopolitics can be conceived as an ambiguous, political and cultural force within civil society – i.e. those institutions and organizations which are neither part of the processes of material production in the economy, nor part of state-funded or state-controlled organizations (e.g. religious institutions, the media, voluntary organizations, educational institutions and trade unions) – that articulates two interrelated forms of counter-hegemonic struggle. First, it challenges the *material* (economic and military) geopolitical power of states and global institutions, and second, it challenges the *representations* imposed by political elites upon the world and its different peoples, that are deployed to serve their geopolitical interests.

Anti-geopolitics can take myriad resistant forms, from the oppositional discourses of dissident intellectuals, the strategies and tactics of social movements to armed insurrection and terrorism. Forms of resistance can articulate liberationary and/or anti-liberationary goals and practices, and thus "resistance" is an ambiguous practice for several reasons. First, resistance may occur *within* regimes of geopolitical power, for example, when the state upholds certain democratic rights within a society that are being challenged by reactionary anti-democratic resistances. The rights of African Americans to full citizenship rights or women to full reproductive choice in the United States, for example, have been upheld by federal institutions in the face of reactionary "Jim Crow" state laws and violent actions like the fire-bombing of reproductive heath clinics or the murder of doctors who provide abortion services. Second, various forms of domination may occur within resistance practices themselves, such as through the creation of internal hierarchies; the silencing of dissent; peer pressure; or how various forces of hegemony are internalized and reproduced within such practices. For example, Rubin (1998) recounts how the Coalition of Workers, Peasants and Students of the Isthmus in southern Mexico, while struggling against the Mexican state, continued to enact violence toward, and exclusion of, women, as well as more widely repressing internal democracy within its practices. Nationalist movements are often deeply patriarchal and repressive of minorities. Third, certain forms of resistance are reproductions or extensions of dominating state power, nurtured by states to further domestic or foreign policy agendas. For example, in the 1980s the white minority government in South Africa organized and funded an anti-government guerrilla movement in Mozambique, Renamo, whose war against peasant communities formed part of the apartheid regime's campaign of destabilization against oppositional neighboring states. Finally, forms of resistance may be violent and/or non-violent in character (Sharp et al., 2000).

When thinking about the use of violence, it is important to distinguish between "violence" against property (which does not involve physical harm to people) which might be more accurately described as non-violent sabotage, forms of violence that are aimed at military targets (personnel and materiel) and forms of violence conducted against civilians or non-combatants, which would be included in definitions of terrorism (to be discussed below). However, even the discourse of "resistant" violence is ambiguous, dependent upon the context in which it takes place, and who the perpetrators are. For example, colonial occupiers frequently referred to armed anti-colonial resistance as "terrorism," even when it was focused upon military targets. More recently, the targeting of civilians by state-sponsored groups – such as the US-funded *contras* in Nicaragua in the 1980s – was interpreted by the United States as "freedom fighting." In reality, armed struggle against a perceived oppressor frequently involves violence against both military and civilian/non-combatant targets whether this is intentional or not.

While anti-geopolitical practices are usually located within the political boundaries of a state, with the state frequently being the principal opponent, this is not to suggest that anti-geopolitics is necessarily localized. For example, with the intensity of the processes of globalization, social movements and other resistance formations are increasingly operating across regional, national and international scales, both independently and as part of broader oppositional networks, as they challenge elite international institutions and global structures of domination. Let us briefly consider the anti-geopolitical resistance to the various forms of geopolitics we have discussed in this volume.

COLONIAL ANTI-GEOPOLITICS

Anti-geopolitical resistance to colonialism was premised upon securing independence for people under imperial control, and frequently involved the organization of powerful nationalist movements, whose primary aim was to secure the independence of their countries and their effective sovereignty in world affairs. The process of decolonization, while linked historically to the character of the international political and economic system, was also place-specific in character and outcome, and highlights some of the ambiguities inherent within the practice of resistance. For example, while US independence from British colonial rule was established after a revolutionary war in 1776, African Americans and Native Americans in the US state long remained subject to slavery, conquest and, at times, extermination. While Haiti achieved independence in 1804 from France after a successful slave revolution, the country was subsequently ruled by an alliance of merchants and the military, before being occupied by the United States in 1915. During the Boer War in South Africa, from 1899 to 1902 descendants of an earlier phase of Dutch colonialism in South Africa (the Boers) struggled for independence against British occupation. Both groups of colonizers subjugated the indigenous population and shared similar white supremacist views. The Boers used child soldiers (Penkiops) and waged an armed guerrilla struggle against the British. In response, the British waged a "dirty war," burning Boer farms and establishing concentration camps where an estimated 20,000 people, mostly women and children, died. Finally, Irish independence was the outcome of a prolonged and mainly constitutional agitation for greater autonomy from the United Kingdom (1870–1914) followed by a short guerrilla struggle aimed at outright secession from 1916 to 1921. While this was achieved in 1921, some Irish republican nationalists claim decolonization remains partial since the six northeastern counties remained in the United Kingdom as the province of Northern Ireland (Krieger, 1993).

At the end of the World War II, the world became reconfigured into a bipolar political order characterized by the geopolitical, military and ideological competition between the United States and the Soviet Union. The "Third World" soon became a battleground upon which this competition was played out as the superpowers waged an ideological struggle for the hearts and minds of non-Western peoples who were liberating themselves from colonialism. During the Cold War, economic aid and development served as a means to encourage capitalist market economies, thereby providing conditions under which Western-style "democracy" could flourish. Economic development programmes were constructed as a strategy to bring "Third World" states into the geopolitical orbit of the United States and its allies. It was a process by which the "colonial world" was reconfigured into the "developing world" (Peet and Watts, 1993; Sachs, 1992). Concurrently, the Soviet Union also provided aid to those states which had nascent revolutionary or communist governments such as Cuba and Vietnam.

Independence from colonialism was aided by the emergence of the United States and the Soviet Union as two self-proclaimed anti-imperialist superpowers, the weakening of the economic and military strength of the imperial powers (such as Britain and France) due to the debilitating effects of World War II, and the development of powerful nationalist political movements within the colonized countries. Although the character of these nationalist movements varied widely according to local contexts, armed struggle and guerrilla warfare were frequently employed against the colonizing forces in order to defeat and remove them as a force of occupation (e.g. the Mau Mau rebellion in Kenya against the British, and the Algerian resistance against the French).

However, what frequently defined and decided the successful outcome of independence struggles was the social struggle of the population at large in combination with the armed struggle. For, armed struggle by a guerrilla army was frequently used in concert with non-violent sanctions such as strikes and civil disobedience conducted by the population at large (see Sharp, 1973). Together, these sanctions effectively withdrew popular consent from the colonizing ruling power. The power of the guerrilla army, then, lay both in its military capabilities and in the fact that it represented the embodiment of the collective will of a colonized people to resist. The use of non-violent sanctions enabled widespread popular participation in the struggle against colonialism which, in turn, frequently enabled the development of a national consciousness to develop among the colonized. The strategy of non-violent resistance was probably most effective in the Indian independence movement led by Mohandas K. Gandhi and the Indian National Congress against the British.

One of the principle legitimizations of colonial exploitation had been that empire building was, in part, an unselfish, even noble act. Colonialism was further legitimized through the (mis)representation of other cultures and places as primitive, savage, and uneducated, in need of Western civilization and enlightenment. Edward Said – late Professor of Comparative Literature at Columbia University and former member of the Palestinian National Council – argues that such representations were "imaginative geographies," or fictional realities, that shaped the West's perception and experience of other places and cultures.

As noted in the general introduction to the Reader, the geopolitical culture of a state is shaped by the primary forms of identification and boundary-formation that characterize its social, cultural and political life. Such geographical imaginations influence a state's behaviour towards other states and the international community. In the context of colonialism such representations designated geographical space into familiar and unfamiliar spaces, dramatizing the distance and difference between the West and its others, separating the occident from the orient, the colonizers from the colonized, and the developed from the underdeveloped (Reading 29). These representations were constructed around essentialist conceptions of (non-Western) *others* that equated difference with inferiority, and served to inform and legitimize geopolitical strategies of control and colonization by the Western countries, as they subjected other territories to military conquest and commercial exploitation. Said exposes and articulates the ideological and political purposes of imaginative geographies for the purposes of imperialism and notes that such distortions are not confined to the colonial era, they are continually deployed to this day because they serve geopolitical ends. Hence in the Cold War, American policy makers such as Kennan could refer to the inherent deception of the Russian mind when explaining Soviet foreign policy, while more recently the demonization of Islam in general, and the Palestinian cause in particular, have served to legitimize the Israeli occupation of the West Bank. Said's work poses a challenge to the representation of others by Western "experts" articulating an intellectual project against such deformed and self-interested representations of the world. For Said the role of the dissident intellectual is to articulate an "oppositional consciousness" to dominant (Western or elitist) representations of others.

Colonialism was invariably a violent process, constructed upon and maintained by a profound alienation of colonized peoples, and premised upon a geographical, political and cultural division of the colonized world. This is described by Frantz Fanon – a medical doctor and psychiatrist who worked for the Front de Liberation Nationale (FLN), the principal nationalist organization involved in the anti-colonial war against French occupation of Algeria (Reading 30). Fanon articulates how the indigenous inhabitants of colonial societies are "othered" by the colonizing culture, which constructs a Manichean world of colonizer and colonized. Such a division occurs on both physical and representational levels as the colonized are spatially separated from the colonizers, and their culture depicted in negative terms relative to the colonizer. As Fanon explains, the colonized are dehumanized by the colonizers in order to legitimize their control and exploitation. Writing from within the turmoil of decolonization, Fanon speaks with the voice of the heterogeneous peoples who comprised the colonized: those who were silenced and (mis)represented by the West. The constant state of emergency that exists within the colonized world, Fanon explains, also becomes a state of emergence for the colonized, to throw off both the colonizer's material occupation – frequently through violent struggle – and their appropriation of the colonized's right to speak for themselves and to represent themselves.

Moreover, as Fanon notes, the decolonization process was a global phenomenon, influenced by both other anti-colonial struggles (e.g. the French army's defeat at Dien Bien Phu in Vietnam) and the geopolitics of the Cold War. Both the United States and the Soviet Union attempted to support and control independence movements as part of broader geopolitical strategies against one another. For Fanon, decolonization entails both the physical removal of the occupier from one's territory and what Ngugi wa Thiong'o (1986) has termed a "decolonization of the mind." This involves opposition to Western ways of representing and organizing the world and the peoples in it – a struggle over who decides and controls how different cultures are interpreted and represented.

Decolonization had mixed results. For example, independence left many states with borders that had been established arbitrarily by colonial rulers. As a result, many newly independent states were forced to manage a poisonous legacy of colonialism as certain ethnic groups were divided by states while other ethnic groups were empowered by state geography to divide and rule competing ethnic groups. The disjuncture between state geography and ethnic geography has been a major source of tension and conflict in post-colonial Africa. Moreover, independence often resulted in the replacement of white colonial rule by indigenous elites dedicated to perpetuating ethnic divisions as a means of staying in power and continuing the exploitation of the country's peoples and natural resources. This process often took place with the active support of Western states and institutions, who, through so-called "development programs", helped establish what Kwame Nkrumah of Ghana, a leading advocate of pan-Africanism, called *neocolonialism*. An example of this process was the regime of General Joseph Mobutu established in Zaire. Covertly supported by the Western powers, Mobutu (who was on the CIA's payroll) came to power in 1965 by means of a military coup d'état. With the goodwill and support of states like France and the United States, who viewed him as a stabilizing force and anti-communist ally, Mobutu established a corrupt and repressive regime which plundered the country's resources and impoverished the country's people. As the living standards of average Zaireans fell, Mubuto enriched himself and his entourage, purchasing luxury homes in the south of France and in Switzerland.

COLD WAR ANTI-GEOPOLITICS

Decolonization took place within a global geopolitical map of satellites and spheres of influence, dominated by the two superpowers. In addition to Western Europe, where it had deployed troops and military bases during the course of World War II, the United States wielded considerable influence in Central and Latin America. Over the past century and a half, in order to protect its geopolitical and geoeconomic interests in its "backyard" – maintaining its southern neighbors as a rich source of resources and, at times, profits – the United States has engaged in direct military intervention (e.g. Dominican Republic 1965), the threat of force, the use of surrogate troops (e.g. the *contras* in Nicaragua, 1981–1989), clandestine "destabilizing" operations against radical regimes (e.g. Chile 1973) and economic blockades and sanctions (e.g. Cuba 1962 to the present).

Although American foreign policy was constantly cloaked in the rhetoric of anti-communism, it was, nevertheless, intimately connected with its economic interests in the region. As Jenny Pearce (1981) argues:

> The U.S. dominates the economies of the region, shaping them to its needs through investment and trading policies in a way that has left a lasting legacy of dependency and underdevelopment. The history of United States foreign policy in the region is also the history of its support for local elites favourable to its interests. The close alliance between those who control political and economic power within the region and the military and economic might of the U.S. has resulted in some of the most extreme forms of exploitation and repression anywhere in the world. Such attempts by one country to dominate others are usually called "imperialism".
>
> (Pearce 1981: 2)

There were both intellectual and material challenges to this state of affairs. Prominent among the intellectual challenges were a group of scholars known as "dependency theorists," who focused on the unequal economic and political exchanges that took place between and within the advanced capitalist countries (the "core") and the developing countries (the "periphery") (see Amin, 1976; Cardoso and Faletto, 1979; Emmanuel, 1972; Frank, 1967, 1978).

The most important material challenge was the Cuban Revolution of 1959, led by Fidel Castro and Che Guevara, which overthrew the US-supported Batista regime. Supported by Soviet aid, Cuba withstood repeated attempts by the United States at destabilization, and instituted land reforms, literacy, housing and public health improvements. Inspired by the success of Cuba, numerous peasant guerrilla movements emerged throughout Central and Latin America in attempts to challenge authoritarian regimes and alleviate poverty (e.g. the Farabundo Marti National Liberation Front in El Salvador, and the Sandinista National Liberation Front in Nicaragua, which overthrew the Somoza dictatorship in 1979). The response of the United States to these resistances was to interpret them as examples of communist subversion and threats to US interests and intervene indirectly through the provision of military training, financing, and hardware to its puppet regimes in order to establish brutal counter-insurgency programs, euphemistically termed "low intensity conflict" (Galeano, 1973, 1995).

Outside of the Americas, the United States adopted similar methods when faced with newly independent revolutionary regimes, following the defeat of the colonial powers. Such events were interpreted within the logics of Cold War geopolitics, demanding US intervention to counter potential Soviet or Chinese influence. This was dramatically illustrated by the involvement of the United States in Vietnam, which had been partitioned in 1954, the North remaining under the control of the Vietnam Communist Party, while the South was controlled by the anti-communist government of Ngo Dinh Diem. Framed within the discourse of communist containment, US policy actively sought to prevent a reunification of Vietnam under the control of the North's leader, Ho Chi Minh. Such a geopolitical strategy determined that the United States militarily support an unpopular, authoritarian regime in the South irrespective of the cost in human lives (see Reading 10). However, the United States was unable to defeat the National Liberation Front, who waged an effective guerrilla war, mobilizing the entire population of the North against the joint US and South Vietnamese forces. Moreover, as evidence of US atrocities in the war became known (e.g. mass bombings, the use of napalm against civilian populations, the poisoning of food and water resources, and over 2 million Vietnamese dead) and as the number of US casualties mounted, so public pressure began to mount within the United States to bring an end to the war.

A very different form of resistance to US influence emerged in Iran, where a US and British-supported military coup d'état in 1953 had restored the king (shah) to power. During the late 1970s, the royal dictatorship was opposed by a religious–secular coalition under the leadership of an exiled Shiite Islamic clergyman Ayatollah Khomeini. A popular revolutionary movement grew rapidly and overthrew the monarchy in February 1979. Again, such resistance was ambiguous in character and outcome. Through their control of revolutionary organizations that assumed judicial and security functions, the religious nationalists swiftly sidelined their erstwhile secular partners and established an Islamic republic. Unlike the majority of Cold War revolutionary movements, the Iranian revolution justified itself in religious (Islamic) terms, rejected modern ideas of progress, democracy and material well-being, yet involved relatively little armed conflict (Krieger, 1993).

While the Soviet Union sought to support certain revolutionary movements in the "Third World" for its own geopolitical purposes *contra* the United States, within Eastern Europe, each country within the Soviet bloc was penetrated in varying ways by Soviet mechanisms of control, and subordinated to the interests of Soviet communism. Despite its military dominance within Eastern Europe, popular uprisings against Soviet occupation and control periodically surfaced within its "satellites" – in the German Democratic Republic in 1953, in Hungary in 1956, in Czechoslovakia in 1968 and in Poland in 1981. Although these expressions of opposition proved unsuccessful, they were indicative of broader counter-hegemonic currents within the Soviet bloc.

What first came to the notice of the West as "dissent" – articulated by dissidents such as Andrei Sakharov (in the Soviet Union) and Vaclav Havel (in Czechoslovakia) and organisations such as the Czech-based human rights group Charter 77 – was symptomatic of the development within the Soviet Union and the Warsaw Pact countries of various independent initiatives that emerged "from below." For example, in Poland in 1980, shipbuilding and steel workers agitated for the right to form trade unions independent of state and Communist Party control. General strikes in Gdansk and Szczecin culminated in the creation of an independent trade union "Solidarity" led by Lech Walesa. This resistance was ambiguous in the sense that it represented a variety of interests: it was an amalgam of militant trade union, democratic political opposition and Catholic fundamentalist movement for national revival. Solidarity was outlawed in 1981 when the government declared martial law. However, following years of economic crises and public protests a power sharing deal between government and former union leaders was negotiated, followed by elections in 1989 which saw Solidarity representatives form a government under the populist and somewhat authoritarian leadership of Walesa (Krieger, 1993).

The actions of dissident groups laid the groundwork for increased popular participation in challenging the Soviet bloc regimes as public space for open protests in the late 1980s began with the implementation of *glasnost* in the Soviet Union. Resistance to Soviet control was often rooted in nationalist sentiments. For example, in the Baltic states (Lithuania, Latvia and Estonia), secessionist nationalist popular fronts organized huge protests in favour of sovereignty and overturning the entrenched communist regimes, leading to elections and the emergence of parliamentary democracies in 1989 and 1990. Ethnic Russians in the Baltic states suffered as these national liberation movements secured independence (Smith, 1999). Meanwhile, nationalist sentiment against Soviet control in Azerbaijan was initially in response to efforts by the Armenian population in the area of Nagorno-Karabakh to secede to join Armenia, which resulted in a protracted war between Armenian partisans and Azerbaijani militia troops, and the flow of refugees between the two republics (De Waal, 2003).

Many of the dissident movements in Eastern Europe also forged links with what proved to be the largest popular resistance against the Cold War itself – the peace movement – which opposed the deployment of cruise and Pershing missiles in Europe by NATO and SS20s by the Soviet Union. The movement comprised a variety of anti-nuclear and anti-militarist groups, including the Nuclear Freeze in the United States, the Campaign for Nuclear Disarmament (CND) in Britain and the European Nuclear Disarmament (END) movement (see Reading 14). This movement emerged at a time of heightened tension between the superpowers, and renewed US militarism under the Carter and Reagan presidencies. Although clearly differentiated on a state by state basis, END was non-statal, and sought to evolve mechanisms for transnational solidarity and identity, in an attempt to revitalize democracy within Europe. Intellectuals within the peace movement – such as historian E.P. Thompson (1985) – articulated a theoretical critique of the Cold War, voicing opposition to the superpower arms race and the division of Europe into ideological and militarized blocs.

Thompson argued that the expansionist ideologies of the United States and the Soviet Union were the driving force of the Cold War, each legitimized through the threat of a demonized other (communism and capitalism respectively) that served to define an approved national self-image against that of one's ideological opponent. This ideology permeated both the state and civil society in many areas (e.g. the media) conflating the interests of the state with the public interest, thus compromising the integrity and autonomy of civil society. Thompson argued that the Cold War should be seen as a means by which the dominant states within each bloc controlled and disciplined their own citizens, populations and clients, and by which those who stood to benefit from increased arms production and political anxiety (e.g. financial, commercial, military and political interests) promoted the rivalry. Hence the Cold War served as legitimization for both US and Soviet intervention in other states, their appropriation of vast resources for military purposes, and for keeping powerful elites in both blocs in power. Within this geopolitical regime, nuclear weapons served to suppress the political process, substituting the threat of annihilation for the negotiated resolution of differences.

However, Thompson also noted that in reality the principal threat of the Cold War was not the demonized other but rather was within each of the superpower blocs – i.e. the peace movements of the Western bloc, and the dissident movements of the Eastern bloc. These movements articulated both material challenges to superpower militarism – through direct action, underground organizations, etc. – and also an intellectual challenge to the geopolitical othering that the Cold War was predicated upon. Their calls for international solidarity, rather than antagonism, were seen as a threat to the power of political elites within each bloc to determine geopolitical spheres of influence. Moreover, by attempting to revitalize spaces of public autonomy, these movements challenged each superpower's ability to control public opinion.

George Konrad, a Hungarian writing during the Cold War, terms such resistances "antipolitics" (Reading 31). Antipolitics, as conceived by Konrad, is a moral force within civil society that articulates a distrust and public rejection of the power monopoly of the political class within the state – a power that is wielded against domestic populations through repressive legislation (e.g. censorship) and against others through the threat or prosecution of war. It is a practice that does not seek to overthrow the state, but opposes the political power that is exerted over people. The Cold War articulated a particularly dangerous manifestation of this power since politicians within the NATO and Warsaw Pact blocs had the power to unleash weapons of mass destruction. Konrad critiques Cold War geopolitics as a form of terrorism, since it enables the powerful to keep the masses dominated by the threat of nuclear annihilation. The threat of war, Konrad argues, is synonymous with the absence of democracy. He critiques the complicity of intellectuals with the state, whose role is to manufacture ideological justifications for the prosecution of geopolitical power. In response, the project of antipolitics seeks to expose the propaganda which equates the preparation, threat, or waging of war with patriotism. Moreover, it attempts to develop an internationalist solidarity, premised upon the notions of mutual coexistence, that transgresses state borders in an attempt to undermine the politics of "othering" upon which Cold War discourse is constructed.

CONTEMPORARY ANTI-GEOPOLITICS

With the revolutions of 1989, the demise of the Soviet Union and the Gulf War of 1990–1991, the United States stood as the "sole remaining superpower" in world affairs in the 1990s. Rather than dismantle its Cold War security state, US politicians searched around for new rationales to justify its existence and expansion. NATO was expanded in order to consolidate US influence on the European continent while international institutions promoted neoliberal globalization as an inevitable and necessary successor to the Cold War. The doctrine of neoliberalism articulates an overarching commitment to "free market" principles of unfettered trade, flexible labor and active individualism. It privileges privatization and deregulation (e.g. of state-financed welfare, education, health services and environmental protection) and aggressive intervention by governments around issues such as crime, policing, welfare reform and urban surveillance with the purpose of disciplining and containing those marginalized or dispossessed by neoliberal policies. Alternative development models based upon social redistribution, economic rights or public investment are undermined or foreclosed (Peck and Tickell, 2002).

Neoliberalism entails the centralization of control of the world economy in the hands of transnational corporations (TNCs) and their allies in key government agencies (particularly those of the United States and other members of the G8), large international banks and international institutions such as the International Monetary Fund (IMF), the World Bank and the World Trade Organization (WTO). Although the effects of neoliberalism have been uneven in different countries, global trends are instructive. One billion people live on less than US$1 a day, and the disparities between rich and poor continue to grow: in 1997 the richest 20 percent of the world's population received 90 percent of global income (an increase from 70 percent in 1960), while the poorest 20 percent received 1 percent (a decrease from 2.3 percent in 1960). The combined annual revenues of the largest 200 TNCs are greater than those of 182 countries that contain 80 percent of the world's population (Ellwood, 2001). According to the World Bank, in 1998, the income of the richest 1 percent of the world's population equalled that of the poorest 57 percent (Callinicos, 2003). Overall, neoliberal policies have tended to result in the pauperization and marginalization of indigenous peoples, women, peasant farmers and industrial workers, and a reduction in labor, social and environmental conditions on a global basis.

Responses to neoliberalism have taken different forms. First, a range of actors – leftist guerrillas, social movements, non-government organizations, human rights groups, environmental organizations and indigenous peoples movements – have articulated critiques of neoliberal development ideology and challenged the role of the state in implementing neoliberal policies. Second, coalitions have formed across national borders and across different political ideologies in order to oppose transnational institutions and agreements. Such grassroots globalization networks (Routledge, 2003) are frequently responses to local conditions that are in part the product of global forces, and resistance to these conditions has taken place at the local, national and global level. In contrast to official political discourse about the global economy, these challenges articulate a "globalization from below" – a struggle for inclusive, democratic forms of globalization, using the communicative tools of the global system – and comprise a "geopolitics-from-below" as grassroots globalization networks evolve.

Much of this resistance has been inspired by the Ejercito Zapatista Liberacion National – the EZLN or the Zapatistas – in Chiapas, Mexico, which has articulated resistance to the NAFTA and the Mexican state. Coinciding the emergence of their rebellion with the coming into effect of the NAFTA, the Zapatistas, a predominantly indigenous (Mayan) guerrilla movement, demanded the democratization of Mexican civil society and an end to NAFTA, which they argued was a "death certificate for the ethnic peoples of Mexico." The appearance of an armed insurgency, at a moment when the Mexican economy was entering into a free trade agreement, enabled the Zapatistas to attract national and international media attention. Through their spokesperson, Subcommandante Marcos, the Zapatistas engaged primarily in a war of words, fought with communiqués rather than bullets, giving voice to the victims of neoliberalism. The particular importance of the Zapatista struggle has lain in its ability, with limited resources and personnel, to disrupt international financial markets, and their investments within Mexico, while exposing the inequities on which neoliberal development is predicated. A key moment in the Zapatista struggle came in 1996, when the Zapatistas organized an international encounter in Chiapas, attended by activists, intellectuals and journalists from around the world, who were opposed to neoliberalism. At the encounter, attended by over 3000 activists from over 40 countries, Marcos issued a call for a network of transnational resistance against neoliberalism (Reading 32). This gave both grassroots movements and resistance networks a rallying call. Indeed, several important transnational initiatives were to emerge from this political moment.

First, global days of action have been organized by grassroots globalization networks against neoliberal globalization, typified by international protests against the WTO (Seattle 1999), the World Bank and IMF (Prague 2000) and the G8 (Genoa 2001). Naomi Klein (2002) has argued that such protests represent the convergence of many different movements each with different goals and targets, but with a shared common cause: to resist the neoliberal agenda. Second, two important network initiatives have developed: People's Global Action (PGA), a network of coordination, information sharing and solidarity between grassroots social movements from around the world (see Routledge, 2003) and the World Social Forum, a convergence of a huge diversity of social movements, non-government organizations (NGOs), trade unions and other political forces to protest the World Economic Forum (an annual meeting in Davos, Switzerland where political, business and financial elites meet to determine global economic strategies) and to exchange experiences and pose alternatives to neoliberal capitalism (see Fisher and Ponniah, 2003; Leite, 2005).

In such networks, different groups articulate a variety of potentially conflicting goals (concerning the forms of social change), ideologies (e.g. concerning gender, class and ethnicity) and strategies (e.g. violent and non-violent forms of protest). As a result, a diversity of place-specific solutions to economic and ecological problems are articulated – what Marcos terms "a world made of many worlds." However, certain key areas of agreement have emerged, such as the cancellation of the foreign debt in the developing world (which amounted to US $3000 billion in 1999); the introduction of a tax on international currency transactions and controls on capital flows; the reduction in people's working hours and an end to child labour; the defence of public services; the progressive taxation to finance public services and redistribute wealth and income; the international adoption of enforceable targets for greenhouse emissions and large-scale investment in renewable energy; policies which ensure land, water and food sovereignty for peasant and indigenous people; and the defence of civil liberties (Callinicos, 2003; Fisher and Ponniah, 2003; Leite, 2005).

The geopolitical dimensions of the globalization era have been heavily influenced by the global geopolitical event of 9/11 and the subsequent US-led "war on terror" which has included military invasions of Afghanistan and Iraq. The events of 9/11 have been attributed to Al-Qaeda. The Al-Qaeda network is a continuation of a pre-existing worldwide network of Muslim militias created by Abdullah Azzam, which operated from Pakistan, and recruited men and money for the struggle against the Soviet occupation of Afghanistan. In 1984 Osama Bin Laden set up Al-Qaeda (meaning "the base" and "the way") as another network that operated parallel to the military network originally set up by Azzam. Al-Qaeda's mission was to finance activities beyond the Afghan struggle. Al-Qaeda operated first from Pakistan, and subsequently from Sudan and Saudi Arabia. From 1996 Al-Qaeda was based in Afghanistan, where Osama Bin Laden forged a close relationship with the Taliban.

Al-Qaeda is a network of networks, a "federation" of overlapping relations among not only a financial network comprising Bin Laden's business associates, employees, donors and supporters, but also a terrorist network comprising fundamentalist Islamic groups in several Muslim and non-Muslim countries, some funded by Bin Laden's financial network. Both networks are spatially decentralized, but while power is centralized in the financial network it is decentralized in the terrorist network. Individual groups or cells appear to have a high degree of autonomy, making contact with other groups only when necessary (Ettlinger and Bosco, 2004).

Such groups share a particular interpretation of the Arabic word *jihad*, which means literally "struggle." Islamic scholars have long been divided on how it should be interpreted. For some it means the struggle to defend one's faith and ideals against harmful outside influences. For Al-Qaeda it has come to represent the duty of Muslims to fight to rid the Islamic world of Western influence in the form of corrupt and despotic leaders and occupying armies. This was most clearly articulated in the "declaration of war" against the United States in 1998, issued by Al-Qaeda, claiming that an International Islamic Front had been established to conduct jihad against the United States and Israel.

The "Letter to America" (Reading 33) is purportedly written by Osama Bin Laden. The letter first appeared on the Internet in Arabic and was subsequently translated into English and circulated in Britain by Muhammad al-Mass'ari, secretary-general of the Commission for the Defence of Legitimate Rights, a Saudi opposition group. The "Letter" was subsequently posted on numerous websites around the world, and attributed to Bin Laden. The fact that the precise authorship of this letter remains in doubt is indicative of the uncertainty surrounding Bin Laden. The "Letter" represents a clear and concise statement of the radical Salafi (an orthodox branch of Sunni Islam) worldview. The necessity of jihad is justified by the violence of US foreign policy, and its support of Israel's continued occupation of Palestinian land, and other "un-Islamic" governments of the region. Interestingly, the speech articulates certain anti-capitalist and anti-militarist sentiments that would be shared by many involved in grassroots globalization networks. In its condemnation of US foreign policy hypocrisy (e.g. concerning weapons of mass destruction) and the United States' selective violation of international law (e.g. ignoring United Nations resolutions condemning Israel's occupation of Palestine), the speech also articulates sentiments shared by many who opposed the US invasion of Iraq. However, in another example of the ambiguity of resistance practices, the speech also articulates the supremacy of Islam over all other religions, and is permeated with anti-Semitic, homophobic and sexist discourses. Liberation, here, is selective and implies goals and practices that repress others (e.g. non-Muslims, secular women and gay "others").

Al-Qaeda's organizational form – networks – is similar, in part, to the resistance against neoliberalism. Certain similar tools are deployed in organization (emails, videos, internet, communications technologies); people, money, resources and tactics from different places at different times are pulled together into particular actions without necessarily following any overly centralized strategy; and both kinds of network are relatively discontinuous and decentered (Luke 2003), allowing for fast and flexible movement in response to a rapidly changing environment. Moreover, both kinds of network effectively use symbolic political actions to further their cause. While grassroots globalization networks have utilized mass protests at the sites where neoliberal institutions meet, Al-Qaeda has targeted symbols of US economic and military dominance (e.g. the World Trade Center, the Pentagon, US embassies etc). Of course, such networks differ greatly in both their liberation goals, and in their choice of tactics – grassroots globalization networks tending to be predominantly non-violent, while Al-Qaeda professes the use of armed resistance, particularly in the form of terrorism.

Unlike more traditional place-based movements, terror networks, comprising spatially dispersed groups acting independently and clandestinely, are very difficult for state agencies to penetrate or disenable. They coordinate activities through and around state territories in a manner that eludes border controls. The 9/11 attacks saw the Al-Qaeda network literally and metaphorically fly under the radar of US defensive surveillance in order to launch their attacks. Subsequently there has been a wave of terrorist attacks – e.g. the Bali nightclub bombings (2002), the Istanbul HBSC bank bombing (2003), the Madrid train bombing (2004) and the London Underground and bus bombings (2005).

The core purpose of terrorism is "propaganda by the deed," using the intentional use or threat of violence, against civilians or civilian targets, for political objectives, to influence the "hearts and minds" of government and public opinion. The propagators can be governments (and their agencies and proxies), non-government groups (e.g. Al-Qaeda) and individuals. What this definition excludes are attacks against the military as in military occupations, colonialism and so on. There are many varieties of terrorism: suicide bombings (first used in Sri Lanka by the nationalist Liberation Tigers of Tamil Eelam in 1983, and a common tactic in the occupied territories of Palestine and Iraq); disappearance of government opponents (as occurred in Argentina and Chile in the 1970s and 1980s, a crime for which former Chilean President Pinochet now stands trial), assassinations (e.g. Israeli attacks on Palestinian leaders and Hamas operatives) and state terror (e.g. death squads in El Salvador and Guatemala during the 1980s).

As noted at the beginning of this introduction, how such acts are represented is also important. While the US government might interpret any attack upon its interests as terrorism, for people resisting US hegemony, such acts might be interpreted as freedom fighting (Barker, 2002). United Nations member states have been unable to agree on an anti-terrorism convention including a definition of terrorism. However, in 2004 a UN document argued for a definition that includes any actions that specifically target innocent civilians and non-combatants, are intended to cause death or serious bodily harm to them, and which are designed to intimidate a population or compel a government or an international organization to do, or to abstain from doing, any act (United Nations, 2004). Such a definition would encompass many of the actions of groups such as Al-Qaeda, resistance to foreign occupation that targets civilians and non-combatants as well as the torture and deliberate killing of civilians by any government's military forces. However, the document implies that "terrorism" should be confined to non-state actors since "state terrorism," or the use of armed forces against civilians, is prohibited by international law and better described as a war crime or a crime against humanity (United Nations, 2004: 51–52).

The response of the US and allied governments to 9/11 and subsequent terrorist events has been to declare a "global war on terror" which has included the military attack and occupation of Afghanistan (with approximately 20,000 dead) and Iraq (with approximately 100,000 dead), the detention and torture of terror suspects (such as at Camp X-ray in Guatanamo Bay, Cuba, and the Abu Ghraib prison in Iraq), increased surveillance of citizens and the curtailment of civil liberties (e.g. the Patriot Act in the United States) and increased military spending. The war on terror has enabled the United States and its allies, primarily to focus their challenge to radical Islam around the world, but also, in part, to reassert their ideological coordinates against the threat posed to neoliberalism by grassroots globalization networks (for example through anti-terror legislation that curtails civil liberties to protest). More insidiously the war on terror normalizes a permanent state of global emergency that legitimizes the increasing suspension of legal and other rights, and blurs the distinction between a state of war and a state of peace (Sardar and Davies, 2002; Blum, 2000). Similar to Thompson's arguments about the Cold War discussed earlier, the war on terror serves to discipline domestic populations through the instillation of fear, and legitimize US intervention abroad, and its appropriation of foreign resources such as oil.

Gilbert Achar (Reading 34) argues that the war on terror, rather than representing a clash of civilizations, represents a clash of barbarisms, where the violence of one side engenders violence from the other. After analyzing the events of 9/11, Achar argues that US foreign policy has contributed to the emergence of Islamic fundamentalist terrorism – for example, through the funding of Al-Qaeda as a tool against the Soviet occupation of Afghanistan, and by helping to defeat the socialist and progressive nationalist politics throughout the Islamic world since the mid-1950s, which has freed up space for political Islam to emerge as the only ideological and organizational expression of popular resentment. He argues that, in the war on terror, the United States deploys asymmetric advantages (e.g. it breaks international law at will and enjoys global military dominance) and as a result engenders asymmetric means of retaliation (i.e. unconventional approaches that avoid or undermine US conventional military strengths and exploit their weaknesses), e.g. terrorism. The war on terror, Achar argues, produces terrorism. The United States can respond to attacks such as 9/11 only by asymmetric means, i.e. by increasing the asymmetry of its military dominance still further, and extending its use in more areas in the name of preemptive attacks. This will provoke new asymmetric attacks on its troops and civilian population in the form of terrorism.

Each civilization, Achar argues, has its own barbarisms: some cut throats on planes (such as in the 9/11 attacks), while others use "daisy cutters" (the 15,000 pound bombs of the US Air Force). On both sides, there is absolute hostility toward an absolute enemy. But, Achar cautions, the barbarism of the weak is usually a reaction to the barbarism of the strong, which itself is inherently racist, privileging predominantly ethnically Caucasian countries at the expense of the "rest." He forcefully argues for the need to create a credible alternative to neoliberal capitalism in order to weaken support for reactionary identity politics (e.g. Islamic fundamentalism), in order to channel social discontent toward democracy and social justice. Such alternatives, however, are faced by the Manichean logic of US foreign policy that, in the wake of the 9/11 attacks, is premised upon the logic that "either you are with us or you are with the terrorists". While opposing terrorism is important, this logic contains certain problems. First, it abrogates to itself the right to determine what constitutes "terrorism." In practice this is selective. For example, the United States neither fully condemns nor intervenes in Israeli intimidation and violence against Palestinian civilian populations, or Russian intimidation and violence against Chechen civilian populations. Second, it forecloses criticism of whatever actions are taken to combat terrorism. For example, in 2004 a UN report noted how many governments and civil society organizations have voiced concerns that the "war on terror" has in some instances corroded the very values that terrorists target: human rights and the rule of law. In addition, the focus on military, police and intelligence measures risks undermining efforts to promote good governance and human rights, and has alienated large parts of the world's population, thereby weakening the potential for collective action against terrorism (United Nations, 2004).

The article by Jennifer Hyndman (Reading 35) addresses this problem, advocating a feminist geopolitical viewpoint that attempts to develop an embodied politics of security at multiple scales that includes both states and the people who comprise their citizenry. She argues that geopolitical narratives of the United States and its allies are dominated (as they were during the Cold War) by a binary logic of either/or that leaves little space for those who fail to identify with either the position of the United States or that of radical Islam (or other future opponents). The result of this logic is that the dominant geopolitical narrative of the United States (e.g. the denotation of countries as "rogue states," the definition of something and somewhere to define one's policies against, such as Afghanistan) is presented as the only political option available, and serves to legitimize violence against innocent civilians.

The binary logic of either/or creates mutually exclusive places of "here" and "there" and political relationships of "us" and "them." Focusing upon the US war in Afghanistan, Hyndman shows that rather than being separate, there are many connections between Afghanistan and the United States, and between Bin Laden, the Taliban and US government policy makers, born out of earlier Cold War concerns and US oil and natural gas interests. Dominant geopolitical narratives of either/or, supported and disseminated by the mainstream media, enable the mobilizing of consent for military violence against demonised others. However, Hyndman contends, neither the killings of thousands of innocent civilians in New York and Washington on September 11, nor the killings of thousands of innocent civilians in Afghanistan, is justified – both constitute acts of terrorism.

The final article (Reading 36), by Indian author Arundhati Roy, represents an example of discursive anti-geopolitical resistance, from a woman who has written on behalf of Indian social movements and participated in the World Social Forum. The article represents a critique of both neoliberalism and the "war on terror." Written during the US-led invasion and occupation of Iraq, Roy argues that current US foreign policy is imperialistic and unjust, financed by the United States' poor in the interests of its neoliberal elites. Roy argues that the reasons for war against Iraq given by US elites (and supported by the mainstream media) were based upon falsehoods, elided the historical support given to Saddam Hussein by the US government (which included the provision of weapons of mass destruction) and was conducted in the face of worldwide public opposition.

Democracy in the United States and abroad is in crisis, Roy contends, subverted by neoliberal elites and corporate media. She argues that US "democracy" itself is racist (since African Americans have a lower life expectancy than people born in many developing countries) and is being undermined by anti-terrorism legislation, which curtails civil liberties and criminalizes dissent. As a result US wars of "liberation" are being paid for with the freedoms held dear by US citizens. Roy argues that resistance to the US empire should preclude terrorism, and utilize non-violent methods of civil disobedience such as economic boycotts, targeted at US (and its allies') corporate interests. Such a resistance must eschew the seduction of equating war with patriotism, and be international in character, ignoring national boundaries and allegiances. However, for it to succeed it must begin in the heart of empire, in the United States.

REFERENCES AND FURTHER READING

Foucault, M. (1980) *Power/Knowledge*. New York: Pantheon.
Rubin, J. (1998) "Ambiguity and contradiction in a radical popular movement", in *Cultures of Politics, Politics of Cultures*, eds. S.E. Alvarez, E. Dagnino and A. Escobar. Oxford: Westview.
Sharp, J., Routledge, P., Philo, C. and Paddison, R. (eds) (2000) *Entanglements of Power: Geographies of Domination/Resistance*. London: Routledge.
Zinn, H. (1980) *A People's History of the U.S*. New York: Harper & Row.

On colonial anti-geopolitics

Fanon, F. (1965) *A Dying Colonialism*. New York: Grove.
Guha, R. and Spivak, G.C. (1988) *Selected Subaltern Studies*. New York: Oxford University Press.
Krieger, J. (1993) *The Oxford Companion to Politics of the World*. Oxford: Oxford University Press.
Memmi, A. (1965) *The Colonizer and the Colonized*. Boston, MA: Beacon Press.
Peet, R. and Watts, M. (1993) "Development Theory and Environment in an Age of Market Triumphalism," *Economic Geography*, 69(3): 227–253.
Sachs, W. (ed.) (1992) *The Development Dictionary*. London: Zed.
Said, E. (1978) *Orientalism*. New York: Vintage.
—— (1993) *Culture and Imperialism*. London: Vintage.
Scott, J.C. (1985) *Weapons of the Weak*. New Haven, CT: Yale University Press.
—— (1990) *Domination and the Arts of Resistance*. New Haven, CT: Yale University Press.
Sharp, G. (1973) *The Politics of Non-violent Action*, 3 vols. Boston, MA: Porter Sargent.
Ngugi wa Thiong'o (1986) *Decolonising the Mind*. London: Heinemann.
Wolf, E.R. (1969) *Peasant Wars of the Twentieth Century*. New York: Harper & Row.

On Cold War anti-Geopolitics

Amin, S. (1976) *Unequal Development: An Essay on the Social Formations of Peripheral Capitalism*. New York: Monthly Review Press.

Cardoso, F.H. & Faletto, E. (1979) *Dependency and Development.* Berkeley, CA: University of California Press.

Chailand, G. (1982) *Guerrilla Strategies.* London: Penguin.

—— (1989) *Revolution in the Third World.* London: Penguin.

De Waal, T. (2003) *Black Garden: Armenia and Azerbaijan through War and Peace.* New York: New York University Press.

Emmanuel, A. (1972) *Unequal Exchange: A Study of the Imperialism of Trade.* New York: Monthly Review Press.

Frank, A.G. (1967) *Capitalism and Underdevelopment in Latin America.* New York: Monthly Review Press.

—— (1978) *Dependent Accumulation and Underdevelopment.* London: Macmillan.

Galeano, E. (1973) *The Open Veins of Latin America.* New York: Monthly Review Press.

—— (1995) *Memory of Fire.* London: Quartet.

Katziaficas, G. (1987) *The Imagination of the New Left: A Global Analysis of 1968.* Boston, MA: South End Press.

Krieger, J. (1993) *The Oxford Companion to Politics of the World.* Oxford: Oxford University Press.

Pearce, J. (1981) *Under the Eagle.* London: Latin American Bureau.

Smith, D. and Thompson, E.P. (eds) (1987) *Prospectus for a Habitable Planet.* London: Penguin.

Smith, G. (1999) *The Post-Soviet States: Mapping the Politics of Transition.* London: Arnold.

Thompson, E.P. (1985) *The Heavy Dancers.* New York: Pantheon.

On contemporary anti-geopolitics

Arquilla, J. and Ronfeldt, D. (eds) (2001) *Networks and Netwars: The Future of Terror, Crime and Militancy,* http://www.rand.org/publications/MR/MR1382/

Barker, J. (2002) *The No-Nonsense Guide to Terrorism.* London: Verso.

Bircham, E. and Charlton, J. (eds) (2001) *Anti-Capitalism: A Guide to the Movement.* London: Bookmarks.

Blum, W. (2000) *Rogue State: A Guide to the World's Only Superpower.* Monroe, ME: Common Courage Press.

Burke, J. (2004) *Al Qaeda: The True Story of Radical Islam.* London: Penguin.

Callinicos, A. (2003) *An Anti-Capitalist Manifesto.* Cambridge: Polity Press.

Carter, J. and Morland, D. (eds) (2004) *Anti-Capitalist Britain.* Cheltenham: New Clarion Press.

Castells, M. (1997) *The Power of Identity.* Oxford: Blackwell.

Ellwood, W. (2001) *The Non-Nonsense Guide to Globalization.* London: Verso.

Ettlinger, N. and Bosco, F. (2004) "Thinking through Networks and their Spatiality: A Critique of the US (Public) War on Terrorism and its Geographic Discourse," *Antipode,* 36(2): 249–271.

Fisher, W.F. and Ponniah, T. (eds) (2003) *Another World is Possible.* London: Zed.

Gedicks, A. (1993) *The New Resource Wars.* Boston, MA: South End Press.

Glassman, J. (2002) "From Seattle (and Ubon) to Bangkok: The Scales of Resistance to Corporate Globalization," *Environment and Planning D: Society and Space,* 19: 513–533.

Gray, J. (2004) *Al Qaeda and What It Means to be Modern.* London: Faber & Faber.

Gregory, D. (2004) *The Colonial Present.* Oxford: Blackwell.

Hardt, M. and Negri, A. (2000) *Empire.* Cambridge, MA: Harvard University Press.

Harvey, D. (2003) *The New Imperialism.* Oxford: Oxford University Press.

Keck, M.E. and Sikkink, K. (1998) *Activists Beyond Borders.* Ithaca, NY: Cornell University Press.

Klein, N. (2000) *No Logo.* London: Flamingo.

—— (2002) *Fences and Windows.* London: Flamingo.

Leite, J.C. (2005) *The World Social Forum: Strategies of Resistance.* Chicago, IL: Haymarket.

Luke, T.W. (2003) "Postmodern Geopolitics: The Case of the 9.11 Terrorist Attacks", in *A Companion to Political Geography*, eds. J. Agnew, K. Mitchell and G. Toal. London: Blackwell.

Notes from Nowhere (2003) *We Are Everywhere*. London: Verso.

Olesen, T. (2004) *International Zapatismo*. London: Zed.

Peck, J. and Tickell, A. (2002) "Neoliberalizing Space," *Antipode*, 34(3): 380–404.

Ross, J. (1995) *Rebellion from the Roots: Indian Uprising in Chiapas*. Monroe, ME: Common Courage Press.

Routledge, P. (2003) "Convergence Space: Process Geographies of Grassroots Globalisation Networks," *Transactions of the Institute of British Geographers*, 28(3): 333–349.

Roy, A. (2004) *The Ordinary Person's Guide to Empire*. London: Flamingo.

Safaoui, M. (2003) *Inside Al Qaeda*. London: Granta.

Sardar, Z. and Davies, M.W. (2002) *Why Do People Hate America?* Cambridge: Icon.

—— (2004) *American Dream Global Nightmare*. Cambridge: Icon.

Schalit, J. (ed.) (2002) *The Anti-Capitalism Reader*. New York: Akashic.

Smith, N. (2005) *The Endgame of Globalization*. London: Routledge.

Starr, A. (2000) *Naming the Enemy: Anti-corporate Movements Against Globalization*. London: Zed.

United Nations (2004) *A More Secure World: Our Shared Responsibility*. New York: United Nations Foundation.

Tormey, S. (2004) *Anti-Capitalism: A Beginner's Guide*. Oxford: Oneworld.

Zeite, J.C. (2005) *The World Social Forum: Strategies of Resistance*. Chicago, IL: Haymarket.

Websites

Adbusters (culture jamming): www.adbusters.org

Corporate Watch (campaigns against TNCs): www.corporatewatch.org.uk

Indymedia (alternative global media source): www.indymedia.org

International Forum on Globalization: www.ifg.org

International Rivers Network (NGO): www.irn.org

International Solidarity Movement (for Palestine): www.palsolidarity.org

Landless People's Movement, Brazil: www.mstbrazil.org

Latin American struggles (info): www.rebelion.org

Narmada Bachao Andolan: www.nba.org (Indian anti-dam struggle)

New Internationalist (magazine on struggles and issues): www.newint.org

No Logo website (news, information): www.nologo.org

People's Global Action: www.agp.org

Via Campesina (international peasant farmer's network): www.viacampesina.org

World Bank Boycott Campaign: www.worldbankboycott.org

World Social Forum: www.forumsocialmundial.org.br/home

The Zapatistas: www.eco.utexas.edu/faculty/Cleaver/chiapas95.html and www.flag.blackened.net/revo

ZNet (excellent website for articles, commentaries on many issues): www.zmag.org

Cartoon 11 Principles of International Relations

From the Cold War to the present, US foreign policy has been determined by a set of supposedly universal principles of international relations that have been selectively interpreted and applied in the United States' self-interest.

Source: M. Wuerker

29

Orientalism Reconsidered

Edward W. Said

from *Europe and Its Others* (1984)

There are two sets of problems that I'd like to take up, each of them deriving from the general issues addressed in *Orientalism*.

As a department of thought and expertise Orientalism of course refers to several overlapping domains: firstly, the changing historical and cultural relationship between Europe and Asia, a relationship with a 4000 year old history; secondly, the scientific discipline in the West according to which beginning in the early nineteenth century one specialised in the study of various Oriental cultures and traditions; and, thirdly, the ideological suppositions, images and fantasies about a currently important and politically urgent region of the world called the Orient. The relatively common denominator between these three aspects of Orientalism is the line separating Occident from Orient and this, I have argued, is less a fact of nature than it is a fact of human production, which I have called imaginative geography.

This is, however, neither to say that the division between Orient and Occident is unchanging nor is it to say that it is simply fictional. It is to say – emphatically – that . . . the Orient and the Occident are facts produced by human beings, and as such must be studied as integral components of the social, and not the divine or natural, world. And because the social world includes the person or subject doing the studying as well as the object or realm being studied, it is imperative to include them both in any consideration of Orientalism for, obviously enough, there could be no Orientalism without, on the one hand, the Orientalists, and on the other, the Orientals.

Yet, and this is the first set of problems I want to consider, there is still a remarkable unwillingness to discuss the problems of Orientalism in the political or ethical or even epistemological contexts proper to it. This is as true of professional literary critics who have written about my book, as it is of course of the Orientalists themselves. Since it seems to me patently impossible to dismiss the truth of Orientalism's political origin and its continuing political actuality, we are obliged on intellectual as well as political grounds to investigate the resistance to the politics of Orientalism, a resistance that is richly symptomatic of precisely what is denied.

If the first set of problems is concerned with the problems of Orientalism reconsidered from the standpoint of local issues like who writes or studies the Orient, in what institutional or discursive setting, for what audience, and with what ends in mind, the second set of problems takes us to a wider circle of issues. These are the issues raised initially by methodology and then considerably sharpened by questions as to how the production of knowledge best serves communal, as opposed to factional, ends, how knowledge that is non-dominative and non-coercive can be produced in a setting that is deeply inscribed with the politics, the considerations, the positions, and the strategies of power. In these methodological and moral re-considerations of Orientalism I shall quite consciously be alluding to similar issues raised by the experiences of feminism or women's studies, black or ethnic studies, socialist and anti-imperialist studies, all of which take for their point of departure the right of formerly un- or mis-represented human groups to speak for and represent themselves in domains defined, politically and intellectually as normally excluding them, usurping their signifying and representing functions, overriding their historical reality. In short, Orientalism reconsidered in this wider and libertarian optic entails nothing less than the creation of new objects for a new kind of knowledge.

Certainly there can be no doubt that – in my own rather limited case – the consciousness of being an Oriental goes back to my youth in colonial Palestine and Egypt, although the impulse to resist its accompanying impingements was nurtured in the heady atmosphere of the post-World War II period of independence when Arab nationalism, Nasserism, the 1967 War, the rise of the Palestine national movement, the 1973 War, the Lebanese Civil War, the Iranian Revolution and its horrific aftermath, produced that extraordinary series of highs and lows which has neither ended nor allowed us a full understanding of its remarkable revolutionary impact.

The interesting point here is how difficult it is to try to understand a region of the world whose principal features seem to be, first, that it is in perpetual flux, and second, that no one trying to grasp it can by an act of pure will or of sovereign understanding stand at some Archimedean point outside the flux. That is, the very reason for understanding the Orient generally and the Arab world in particular, was first that it prevailed upon one, beseeched one's attention urgently, whether for economic, political, cultural, or religious reasons, and second, that it defied neutral, disinterested, or stable definition [. . .] even so relatively inert an object as a literary text is commonly supposed to gain some of its identity from its historical moment interacting with the attentions, judgements, scholarship and performances of its readers. But, I discovered, this privilege was rarely allowed the Orient, the Arabs, or Islam, which separately or together were supposed by mainstream academic thought to be confined to the fixed status of an object frozen once and for all in time by the gaze of Western percipients.

Far from being a defence either of the Arabs or Islam – as my book was taken by many to be – my argument was that neither existed except as "communities of interpretation" which gave them existence, and that, like the Orient itself, each designation represented interests, claims, projects, ambitions and rhetorics that were not only in violent disagreement, but were in a situation of open warfare. So saturated with meanings, so overdetermined by history, religion and politics are labels like "Arab" or "Muslim" as subdivisions of "The Orient" that no one today can use them without some attention to the formidable polemical mediations that screen the objects, if they exist at all, that the labels designate.

I do not think it is too much to say that the more these observations have been made by one party, the more routinely they are denied by the other; this is true whether it is Arabs or Muslims discussing the meaning of Arabism or Islam, or whether an Arab or Muslim disputes these designations with a Western scholar. Anyone who tries to suggest that nothing, not even a simple descriptive label, is beyond or outside the realm of interpretation, is almost certain to find an opponent saying that science and learning are designed to transcend the vagaries of interpretation, and that objective truth is in fact attainable. This claim was more than a little political when used against Orientals who disputed the authority and objectivity of an Orientalism intimately allied with the great mass of European settlements in the Orient.

The challenge to Orientalism and the colonial era of which it is so organically a part, was a challenge to the muteness imposed upon the Orient as object. Insofar as it was a science of incorporation and inclusion by virtue of which the Orient was constituted and then introduced into Europe, Orientalism was a scientific movement whose analogue in the world of empirical politics was the Orient's colonial accumulation and acquisition by Europe. The Orient was therefore not Europe's interlocutor, but its silent Other. From roughly the end of the eighteenth century, when in its age, distance and richness the Orient was re-discovered by Europe, its history had been a paradigm of antiquity and originality, functions that drew Europe's interests in acts of recognition or acknowledgement but *from* which Europe moved as its own industrial, economic and cultural development seemed to leave the Orient far behind. Oriental history – for Hegel, for Marx, later for Burkhardt, Nietzsche, Spengler, and other major philosophers of history – was useful in protraying a region of great age, and what had to be left behind.

Here, of course, is perhaps the most familiar of Orientalism's themes – they cannot represent themselves, they must therefore be represented by others who know more about Islam than Islam knows about itself. Now it is often the case that you can be known by others in different ways than you know yourself, and that valuable insights might be generated accordingly. But that is quite a different thing than pronouncing it as immutable law that outsiders *ipso facto* have a better sense of you as an insider than you do of yourself. Note that there is no question of an *exchange* between Islam's views and an outsider's: no dialogue, no discussion, no mutual recognition. There is a flat assertion of quality, which the Western policy-maker, or his faithful servant possesses by virtue of his being Western, white, non-Muslim.

Now this, I submit, is neither science, nor knowledge, nor understanding: it is a statement of power and a claim for relatively absolute authority. It is constituted out of racism, and it is made comparatively acceptable to an audience prepared in advance to listen to its muscular truths [. . .] for whom Islam is not a culture, but a nuisance [. . .] associate[d] with the other nuisances of the 60's and the 70's – blacks, women, post-colonial Third World nations that have tipped the balance against the U.S. in such places as UNESCO and the U.N.

[. . .] Orientalism's large political setting, which is routinely denied and suppressed [. . .] comprises two other elements, about which I'd like to speak very briefly, namely the recent (but at present uncertain) prominence of the Palestinian movement, and secondly, the demonstrated resistance of Arabs in the United States and elsewhere against their portrayal in the public realm.

As for the Palestinian issue [. . .] the Israeli occupation of the West Bank and Gaza, the destruction of Palestinian society, and the sustained Zionist assault upon Palestinian nationalism has quite literally been led and staffed by Orientalists. Whereas in the past it was European Christian Orientalists who supplied European culture with arguments for colonising and suppressing Islam, as well as for despising Jews, it is now the Jewish national movement that produces a cadre of colonial officials whose ideological theses about the Islamic or Arab mind are implemented in the administration of the Palestinian Arabs, an oppressed minority within the white-European-democracy that is Israel. [. . .] Hebrew University's Islamic studies department has produced every one of the colonial officials and Arab experts who run the Occupied Territories.

Underlying much of the discussion of Orientalism is a disquieting realisation that the relationship between cultures is both uneven and irremediably secular. This brings us to the point I alluded to a moment ago, about recent Arab and Islamic efforts, well-intentioned for the most part, but sometimes motivated by unpopular regimes, who in attracting attention to the shoddiness of the Western media in representing the Arabs or Islam divert scrutiny from the abuses of their rule and therefore make efforts to improve the so-called image of Islam and the Arabs. Parallel developments have

been occurring [. . .] in UNESCO where the controversy surrounding the world information order – and proposals for its reform by various Third World and Socialist governments – has taken on the dimensions of a major international issue. Most of these disputes testify, first of all, to the fact that the production of knowledge, or information, of media images, is unevenly distributed: its locus, and the centers of its greatest force are located in what, on both sides of the divide, has been polemically called the metropolitan West. Second, this unhappy realisation on the part of weaker parties and cultures, has reinforced their grasp of the fact that although there are many divisions within it, there is only one secular and historical world, and that neither nativism, nor divine intervention, nor regionalism, nor ideological smokescreens can hide societies, cultures and peoples from each other, especially not from those with the force and will to penetrate others for political as well as economic ends. But, third, many of these disadvantaged post-colonial states and their loyalist intellectuals have, in my opinion, drawn the wrong set of conclusions, which in practice is that one must either attempt to impose control upon the production of knowledge at the source, or, in the worldwide media economy, to attempt to improve, enhance, ameliorate the images currently in circulation without doing anything to change the political situation from which they emanate and on which to a certain extent they are based.

The failings of these approaches strike me as obvious, and here I don't want to go into such matters as the squandering of immense amounts of petro-dollars for various short-lived public relations scams, or the increasing repression, human-rights abuses, outright gangsterism that has taken place in many formerly colonial countries, all of them occurring in the name of national security and fighting neo-imperialism. What I do want to talk about is the much larger question of what, in the context recently provided by such relatively small efforts as the critique of Orientalism, is to be done, and on the level of politics and criticism how we can speak of intellectual work that isn't merely reactive or negative.

I come finally now to the second and, in my opinion, the more challenging and interesting set of problems that derive from the reconsideration of Orientalism. One of the legacies of Orientalism, and indeed one of its epistemological foundations, is historicism, that is, the view propounded by Vico, Hegel, Marx, Ranke, Dilthey and others, that if humankind has a history it is produced by men and women, and can be understood historically as, at each given period, epoch or moment, possessing a complex, but coherent unity. So far as Orientalism in particular and the European knowledge of other societies in general have been concerned, historicism meant that the one human history uniting humanity either culminated in or was observed from the vantage point of Europe, or the West. What was neither observed by Europe nor documented by it was therefore "lost" until, at some later date, it too could be incorporated by the new sciences of anthropology, political economics, and linguistics. It is out of this later recuperation of what Eric Wolf has called people without history, that a still later disciplinary step was taken, the founding of the science of world history, whose major practitioners include Braudel, Wallerstein, Perry Anderson and Wolf himself.

But along with the greater capacity for dealing with – in Ernst Bloch's phrase – the non-synchronous experiences of Europe's Other, has gone a fairly uniform avoidance of the relationship between European imperialism and these variously constituted, variously formed and articulated knowledges. What, in other words, has never taken place is an epistemological critique at the most fundamental level of the connection between the development of a historicism which has expanded and developed enough to include antithetical attitudes such as ideologies of Western imperialism and critiques of imperialism on the one hand, and on the other, the actual practise of imperialism by which the accumulation of territories and population, the control of economies, and the incorporation and homogenisation of histories are maintained. If we keep this in mind we will remark, for example, that in the methodological assumptions and practice of world history – which is ideologically anti-imperialist – little or no attention is given to those cultural practises like Orientalism or ethnography affiliated with imperialism, which in genealogical fact fathered world history itself; hence the emphasis in world history as a discipline has been on economic and political practices, defined by the processes of world historical writing, as in a sense separate and different from, as well as unaffected by, the knowledge of them which world history produces. The curious result is that the theories of accumulation on a world scale, or the capitalist world state, or lineages of absolutism depend (a) on the same displaced percipient and historicist observer who had been an Orientalist or colonial traveller three generations ago; (b) they depend also

on a homogenising and incorporating world historical scheme that assimilated non-synchronous developments, histories, cultures, and peoples to it; and (c) they block and keep down latent epistemological critiques of the institutional, cultural and disciplinary instruments linking the incorporative practise of world history with partial knowledges like Orientalism on the one hand, and on the other, with continued "Western" hegemony of the non-European, peripheral world.

In fine, the problem is once again historicism and the universalising and self-validating that has been endemic to it [. . .] in a whole series of studies produced in a number of both interrelated and frequently unrelated fields, there has been a general advance in the process of, as it were, breaking up, dissolving and methodologically as well as critically re-conceiving the unitary field ruled hitherto by Orientalism, historicism, and what could be called essentialist universalism.

I shall be giving examples of this dissolving and decentering process in a moment. What needs to be said about it immediately is that it is neither purely methodological nor purely reactive in intent. You do not respond, for example, to the tyrannical conjuncture of colonial power with scholarly Orientalism simply by proposing an alliance between nativist sentiment buttressed by some variety of native ideology to combat them. This, it seems to me, has been the trap into which many Third World and anti-imperialist activists fell in supporting the Iranian and Palestinian struggles, and who found themselves either with nothing to say about the abominations of Khomeini's regime or resorting, in the Palestine case, to the time-worn clichés of revolutionism and, if I might coin a deliberately barbaric phrase, rejectionary armed-strugglism after the Lebanese debacle. Nor can it be a matter simply of re-cycling the old Marxist or world-historical rhetoric which only accomplishes the dubiously valuable task of re-establishing intellectual and theoretical ascendancy of the old, by now impertinent and genealogically flawed, conceptual models. No: we must, I believe, think both in political and above all theoretical terms, locating the main problems in what Frankfurt theory identified as domination and division of labor, and along with those, the problem of the absence of a theoretical and utopian as well as libertarian dimension in analysis. We cannot proceed unless therefore we dissipate and re-dispose the material of historicism into radically different objects and pursuits of knowledge, and we cannot do that until we are aware clearly that no new projects of knowledge can be constituted unless they fight to remain free of the dominance and professionalized particularism that comes with historicist systems and reductive, pragmatic, or functionalist theories.

These goals are less grand and difficult than my description sounds. For the reconsideration of Orientalism has been intimately connected with many other activities of the sort I referred to earlier, and which it now becomes imperative to articulate in more detail. Thus, for example, we can now see that Orientalism is a praxis of the same sort, albeit in different territories, as male gender dominance, or patriarchy, in metropolitan societies: the Orient was routinely described as feminine, its riches as fertile, its main symbols the sensual woman, the harem, and the despotic – but curiously attractive – ruler. Moreover, Orientals like Victorian housewives were confined to silence and to unlimited enriching production. Now much of this material is manifestly connected to the configurations of sexual, racial and political asymmetry underlying main stream modern Western culture, as adumbrated and illuminated respectively by feminists, by black studies critics, and by anti-imperialist activists.

What I want to do in conclusion is to try to draw [. . .] the larger enterprise of which the critique of Orientalism is a part. First, we note a plurality of audiences and constituencies; none of the works and workers I have cited claims to be working on behalf of One audience which is the only one that counts, or for one supervening, over-coming Truth, a truth allied to Western (or for that matter Eastern) reason, objectivity, science. On the contrary, we note here a plurality of terrains, multiple experiences and different constituencies, each with its admitted (as opposed to denied) interest, political desiderata, disciplinary goals. All these efforts work out of what might be called a decentered consciousness, not less reflective and critical for being decentered, for the most part non- and in some cases anti-totalizing and anti-systematic. The result is that instead of seeking common unity by appeals to a centre of sovereign authority, methodological consistency, canonicity, and science, they offer the possibility of common grounds of assembly between them. They are therefore planes of activity and praxis, rather then one topography commanded by a geographical and historical vision locatable in a known centre of metropolitan power. Second, these activities and praxes are consciously secular, marginal and oppositional with reference to the mainstream, generally authoritarian systems from which they

emanate, and against which they now agitate. Thirdly, they are political and practical in as much as they intend – without necessarily succeeding in implementing the end of dominating, coercive systems of knowledge. I do not think it too much to say that the political meaning of analysis, as carried out in all these fields, is uniformly and programmatically libertarian by virtue of the fact that, unlike Orientalism, it is not based on the finality and closure of antiquarian or curatorial knowledge, but on investigative open models of analysis, even though it might seem that analyses of this sort – frequently difficult and abstruse – are in the final count paradoxically quietistic. I think we must remember the lesson provided by Adorno's negative dialectics, and regard analysis as in the fullest sense being *against* the grain, deconstructive, utopian.

But there remains the one problem haunting all intense, self convicted and local intellectual work, the problem of the division of labor, which is a necessary consequence of that reification and commodification first and most powerfully analysed in this century by George Lukacs. This is the problem sensitively and intelligently put by Myra Jehlen for women's studies, whether in identifying and working through antidominant critiques, subaltern groups – women, blacks, and so on can resolve the dilemma of autonomous fields of experience and knowledge that are created as a consequence. A double kind of possessive exclusivism could set in: the sense of being an excluding insider by virtue of experience (only women can write for and about women, and only literature that treats women or Orientals well is good literature), and second, being an excluding insider by virtue of method (only Marxists, anti-Orientalists, feminists can write about economics, Orientalism, women's literature).

This is where we are now, at the threshold of fragmentation and specialisation, which impose their own parochial dominations and fussy defensiveness, or on the verge of some grand synthesis which I for one believe could very easily wipe out both the gains and the oppositional consciousness provided by these counter-knowledges hitherto. Several possibilities propose themselves, and I shall conclude simply by listing them. A need for greater crossing of boundaries, for greater interventionism in cross-disciplinary activity, a concentrated awareness of the situation – political, methodological, social, historical – in which intellectual and cultural work is carried out. A clarified political and methodological commitment to the dismantling of systems of domination which since they are collectively maintained must, to adopt and transform some of Gramsci's phrases, be collectively fought, by mutual siege, war of manoeuvre and war of position. Lastly, a much sharpened sense of the intellectual's role both in the defining of a context and in changing it, for without that, I believe, the critique of Orientalism is simply an ephemeral pastime.

F
I
V
E

30

Concerning Violence

Frantz Fanon
from *The Wretched of the Earth* (1963)

Decolonization is the meeting of two forces, opposed to each other by their very nature, which in fact owe their originality to that sort of substantification which results from and is nourished by the situation in the colonies. Their first encounter was marked by violence and their existence together – that is to say the exploitation of the native by the settler – was carried on by dint of a great array of bayonets and cannons.

In decolonization, there is therefore the need of a complete calling in question of the colonial situation. If we wish to describe it precisely, we might find it in the well known words: "The last shall be first and the first last." Decolonization is the putting into practice of this sentence. The naked truth of decolonization evokes for us the searing bullets and bloodstained knives which ernanate from it. For if the last shall be first, this will only come to pass after a murderous and decisive struggle between the two protagonists.

The colonial world is a world cut in two. The dividing line, the frontiers are shown by barracks and police stations. In the colonies it is the policeman and the soldier who are the official, instituted go-betweens, the spokesmen of the settler and his rule of oppression. In capitalist societies the educational system, whether lay or clerical, the structure of moral reflexes handed down from father to son, the exemplary honesty of workers who are given a medal after fifty years of good and loyal service, and the affection which springs from harmonious relations and good behavior – all these aesthetic expressions of respect for the established order serve to create around the exploited person an atmosphere of submission and of inhibition which lightens the task of policing considerably. In the capitalist countries a multitude of moral teachers, counselors and "bewilderers" separate the exploited from those in power. In the colonial countries, on the contrary, the policeman and the soldier, by their immediate presence and their frequent and direct action maintain contact with the native and advise him by means of rifle butts and napalm not to budge. It is obvious here that the agents of government speak the language of pure force. The intermediary does not lighten the oppression, nor seek to hide the domination; he shows them up and puts them into practice with the clear conscience of an upholder of the peace; yet he is the bringer of violence into the home and into the mind of the native.

The zone where the natives live is not complementary to the zone inhabited by the settlers. The two zones are opposed, [. . .] they both follow the principle of reciprocal exclusivity.

The settler's town is a strongly built town, all made of stone and steel. It is a brightly lit town; the streets are covered with asphalt, and the garbage cans swallow all the leavings, unseen, unknown and hardly thought about. The settler's feet are never visible, except perhaps in the sea; but there you're never close enough to see them. His feet are protected by strong shoes although the streets of his town are clean and even, with no holes or stones. The settler's town is a well-fed town, an easygoing town; its belly is always full of good things. The settler's town is a town of white people, of foreigners.

The town belonging to the colonized people, or at least the native town, the Negro village, the *medina*, the reservation, is a place of ill fame, peopled by men of evil repute. They are born there, it matters little where or how; they die there, it matters not where, nor how.

It is a world without spaciousness; men live there on top of each other, and their huts are built one on top of the other. The native town is a hungry town, starved of bread, of meat, of shoes, of coal, of light. The native town is a crouching village, a town on its knees, a town wallowing in the mire. It is a town of niggers and dirty Arabs.

In the colonies, the foreigner coming from another country imposed his rule by means of guns and machines. In defiance of his successful transplantation, in spite of his appropriation, the settler still remains a foreigner. It is neither the act of owning factories, nor estates, nor a bank balance which distinguishes the governing classes. The governing race is first and foremost those who come from elsewhere, those who are unlike the original inhabitants, "the others."

To break up the colonial world does not mean that after the frontiers have been abolished lines of communication will be set up between the two zones. The destruction of the colonial world is no more and no less that the abolition of one zone, its burial in the depths of the earth or its expulsion from the country.

The colonial world is a Manichean world. It is not enough for the settler to delimit physically, that is to say with the help of the army and the police force, the place of the native. As if to show the totalitarian character of colonial exploitation the settler paints the native as a sort of quintessence of evil. Native society is not simply described as a society lacking in values. It is not enough for the colonist to affirm that those values have disappeared from, or still better never existed in, the colonial world. The native is declared insensible to ethics; he represents not only the absence of values, but also the negation of values. He is, let us dare to admit, the enemy of values, and in this sense he is the absolute evil. He is the corrosive element, destroying all that comes near him; he is the deforming element, disfiguring all that has to do with beauty or morality, he is the depository of maleficent powers, the unconscious and irretrievable instrument of blind forces.

At times this Manicheism goes to its logical conclusion and dehumanizes the native, or to speak plainly, it turns him into an animal. In fact, the terms the settler uses when he mentions the native are zoological terms. He speaks of the yellow man's reptilian motions, of the stink of the native quarter, of breeding swarms, of foulness, of spawn, of gesticulations.

The native knows all this, and laughs to himself every time he spots an allusion to the animal world in the other's words. For he knows that he is not an animal; and it is precisely at the moment he realizes his humanity that he begins to sharpen the weapons with which he will secure its victory.

In the colonial context the settler only ends his work of breaking in the native when the latter admits loudly and intelligibly the supremacy of the white man's values. In the period of decolonization, the colonized masses mock at these very values, insult them and vomit them up.

The immobility to which the native is condemned can only be called in question if the native decides to put an end to the history of colonization – the history of pillage – and to bring into existence the history of the nation – the history of decolonization.

The uprising of the new nation and the breaking down of colonial structures are the result of one of two causes: either of a violent struggle of the people in their own right, or of action on the part of surrounding colonized peoples which acts as a brake on the colonial regime in question.

A colonized people is not alone. In spite of all that colonialism can do, its frontiers remain open to new ideas and echoes from the world outside. It discovers that violence is in the atmosphere, that it here and there bursts out, and here and there sweeps away the colonial regime – that same violence which fulfills for the native a role that is not simply informatory, but also operative. The great victory of the Vietnamese people at Dien Bien Phu is no longer, strictly speaking, a Vietnamese victory. Since July, 1954, the question which the colonized peoples have asked themselves has been, "What must be done to bring about another Dien Bien Phu? How can we manage it?" Not a single colonized individual could ever again doubt the possibility of a Dien Bien Phu; the only problem was how best to use the forces at their disposal, how to organize them, and when to bring them into action. This encompassing violence does not work upon the colonized people only; it modifies the attitude of the colonialists who become aware of manifold Dien Bien Phus. This is why a veritable panic takes hold of the colonialist governments in turn. Their purpose is to capture the vanguard, to turn the movement of liberation toward the fight, and to disarm the people: quick, quick, let's decolonize. Decolonize the Congo before it turns into another Algeria. [. . .] To the strategy of Dien Bien Phu, defined by the colonized peoples, the colonialist replies by the strategy of encirclement – based on the respect of the sovereignty of states.

[. . .] The reconstruction of the nation continues within the framework of cutthroat competition between capitalism and socialism.

This competition gives an almost universal dimension to even the most localized demands. Every meeting held, every act of repression committed, reverberates in the international arena. [. . .] Each act of sedition in the Third World makes up part of a picture framed by the Cold War. Two men are beaten up in Salisbury, and at once the whole of a bloc goes into action, talks about those two men, and uses the beating-up incident to bring up the particular problem of Rhodesia, linking it, moreover, with the whole African question and with the whole question of colonized people. The other bloc however is equally concerned in measuring by the magnitude of the campaign the local weaknesses of its system. Thus the colonized peoples realize that neither clan remains outside local incidents. They no longer limit themselves to regional horizons, for they have caught on to the fact that they live in an atmosphere of international stress.

When every three months or so we hear that the Sixth or Seventh Fleet is moving toward such-and-such a coast; when Khrushchev threatens to come to Castro's aid with rockets; when Kennedy decides upon some desperate solution for the Laos question, the colonized person or the newly independent native has the impression that whether he wills it or not he is being carried away in a kind of frantic cavalcade. In fact, he is marching in it already.

Strengthened by the unconditional support of the socialist countries, the colonized peoples fling themselves with whatever arms they have against the impregnable citadel of colonialism. If this citadel is invulnerable to knives and naked fists, it is no longer so when we decide to take into account the context of the Cold War. In this fresh juncture, the Americans take their role of patron of international capitalism very seriously. Early on, they advise the European countries to decolonize in a friendly fashion. Later on, they do not hesitate to proclaim first the respect for and then the support of the principle of "Africa for the Africans." The United States is not afraid today of stating officially that they are the defenders of the right of all peoples to self-determination.

[. . .] We understand why the violence of the native is only hopeless if we compare it in the abstract to the military machine of the oppressor. On the other hand, if we situate that violence in the dynamics of the international situation, we see at once that it constitutes a terrible menace for the oppressor. Persistent *jacqueries* and Mau-Mau disturbance unbalance the colony's economic life but do not endanger the mother country. What is more important in the eyes of imperialism is the opportunity for socialist propaganda to infiltrate among the masses and to contaminate them. This is already a serious danger in the Cold War; but what would happen to that colony in case of real war, riddled as it is by murderous guerrillas? Thus capitalism realizes that its military strategy has everything to lose by the outbreak of nationalist wars.

Again, within the framework of peaceful coexistence, colonies are destined to disappear, and in the long run neutralism is destined to be respected by capitalism. What must at all costs be avoided is strategic insecurity: the breakthrough of enemy doctrine into the masses – and the deep-rooted hatred of millions of men. The colonized peoples are very well aware of these imperatives which rule international political life; for this reason even those who thunder denunciations of violence take their decisions and act in terms of this universal violence.

The mobilization of the masses, when it arises out of the war of liberation, introduces into each man's consciousness the ideas of a common cause, of a national destiny, and of a collective history. In the same way the second phase, that of the building-up of the nation, is helped on by the existence of this cement which has been mixed with blood and anger. Thus we come to a fuller appreciation of the originality of the words used in these underdeveloped countries. During the colonial period the people are called upon to fight against oppression; after national liberation, they are called upon to fight against poverty, illiteracy, and underdevelopment. The struggle, they say, goes on. The people realize that life is an unending contest.

31

Antipolitics

A Moral Force

George Konrad

from *Antipolitics* (1984)

ANTIPOLITICS: A MORAL FORCE

The political leadership elites of our world don't all subscribe equally to the philosophy of a nuclear *ultima ratio*, but they have no conceptual alternative to it. They have none because they are professionals of power. Why should they choose values that are in direct opposition to physical force? Is there, can there be, a political philosophy – a set of proposals for winning and holding power – that renounces a priori any physical guarantees of power? Only antipolitics offers a radical alternative to the philosophy of a nuclear *ultima ratio*.

Antipolitics strives to put politics in its place and make sure it stays there, never overstepping its proper office of defending and refining the rules of the game of civil society. Antipolitics is the ethos of civil society, and civil society is the antithesis of military society. There are more or less militarized societies – societies under the sway of nation states whose officials consider total war one of the possible moves in the game. Thus military society is the reality, civil society is a utopia.

Antipolitics means refusing to consider nuclear war a satisfactory answer in any way. Antipolitics regards it as impossible in principle that any historical misfortune could be worse than the death of one to two billion people. Antipolitics bases politics on the conscious fear of death. It recognizes that we are a homicidal and suicidal species, capable of thinking up innumerable moral explanations to justify our homicidal and suicidal tendencies.

We shouldn't shy away from the suspicion that the generals think of war with something other than pure horror. It's inconceivable that the American President or the Russian President is not pleased at the thought of being the most powerful man in the world. His pleasure is only disturbed, perhaps, by the fact that he can't be quite sure about it, since his opposite number may very well think the same thing. [. . .]

The career of Adolf Hitler was an extreme paradigm of the politician's trade. He rose from the ranks of the feckless lumpen-intelligentsia to become Gotterdammerung incarnate over the bodies of fifty million people, like a wayside angel of death. When he addressed his followers, a veritable frenzy of verbal aggression gripped the speaker himself and suffused the glowing faces of his listeners as they breathed "Sieg Heil!" in response.

I am afraid of a third world war because, to my mind, there lives in every politician more or less of the delirium that was Hitler's demon. More exalted than the others, he could find exhilaration in pure unbridled power abstracted from all other considerations, from economic rationality or cultural values. We have to be wary of them because in all politicians worth their salt there is present, albeit in more sober form, some of the dynamite that came out in Hitler with such savage brutality; if there were not, they would not have chosen the politician's trade.

No matter what ideology a politician may appeal to, what he says is only a means of gaining and keeping power. A politician for whom the exercise of power is not an end in itself is a contradiction in terms. In culturally stable societies this kind of cynicism runs up against strong social and ethical inhibitions, and any observer who discerns this cynicism behind the inhibitions is himself called cynical. In less well-balanced societies the relentless instrumentalism of political power may come to the fore in hysterical crises of identity, exacerbating hidden suicidal tendencies by hazarding the greatest risk of all, the doomsday gamble.

Politicians have to be guarded against because the peculiarity of their function and mentality lies in the fact that they are at times capable of pushing the button for atomic war. There is in them some of that mysterious hubris that would like to elevate the frail and mortal "I" into a simulacrum of the Almighty. If this psychic dynamite should go off, it could draw all mankind into a global Auschwitz.

Why should I as a writer stick my nose into political matters? Because they frighten me. I feel mortally threatened by them, because there is more and more talk in political circles of rearmament and the likelihood of war. If the other side doesn't back down, they say, there will be war, and the responsibility will be wholly theirs.

All right, I look at the other side: they don't back down, and they say exactly the same thing. All right, I say, neither one will back down, now what? Are they going to have it out, or are they bluffing? Are they just trying to scare us, or are they serious after all?

I am speaking out because I feel confined by the Iron Curtain and the web of censorship restrictions that has grown up along with it. I know I may be locked up if tensions mount and the regime becomes more stringent. Most of the world is poor and military waste infuriates me. I loathe a culture that represents preparations to kill millions of people as a patriotic obligation. Thanks to the whims of politicians, I have more than once been in a fair way to depart violently from this scene, where otherwise it is still possible to live a good life. On the basis of their public statements, I suspect that politicians still think of war, even in the nuclear age, as a possible political action – "politics with bloodshed," to quote Mao Tse-tung's more graphic version of Clausewitz's aphorism.

I don't like it when they want to kill me. I don't like it when the agents of the politicians hold a gun on me. To me it doesn't matter much whether a bomb kills me or a death squad. To die by war is no better than to die by terror. War is terror too; the possibility of war is terror, and those who prepare for war are terrorists. The prospect of war and the absence of democracy are two sides of the same reality: politicians threatening defenseless people. If reality means people working at

their own deaths for fatuous reasons, then I am bound to think reality even more absurd than deadly.

Escalation is the rule when weapons are put to use, yet no manner of social conflict can be solved by atom bombs. Our entire mythology of revolution and counterrevolution is an anachronistic shadow from the days of simple firearms. I am convinced that the redeeming doctrine of war as a continuation of power politics – the doctrine of the balance of terror – doesn't work any longer.

The abject stupidity of the flower of our intellectuals has contributed to the killing of millions in the big and little wars of this century. The ideologue is responsible because it is possible to kill with ideologies. In order to make war, drop bombs, build concentration camps, and dispose efficiently of the bodies, the skills of intellectuals are required.

I am repelled by men of ideas who chatter in tune with the military's propaganda machinery, who never lifted a finger against the butchery, who are left with only the sad excuse of declaring afterward that they were not in agreement with the terror to which they paid homage. A disturbingly large proportion of our thinkers have become experts in the service of our leaders. They are at pains to depict in rational colors something that is deathly irrational. The intellectual specialists in the logic of atomic and ideological war get their money for deceiving others, for leading them like lambs to the slaughter.

I was in a slaughterhouse once – I saw the lambs. A sly faced black ram led them. Just before reaching the block he slipped to one side, escaping from that corridor of death through a trapdoor. The others, following in his tracks, kept on going – right up to the block. They called the black ram Miska. After each of these performances he would go up to the canteen, where he was given a roll with salami and some cake, and he would eat. For me, the scholars of ideological war are so many Miskas, except that they themselves have no way of slipping through any trapdoor to safety. [. . .]

THE POWER OF THE STATE AND THE POWER OF THE SPIRIT: POLITICS AND ANTIPOLITICS

Antipolitics is the political activity of those who don't want to be politicians and who refuse to share in power. Antipolitics is the emergence of independent forums that can be appealed to against political power; it is a counter power that cannot take power and does not wish to. Power it has already, here and now, by reason of its moral and cultural weight. If a notable scholar or writer takes a ministerial post in a government, he thereby puts his previous work aside. Henceforth he must stand his ground as a representative of his government, and in upholding his actions against the criticisms of democratic antipolitics he may not use his scholarly or literary distinction as either a defense or an excuse.

Antipolitics and government work in two different dimensions, two separate spheres. Antipolitics neither supports nor opposes governments; it is something different. Its people are fine right where they are; they form a network that keeps watch on political power, exerting pressure on the basis of their cultural and moral stature alone, not through any electoral legitimacy. That is their right and their obligation, but above all it is their self-defense. A rich historical tradition helps them exercise their right.

Antipolitics is the rejection of the power monopoly of the political class. The relationship between politics and antipolitics is like the relationship between two mountains: neither one tries to usurp the other's place; neither one can eliminate or replace the other. If the political opposition comes to power, antipolitics keeps at the same distance from, and shows the same independence of, the new government. It will do so even if the new government is made up of sympathetic individuals, friends perhaps; indeed, in such cases it will have the greatest need for independence and distance.

In his thinking, the antipolitician is not politic. He doesn't ask himself whether it is a practical, useful, politic thing to express his opinion openly. In contrast with the secrecy of the leadership, antipolitics means publicity; it is a power exercised directly over society, through civil courage, and one that differs by definition from any present or future power of the state.

Antipolitics means perspicacity; it means ineradicable suspicion toward the mass of political judgments that surround us. Often these judgments are simply aggression in another form. We shouldn't forget that older men whose physical and nervous energies are failing are especially prone to intellectual aggression of the most savage and relentless kind, though always in the name of noble ideals. Spiritual authority is the practice of this kind of antipolitical understanding.

But what does spiritual authority have to offer that is positive? How is it anything more than sheer negativity? It asserts the worth of human life as a value in itself, not requiring further justification. It respects human beings' fear of death. It views the lives of people of other countries and cultures as equal in value to those of our countrymen. It refuses to license killing on any political grounds whatever. I regard the commandment "Thou shalt not kill" as an absolute command. I have never killed, I want to avoid killing, yet it's not impossible that situations may arise in which I will kill. If I do, I will be a murderer and will consider myself one. Murderers must expiate their crimes.

Antipolitics looks kindly on the ecumenical variety of religions and styles and doesn't believe that the condition for the existence of one cultural reality is the extinction of an other.

Antipolitics prefers qualitative competition to silly quantitative questions about who is stronger. Who is stronger is really of no interest. For the antipolitician, it is more interesting to know whether a community produces an intelligent and honest portrait of itself, not how much technical power it commands.

Antipolitics asserts the right of every community to defend itself, with adequate defensive weapons, against occupiers. It is a great misfortune to have to fire on occupiers. We would become murderers ourselves in so doing, but it may happen that we will decide we have to be murderers.

Cartoon 12 Globalization: It's Good for Your Family Too!
Neoliberal globalization has seen transnational corporations outsource their operations to countries in the global south in order to take advantage of lower wage costs, and to evade various regulations (e.g. labor conditions) in their "home" countries. This cartoon satirizes such practices by applying such corporate logic to parenting.

Source: M. Wuerker

32

Tomorrow Begins Today

Invitation to an Insurrection

Subcommandante Marcos
from *We Are Everywhere* (2003)

Welcome to the Zapatista reality. Welcome to this territory in struggle for humanity. Welcome to this territory in rebellion against neoliberalism.

When this dream that awakens today in La Realidad began to be dreamed by us, we thought it would be a failure. We thought that, maybe, we could gather here a few dozen people from a handful of continents. We were wrong. As always, we were wrong. It wasn't a few dozen, but thousands of human beings, those who came from the five continents to find themselves in the reality at the close of the twentieth century.

The word born within these mountains, these Zapatista mountains, found the ears of those who could listen, care for and launch it anew, so that it might travel far away and circle the world. The sheer lunacy of calling to the five continents to reflect clearly on our past, our present, and our future, found that it wasn't alone in its delirium. Soon lunacies from the whole planet began to work on bringing the dream to rest in La Realidad.

Who are they who dare to let their dreams meet with all the dreams of the world? What is happening in the mountains of the Mexican Southeast that finds an echo and a mirror in the streets of Europe, the suburbs of Asia, the countryside of America, the townships of Africa, and the houses of Oceania? What is it that is happening with the peoples of these five continents who, so we are all told, only encounter each other to compete or make war? Wasn't this turn of the century synonymous with despair, bitterness, and cynicism? From where and how did all these dreams come to La Realidad?

May Europe speak and recount the long bridge of its gaze, crossing the Atlantic and history in order to rediscover itself in La Realidad. May Asia speak and explain the gigantic leap of its heart to arrive and beat in La Realidad. May Africa speak and describe the long sailing of its restless image to come to reflect upon itself in La Realidad. May Oceania speak and tell of the multiple flight of its thought to come to rest in La Realidad. May America speak and remember its swelling hope to come to renew itself in La Realidad.

May the five continents speak and everyone listen. May humanity suspend for a moment its silence of shame and anguish.

May humanity speak.
May humanity listen [. . .]
Each country,
each city,
each countryside,
each house,
each person,
each is a large or small battleground.

On the one side is neoliberalism with all its repressive power and all its machinery of death; on the other side is the human being.

In any place in the world, anytime, any man or woman rebels to the point of tearing off the clothes that resignation has woven for them and cynicism has dyed grey. Any man or woman, of whatever colour, in whatever tongue, speaks and says to himself, to herself: "Enough is enough! —¡Ya Basta!"

For struggling for a better world all of us are fenced in, threatened with death. The fence is reproduced globally. In every continent, every city, every countryside, every house. Power's fence of war closes in on the rebels, for whom humanity is always grateful.

> But fences are broken.
> In every house,
> in every countryside,
> in every city,
> in every state,
> in every country,
> on every continent,
> the rebels, whom history repeatedly has given
>> the length of its long trajectory, struggle and
>> the fence is broken.
> The rebels search each other out. They walk
>> toward one another.
> They find each other and together break other
>> fences.

In the countrysides and cities, in the states, in the nations, on the continents, the rebels begin to recognize each other, to know themselves as equals and different. They continue on their fatiguing walk, walking as it is now necessary to walk, that is to say, struggling. [. . .]

A reality spoke to them then. Rebels from the five continents heard it and set off walking.

Some of the best rebels from the five continents arrived in the mountains of the Mexican Southeast. All of them brought their ideas, their hearts, their worlds. They came to La Realidad to find themselves in others' ideas, in others' reasons, in others' worlds.

A world made of many worlds found itself these days in the mountains of the Mexican Southeast.

A world made of many worlds opened a space and established its right to exist, raised the banner of being necessary, stuck itself in the middle of earth's reality to announce a better future.

But what next?

A new number in the useless enumeration of the numerous international orders?

A new scheme that calms and alleviates the anguish of having no solution?

A global program for world revolution?

A utopian theory so that it can maintain a prudent distance from the reality that anguishes us?

A scheme that assures each of us a position, a task, a title, and no work?

The echo goes, a reflected image of the possible and forgotten: the possibility and necessity of speaking and listening; not an echo that fades away, or a force that decreases after reaching its apogee.

Let it be an echo that breaks barriers and re-echoes.

Let it be an echo of our own smallness, of the local and particular, which reverberates in an echo of our own greatness, the intercontinental and galactic.

An echo that recognizes the existence of the other and does not overpower or attempt to silence it.

An echo of this rebel voice transforming itself and renewing itself in other voices.

An echo that turns itself into many voices, into a network of voices that, before Power's deafness, opts to speak to itself, knowing itself to be one and many.

Let it be a network of voices that resist the war that the Power wages on them.

A network of voices that not only speak, but also struggle and resist for humanity and against neo-liberalism.

The world, with the many worlds that the world needs, continues.

Humanity, recognizing itself to be plural, different, inclusive, tolerant of itself, full of hope, continues.

The human and rebel voice, consulted on the five continents in order to become a network of voices and of resistances continues.

We declare:

That we will make a collective network of all our particular struggles and resistances. An intercontinental network of resistance against neoliberalism, an intercontinental network of resistance for humanity.

This intercontinental network of resistance, recognizing differences and acknowledging similarities, will search to find itself with other resistances around the world.

This intercontinental network of resistance is not an organizing structure; it doesn't have a central head or decision maker; it has no central command or hierarchies. We are the network, all of us who resist.

33

Letter to America

Osama Bin Laden
translated from Arabic document posted on Islamist websites (2002)

In the Name of Allah, the Most Gracious, the Most Merciful, [. . .] Some American writers have published articles under the title 'On what basis are we fighting?' [. . .] While seeking Allah's help, we form our reply based on two questions directed at the Americans:

(Q1) Why are we fighting and opposing you?

(Q2) What are we calling you to, and what do we want from you?

As for the first question: Why are we fighting and opposing you? The answer is very simple:

(1) Because you attacked us and continue to attack us.

(a) You attacked us in Palestine:
Palestine, which has sunk under military occupation for more than 80 years. The British handed over Palestine, with your help and your support, to the Jews, who have occupied it for more than 50 years; years overflowing with oppression, tyranny, crimes, killing, expulsion, destruction and devastation. The creation and continuation of Israel is one of the greatest crimes, and you are the leaders of its criminals. And of course there is no need to explain and prove the degree of American support for Israel. The creation of Israel is a crime which must be erased. Each and every person whose hands have become polluted in the contribution towards this crime must pay its price, and pay for it heavily. [. . .]

(b) You attacked us in Somalia; you supported the Russian atrocities against us in Chechnya, the Indian oppression against us in Kashmir, and the Jewish aggression against us in Lebanon.

(c) Under your supervision, consent and orders, the governments of our countries which act as your agents, attack us on a daily basis;

(i) These governments prevent our people from establishing the Islamic Shariah, using violence and lies to do so.

(ii) These governments give us a taste of humiliation, and places us in a large prison of fear and subdual.

(iii) These governments steal our Ummah's wealth and sell them to you at a paltry price.

(iv) These governments have surrendered to the Jews, and handed them most of Palestine, acknowledging the existence of their state over the dismembered limbs of their own people.

(v) The removal of these governments is an obligation upon us, and a necessary step to free the Ummah, to make the Shariah the supreme law and to regain Palestine. And our fight against these governments is not separate from out fight against you.

(d) You steal our wealth and oil at paltry prices because of you[r] international influence and military threats. This theft is indeed the biggest theft ever witnessed by mankind in the history of the world.

(e) Your forces occupy our countries; you spread your military bases throughout them; you corrupt our lands, and you besiege our sanctities, to protect the security of the Jews and to ensure the continuity of your pillage of our treasures.

(f) You have starved the Muslims of Iraq, where children die every day. It is a wonder that more than 1.5 million Iraqi children have died as a result of your sanctions, and you did not show concern. Yet when 3000 of your people died, the entire world rises and has not yet sat down. [. . .]

(2) These tragedies and calamities are only a few examples of your oppression and aggression against us. It is commanded by our religion and intellect that the oppressed have a right to return the aggression. Do not await anything from us but Jihad, resistance and revenge. Is it in any way rational to expect that after America has attacked us for more than half a century, that we will then leave her to live in security and peace?!!

(3) You may then dispute that all the above does not justify aggression against civilians, for crimes they did not commit and offenses in which they did not partake:

(a) This argument contradicts your continuous repetition that America is the land of freedom, and its leaders in this world. Therefore, the American people are the ones who choose their government by way of their own free will; a choice which stems from their agreement to its policies. Thus the American people have chosen, consented to, and affirmed their support for the Israeli oppression of the Palestinians, the occupation and usurpation of their land, and its continuous killing, torture, punishment and expulsion of the Palestinians. The American people have the ability and choice to refuse the policies of their Government and even to change it if they want.

(b) The American people are the ones who pay the taxes which fund the planes that bomb us in Afghanistan, the tanks that strike and destroy our homes in Palestine, the armies which occupy our lands in the Arabian Gulf, and the fleets which ensure the blockade of Iraq. These tax dollars are given to Israel for it to continue to attack us and penetrate our lands. So the American people are the ones who fund the attacks against us, and they are the ones who oversee the expenditure of these monies in the way they wish, through their elected candidates.

(c) Also the American army is part of the American people. It is this very same people who are shamelessly helping the Jews fight against us.

(d) The American people are the ones who employ both their men and their women in the American Forces which attack us.

(e) This is why the American people cannot be [. . .] innocent of all the crimes committed by the Americans and Jews against us.

(f) Allah, the Almighty, legislated the permission and the option to take revenge. Thus, if we are attacked, then we have the right to attack back. Whoever has destroyed our villages and towns, then we have the right to destroy their villages and towns. Whoever has stolen our wealth, then we have the right to destroy their economy. And whoever has killed our civilians, then we have the right to kill theirs.

The American Government and press still refuses to answer the question:

Why did they attack us in New York and Washington?

If Sharon is a man of peace in the eyes of Bush, then we are also men of peace!!! America does not understand the language of manners and principles, so we are addressing it using the language it understands.

(Q2) As for the second question that we want to answer: What are we calling you to, and what do we want from you?

(1) The first thing that we are calling you to is Islam [. . .]

It is to this religion that we call you; the seal of all the previous religions. It is the religion of Unification of God, sincerity, the best of manners, righteousness, mercy, honour, purity, and piety. It is the religion of showing kindness to others, establishing justice between them, granting them their rights, and defending the oppressed and the persecuted. It is the religion of enjoining the good and forbidding the evil with the hand, tongue and heart. It is the religion of Jihad in the way of Allah so that Allah's Word and religion reign Supreme. And it is the religion of unity and agreement on the obedience to Allah, and total equality between all people, without regarding their colour, sex, or language [. . .]

(2) The second thing we call you to, is to stop your oppression, lies, immorality and debauchery that has spread among you.

(a) We call you to be a people of manners, principles, honour, and purity; to reject the immoral acts of fornication, homosexuality, intoxicants, gambling's, and trading with interest.

We call you to all of this that you may be freed from that which you have become caught up in; that you may be freed from the deceptive lies that you are a great nation, that your leaders spread amongst you to conceal from you the despicable state to which you have reached.

(b) It is saddening to tell you that you are the worst civilization witnessed by the history of mankind:

(i) You are the nation who, rather than ruling by the Shariah of Allah in its Constitution and Laws, choose to invent your own laws as you will and desire [. . .]

(ii) You are the nation that permits Usury, which has been forbidden by all the religions. Yet you build your economy and investments on Usury. As a result of this, in all its different forms and guises, the Jews have taken control of your economy, through which they have then taken control of your media, and now control all aspects of your life making you their servants and achieving their aims at your expense; precisely what Benjamin Franklin warned you against.

(iii) You are a nation that permits the production, trading and usage of intoxicants. You also permit drugs, and only forbid the trade of them, even though your nation is the largest consumer of them.

(iv) You are a nation that permits acts of immorality, and you consider them to be pillars of personal freedom. You have continued to sink down this abyss from level to level until incest has spread amongst you, in the face of which neither your sense of honour nor your laws object [. . .]

(v) You are a nation that permits gambling in its all forms. The companies practice this as well, resulting in the investments becoming active and the criminals becoming rich.

(vi) You are a nation that exploits women like consumer products or advertising tools calling upon customers to purchase them. You use women to serve passengers, visitors, and strangers to increase your profit margins. You then rant that you support the liberation of women.

(vii) You are a nation that practices the trade of sex in all its forms, directly and indirectly. Giant corporations and establishments are established on this, under the name of art, entertainment, tourism and freedom, and other deceptive names you attribute to it.

(viii) And because of all this, you have been described in history as a nation that spreads diseases that were unknown to man in the past. Go ahead and boast to the nations of man, that you brought them AIDS as a Satanic American Invention.

(ix) You have destroyed nature with your industrial waste and gases more than any other nation in history. Despite this, you refuse to sign the Kyoto agreement so that you can secure the profit of your greedy companies and industries.

(x) Your law is the law of the rich and wealthy people, who hold sway in their political parties, and fund their election campaigns with their gifts. Behind them stand the Jews, who control your policies, media and economy.

(xi) That which you are singled out for in the history of mankind, is that you have used your force to destroy mankind more than any other nation in history; not to defend principles and values, but to hasten to secure your interests and profits. You who dropped a nuclear bomb on Japan, even though Japan was ready to negotiate an end to the war. How many acts of oppression, tyranny and injustice have you carried out, O callers to freedom?

(xii) Let us not forget one of your major characteristics: your duality in both manners and values; your hypocrisy in manners and principles. All manners, principles and values have two scales: one for you and one for the others.

(c) The freedom and democracy that you call to is for yourselves and for [the] white race only; as for the rest of the world, you impose upon them your monstrous, destructive policies and Governments, which you call the 'American friends'. Yet you prevent them from establishing democracies. When the Islamic party in Algeria wanted to practice democracy and they won the election, you unleashed your agents in the Algerian army onto them, and to attack them with tanks and guns, to imprison them and torture them – a new lesson from the 'American book of democracy'!!!

(d) Your policy on prohibiting and forcibly removing weapons of mass destruction to ensure world peace: it only applies to those countries which you do not permit to possess such weapons. As for the countries you consent to, such as Israel, then they are allowed to keep and use such weapons to defend their security. Anyone else who you suspect might be manufacturing or keeping these kinds of weapons, you call them criminals and you take military action against them.

(e) You are the last ones to respect the resolutions and policies of International Law, yet you claim to want to selectively punish anyone else who does the same. Israel has for more than 50 years been pushing UN resolutions and rules against the wall with the full support of America.

(f) As for the war criminals which you censure and form criminal courts for – you shamelessly ask that your own are granted immunity!! However, history will not forget the war crimes that you committed against the Muslims and the rest of the world; those you have killed in Japan, Afghanistan, Somalia, Lebanon and Iraq will remain a shame that you will never be able to escape. It will suffice to remind you of your latest war crimes in Afghanistan, in which densely populated innocent civilian villages were destroyed, bombs were dropped on mosques causing the roof of the mosque to come crashing down on the heads of the Muslims praying inside [. . .]

(g) You have claimed to be the vanguards of Human Rights, and your Ministry of Foreign affairs issues annual reports containing statistics of those countries that violate any Human Rights. However, all these things vanished when the Mujahideen hit you, and you then implemented the methods of the same documented governments that you used to curse. In America, you captured thousands [of] Muslims and Arabs, took them into custody with neither reason, court trial, nor even disclosing their names. You issued newer, harsher laws.

What happens in Guatanamo is a historical embarrassment to America and its values, and it screams into your faces – you hypocrites, "What is the value of your signature on any agreement or treaty?"

(3) What we call you to thirdly is to take an honest stance with yourselves – and I doubt you will do so – to discover that you are a nation without principles or manners, and that the values and principles to you are something which you merely demand from others, not that which you yourself must adhere to.

(4) We also advise you to stop supporting Israel, and to end your support of the Indians in Kashmir, the Russians against the Chechens and to also cease supporting the Manila Government against the Muslims in Southern Philippines.

(5) We also advise you to pack your luggage and get out of our lands. We desire for your goodness, guidance, and righteousness, so do not force us to send you back as cargo in coffins.

(6) Sixthly, we call upon you to end your support of the corrupt leaders in our countries. Do not interfere in our politics and method of education. Leave us alone, or else expect us in New York and Washington.

(7) We also call you to deal with us and interact with us on the basis of mutual interests and benefits, rather than the policies of subdual, theft and occupation, and not to continue your policy of supporting the Jews because this will result in more disasters for you.

If you fail to respond to all these conditions, then prepare for fight with the Islamic Nation. The Nation of Monotheism, that puts complete trust on Allah and fears none other than Him. [. . .] The Nation of honour and respect [. . .] The Nation of Martyrdom; the Nation that desires death more than you desire life [. . .] The Nation of victory and success that Allah has promised [. . .] The Islamic Nation that was able to dismiss and destroy the previous evil Empires like yourself; the Nation that rejects your attacks, wishes to remove your evils, and is prepared to fight you. You are well aware that the Islamic Nation, from the very core of its soul, despises your haughtiness and arrogance.

If the Americans refuse to listen to our advice and the goodness, guidance and righteousness that we call them to, then be aware that you will lose this Crusade Bush began, just like the other previous Crusades in which you were humiliated by the hands of the Mujahideen, fleeing to your home in great silence and disgrace. If the Americans do not respond, then their fate will be that of the Soviets who fled from Afghanistan to deal with their military defeat, political breakup, ideological downfall, and economic bankruptcy.

This is our message to the Americans, as an answer to theirs. Do they now know why we fight them and over which form of ignorance, by the permission of Allah, we shall be victorious?

F
I
V
E

34

The Clash of Barbarisms

Gilbert Achar

from *The Clash of Barbarisms* (2002)

In a world in which inequality is increasing inexorably, inside each society as well as among nations, in which the law of the jungle and the principle of "might makes right" reign supreme, the barbarism on one side inevitably engenders barbarism on the other. The clash of these twin barbarisms will not usher in a world at peace. Far from canceling each other out, they reinforce each other, in a spiral of reciprocal escalation. [. . .]

THE UNIQUENESS OF SEPTEMBER 11

[T]he terrorist horror of September 11 has been treated as an absolute [. . .] the event has been buried under a particularly dense layer of superlative epithets. It is thus necessary to put this event in proportion, situating it in the context where it belongs, without giving in to intimidating accusations that any such effort amounts to trivializing the atrocity. No one has a monopoly on moral indignation. Putting a vile act in the context of acts of the same kind does not trivialize it, still less justify it, particularly since its authors or inspirers themselves evoked this same context as their motivation, explicitly and from the beginning. Rather, to put the act in context is to reject selective indignation.

So what was so truly extraordinary about the terrorism of mass destruction that took about 3,300 lives (according to the last adjusted figure) on September 11? On the scale of carnage for which the U.S. government is directly responsible, and has never expressed the least regret for, it was all in all a pretty ordinary massacre. Is it forbidden to mention the 200,000 civilian victims of Hiroshima and Nagasaki, on the pretext that Osama bin Laden himself has made

clever use of the argument? What about the three million Indochinese civilians who were victims of U.S. aggression – whom bin Laden has mentioned much less, by contrast, because as the good anti-Communist fighter he was for so long he had to approve of that war? Do we also need to keep silent, just because bin Laden has constantly referred to them, about the 90,000 people – 40,000 children under five years old and 50,000 other civilians – who according to UN agency estimates have died each year for the last ten years from the effects of the embargo against Iraq? [. . .]

The fact that the September 11 attacks struck New York and Washington, the two capitals of "globalization" – which means first and foremost "Americanization," in the sense of the spread of the U.S. socioeconomic and cultural model – explains not only why Americans were so deeply shocked and moved but also why the rest of the world was to such a degree. Absolute U.S. hegemony over the media universe of fiction and information results in a strong tendency for consumers of images the world over to identify with U.S. citizens. This is also why people identify above all with the metropolises of the U.S. empire, since they are familiar to TV viewers and moviegoers around the planet. [. . .]

THE MEDIA AND THE LOGIC OF WAR

The unavoidable consequence in which the attacks on Washington and New York were unique, due to the very nature of their targets, is the extraordinary media attention they received. This constitutes the second way in which they were unique. Media attention was not just the natural result of the "concentrated," "dramatic" character of the mass murder in Manhattan, as contrasted with the "dispersed," "statistical" character of the scourges that have struck Africa or the Iraqi victims of the UN–US embargo. [...] Over-dramatization of the September 11 attacks was also, and above all, the result of deliberate action by the media in the society of the "world spectacle". [...]

From early on, a political logic – "the logic of war," to use a well-worn phrase – dictated this media over-dramatization. It was necessary to keep imperial atrocities and global poverty under wraps, the better to highlight the "absolute evil" that manifested itself on September 11, along the lines that George W. Bush had laid out. Even after the historic record level of live media coverage devoted to the attacks on New York and Washington, the attacks continued to be referred to and broadcast incessantly, and will be for some time to come, so as to cover up and justify new atrocities committed by the United States and its allies in the guise of reprisals.
[...]

The prevalent code of ethics is more flexible ever since Western war-mongers began to lay claim to "humanitarian" concerns. According to this twisted morality it is thus highly immoral to try to put the crime of September 11 in proportion by referring to the long list of crimes committed by the U.S. government and cited in part by those who planned the attacks. Yet by contrast it is supposed to be a moral imperative, according to the same code of ethics, to put the criminal bombing of Afghanistan in proportion by incessantly referring to the crime that it is supposedly a response to. A double standard is at work here. This is the never-ending iniquity of every form of egocentrism, whether ethnic or social.
[...]

PROPOGATING ISLAMIC FUNDAMENTALISM

The United States [...] contribute[d] directly to propagating Islamic fundamentalism, [...] by helping to defeat and crush the Left and progressive nation-alism through-out the Islamic world, it freed up the space for political Islam as the only ideological and organizational expression of popular resentment.
[...]

Several different factors – the worldwide loss of ideological credibility of socialist values, due to the collapse of the Stalinist system; the specific failure or marginality of all currents of the Left in the Muslim world; Islamic fundamentalism's smooth slide down a chute well oiled by Washington and Riyadh; a context of economic crisis and growing social insecurity against a backdrop of neoliberal deregulation on a world scale and all of this aggravated by the affronts experienced daily by Muslim peoples who identify with the Palestinians or Iraqis – have converged to produce a highly explosive mixture in the form of the most virulent Islamic antagonism to the West.
[...]

HATRED AND STRATEGY

[B]in Laden had no illusions about the possibility of defeating the United States in a head-on confrontation. He understood that, given overwhelming U.S. military superiority, the only way to strike violent, painful blows against it was to resort to what Washington has since 1997 officially called "asymmetric means," defined as "unconventional approaches that avoid or undermine our strengths while exploiting our vulnerabilities."[1]
[...]

Bin Laden seems to have grasped a key aspect of any strategy for resisting the United States: the United States cannot be defeated militarily, and the most effective way to get its government to yield is through U.S. public opinion. But on the other hand, he made two fatal mistakes. First, he underestimated the importance of the Arabian Peninsula, the world's largest reservoir of oil, for the U.S. ruling class.
[...]

Bin Laden's second fatal error was to have thought that by attacking U.S. civilians in such a criminal, devastating way, he would succeed in convincing them to force their rulers to disengage. Hamas's Palestinian suicide bombers make the same mistake. In both cases, terrorist actions against civilians only make the targeted populations stand firmly behind their rulers' most reactionary, brutal policies.
[...]

THE CLASH OF BARBARISMS

Civilization in the singular (which we shall capitalize from now on), in the sense of a refinement of customs, mastery of aggressiveness and pacification of relationships among individuals and states, must be understood as a process.
[. . .]

[E]ach step in the advance of Civilization gives rise to specific modalities of Barbarism. Each civilization produces its own specific forms of barbarism. These are not aberrations in the "civilizing process" – as Enzo Traverso rightly emphasizes in a brilliant short work inspired by Hannah Arendt on the European genealogy of Nazi violence – "but the expression of one of its potentialities, one of its faces, one of the directions in which it can drift off."[2]
[. . .]

Each civilization has its own barbarism. Some people cut throats, a traditional Afghan murder method imported into Algeria by veterans of the anti-Soviet war, and symbolized by the September 11 bombers' box-cutter knives. Others "cut daisies": they kill massively at a distance using "daisy cutters," the most deadly "conventional" (15,000-pound) bombs in existence. Some people hijack airliners so as to use them as missiles to murder civilians. Others launch cruise missiles in "surgical strikes" that are to surgery what chain saws are to scalpels.
[. . .]

Thousands of civilians were killed directly in one morning in New York; tens of thousands of civilians have been killed indirectly each year in Iraq for over ten years. This is the scale of comparison of each barbarism's murders. The richer and more powerful a civilization is, the deadlier its barbarism. A few decades ago in powerful Germany, the Nazis invented industrialized genocide. Today the rich countries are guilty, through industrialized failure to assist, of "biogenocide" at the expense of Black Africans and other miserable people with AIDS.
[. . .]

Rather than a "clash of civilizations," the battle in progress is [. . .] a clash of the barbarisms that civilizations secrete in varying quantities in the course of the long historical and dialectical process of Civilization. The more gluttonous societies are, the greater the barbarism they excrete.
[. . .]

On both sides, "absolute hostility" toward the "absolute enemy," . . . entails the deployment of extreme violence and a logic of extermination:

> People who use such means against other people find themselves constrained to destroy these other people, who are their victims as well as their objects, morally as well. They are forced to declare the entire enemy camp criminal and inhuman, to reduce it to a complete nullity. Otherwise they would be criminals and monsters themselves."[3]

[. . .] No civilized ethic can justify deliberate assassination of noncombatants or children, whether indiscriminate or deliberate, by state or nongovernmental terror. [. . .] The notion of "collateral damage" applied by the Pentagon to civilian victims of its bombings not only cynically reduces the murder of innocents to something banal; it is a hypocritical attempt to excuse the murders that result from repetitive recourse to military force.
[. . .]

The different barbarisms do not carry the same weight in the scales of justice. Admittedly, barbarism can never be an instrument of "legitimate self-defense"; it is always illegitimate by definition. But this does not change the fact that when two barbarisms clash, the strongest, the one that acts as the oppressor, is still the more culpable. Except in cases of manifest irrationality, the barbarism of the weak is most often, logically enough, a reaction to the barbarism of the strong. Otherwise why would the weak provoke the strong, at the risk of being crushed themselves? This is, incidentally, why the strong seek to hide their culpability by portraying their adversaries as demented, demonic, and bestial.

PREVENTING TERRORISM

How can we explain the inexorable escalation in Al-Qaeda's level of violence to more and more destructive forms of terrorism [. . .] ? [. . .] It is at the heart of the concept of "asymmetric means" mentioned above.
[. . .]

In the face of adversaries this determined – and prepared to commit suicide – clearly no preventive security measures could ever be enough. The billions of dollars that the Bush administration is now spending on "homeland security" will not accomplish much – except considerably strengthening the Big Brother watching over the words and deeds of U.S. citizens themselves.

The determination of the U.S. enemies and their "absolute hostility" make deterrence an irrelevant notion. [. . .] It is therefore necessary to rely on political prevention. In other words, the causes of "absolute hostility" must be reduced or eliminated, in such a way that a "common interest" emerges as a possibility. [. . .]

ON ASYMMETRIC DOMINANCE

Even more fundamental is the overall strategic vision that underlies Washington's Middle East policies. These policies are only one illustration among others of a general attitude focused on preserving U.S. "dominance." According to the Pentagon's own analysis, this is what explains the increasing tendency of Washington's adversaries to resort to "asymmetric means," targeting both U.S. missions abroad and U.S. civilians at home. U.S. "dominance," Washington's euphemism for domination, is at the heart of post-Cold War U.S. military planning.
[. . .]

The U.S. sole concern has become to assure itself "full spectrum dominance," that is, dominance in all types of conflict. [. . .] In other words, the United States must be master of the world.
[. . .]

The U.S. military budget [. . .] stabilized in the mid-1990s at a level still worthy of the Cold War years: one-third of world military spending, and equivalent to the combined military budgets of the six next highest-spending countries: not only Russia and China, but also Japan, France, Germany and the United Kingdom.
[. . .]

HEGEMONIC UNILATERALISM

[. . .]
Bush has announced that this "war on terror" will have to last several more years. In this war the United States alone will choose the targets. It will assign to its allies the tasks that it wants them to take on, under U.S. command or supervision: essentially the tasks that the Pentagon balks at taking on itself.
[. . .]

Washington has demonstrated the purely utilitarian relationship that it maintains with the UN [. . .] the UN is nothing more than a postwar management tool for territories ravaged by military interventions decided in Washington.

The United States can get away with this kind of behavior, placing itself above international law and international institutions, thanks precisely to its military dominance over the rest of the world. [. . .] It combines this asymmetric power of the strong over the weak with the example it sets by breaking all the rules and laws. In other words, it is inflicting the reign of its overpowering, arbitrary will on the rest of the world. This is the most effective recipe for making countless people reach the conclusion that the United States' "asymmetric advantages" can only by countered by "asymmetric means" targeting the most vulnerable among the U.S. population, and accordingly turn to terrorism.

The U.S. government itself, through the political and military choices it has made since the end of the Cold War, is inexorably producing the terrorism that it means to fight against. Bush's "war on terror" will inevitably lead to new terrorist attacks on U.S. citizens.
[. . .]
According to Washington's own analysis [. . .] it can only respond to attacks by "asymmetric means" provoked by its military dominance by increasing the asymmetry of its dominance still further and extending the areas in which it is exercised, in the name of "preemption." In so doing it is bound to provoke new asymmetric attacks on its troops and civilian population.
[. . .]

URBAN VIOLENCE AND ANOMIE

Our world seems decidedly more and more dangerous and terrifying. But these apocalyptic scenarios, which reality is fast catching up with if not surpassing, are only the most visible, most spectacular aspects of the rising tide of violence in the last few decades. The various forms of terrorism, governmental and nongovernmental, are part of a generalized increase in urban violence.
[. . .]

According to UN agency figures, the percentage of people who had experienced violence in urban areas with more than 100,000 inhabitants during the years

1988–1994 was 11 percent in Asia and the same in Eastern Europe [. . .] as opposed to 15 percent in Western Europe and 20 percent in North America. Only South America (at 31 percent) and Africa (33 percent) surpassed the North American level of violence.

[. . .]

The universal phenomenon of rising urban violence during the two last decades is not hard to explain. Franz Vanderschueren, technical adviser to Habitat's Urban Management Program, has identified its causes quite well:

Violence is not a spontaneous phenomenon but the product of a society characterized by inequality and exclusion and lack of institutional or social control. However, urban marginality and poverty do not automatically lead to violence, but may favour it in certain circumstances [. . .] In a society that promotes consumption and competition to the detriment of sharing and solidarity, young people with no hope of employment or success look for ways to survive and to gain a sense of recognition at least from their peer group. This often leads to violence and youth gangs.[4]

Habitat's 2001 report on "the state of the world's cities" helps dot the I's and cross the T's about the causes of these phenomena: In the 1970s, the world embarked on a phase of globalization aimed at de-regulating labour markets, privatizing government functions and liberalizing finance. Financial liberalization was supposed to move savings from developed to developing countries, lower the costs of borrowing, reduce risk through new financial instruments, and increase economic growth. Much the opposite materialized: savings have flowed from poor to richer countries, interest rates have generally increased, risk has risen and economic growth throughout the world has slowed for the vast majority of countries, rich and poor.[5]

We need to supplement this observation by noting the steadily deepening inequality among and within countries over the past two decades. This gives us a fairly complete list of the main socioeconomic factors resulting in the great degree of anomie, the disintegration of social norms and reference points, characteristic of our times. [. . .]

[. . .]

The combination of [. . .] two dimensions – socioeconomic anomie together with political and ideological anomie – has inevitably led people to fall back on other factors of social solidarity such as religion, family, and fatherland. [. . .]

But since we cannot transform society right away [. . .] it is necessary for a credible progressive alternative to neoliberal capitalism to emerge again. Only this kind of alternative can pull the rug out from under reactionary identity politics, by channeling social discontent toward transformative action in the pursuit of democracy and justice [. . .]

In reality, the struggle against neoliberal globalization – born in the last years of the dying twentieth century, and growing rapidly among the new generation [. . .] of the twenty-first century – is our best hope for defeating the wave of reaction.

NOTES

1 William Clinton, "A National Security Strategy for a New Century", speech delivered at the White House, May 1997.

2 Enzo Traverso, *La Violence nazie: Une genealogie europeenne* (Paris: La Fabrique, 2002), p. 166.

3 Carl Schmitt, *Theorie des Partisanen: Zwischenbemerkung zum Begriff des Politischen* (Berlin: Duncker & Humblot, 1963), p. 95.

4 Quoted from *Habitat Debate: Towards Safer Cities* 4:1 (March 1998).

5 UNCHS (Nairobi), *The State of the World's Cities Report 2001* (New York: United Nations Publications, 2001), p. 19.

Cartoon 13 Your Guide to Moral Warfare

In the clash of barbarisms, the violence of one side engenders violence from the other. However, as this cartoon suggests, the barbarism of the strong is morally privileged against the barbarism of the weak.

Source: M. Wuerker

35

Beyond Either/Or

A Feminist Analysis of September 11th

Jennifer Hyndman

from *ACME: An International E-Journal for Critical Geographies* (2003)

The events and aftermath of September 11th ineluctably ended the already precarious distinction between domestic space, that within a sovereign state, and more global space where transnational networks, international relations, multilateral institutions, and global corporations operate [. . .] Acts of violence perpetrated by people who entered the country legally from states outside the US, using domestic aircraft, imploded any notion that political borders contain political conflict. Feminists have long argued that private–public distinctions serve to depoliticize the private domestic spaces of 'home' compared to more public domains. The attacks certainly exposed the limitations of 'domestic' space, somehow bounded and separated from the processes and politics of economic, cultural, and political integration. This paper interrogates [. . .] the spatialized imaginary of politics and violence in the context of September 11th.

[. . .]

While I [. . .] aim to analyze the events and aftermath of September 11th, I also want to position myself politically in relation to the violence they embody. To my mind, nothing justified the killing of innocent people on September 11th. Nothing justifies the retaliatory killing of innocent people anywhere else [. . .] Terror in the US on September 11th has been met with more terror in Afghanistan since October 7th, continuing into 2002.

The surge of insecurity experienced by Americans after the attacks has been stoked by fears of anthrax infection and repeat attacks. In this climate of fear, public consent has been mobilized to reconstitute the country as a bounded area that can be fortified against outsiders and other global influences. In this imagining of nation, the US ceases to be a constellation of local, national, international, and global relations, experiences, and meanings that coalesce in places like New York City and Washington DC; rather, it is increasingly defined by a 'security perimeter' and the strict surveillance of borders. Such [. . .] bounded thinking about discrete places has had concrete implications for airline security, immigration and visitor visa regulations (in the US and Canada), and customs control, especially at the land border between the US and Canada. But it has more reactionary, if less tangible, implications for American politics, US immigrants, and questions of how 'civil' society should be. Anti-terrorist legislation, anti-immigrant rhetoric, and public support for a military response are as much an expression of outrage and insecurity as they are evidence of government resolve and a heightened need for increased security in the face of heinous attacks of September 11th. What has been disturbing, however, is the shrinking space and number of venues available for open dialogue about the attacks and responses to them. I attempt to reclaim some of this space by calling for what might be thought of as a feminist geopolitics, a more accountable and embodied notion of politics that analyzes the intersection of power and space at multiple scales, one that eschews violence as a legitimate means to political ends (Hyndman, 2001).

In what follows [. . .] I construct a political space beyond the binary logic of 'either/or' and advocate more embodied ways of seeing and knowing by examining the casualties in the wars of/on terrorism.

FEMINIST GEOPOLITICS

[. . .] I define 'feminist' as analyses and political interventions that address the inequitable and violent relationships of power among people and places based on real or perceived differences [. . .] Feminist geopolitics attempts to develop a politics of security at multiple scales, including that of the (civilian) body. It decentres state security, the conventional subject of geopolitics, and contests the militarization of states and societies [. . .] It seeks embodied ways of seeing and material notions of protection for people on the ground [. . .] It is an analytic and politics that is contingent upon context, place, and time. Just as place is constituted at multiple scales [. . .] so too is geopolitics. [. . .]

Feminist geopolitics [. . .] navigate[s] between nations and across space, cross-cutting the dominant framing of territorial sovereignty, interrogating security as a concern not only for states, but also for the people who comprise the citizenry of states. Moreover, I am concerned about the consequences of 'terror' for unsuspecting citizens on both ends of this continuum of violence.

EITHER/OR – NEITHER/NOR

In his televised address on September 20th, President Bush drew a clear line between the two sides in 'the war against terrorism': 'if you are not with us, you are against us.' On October 7th, Osama Bin Laden stated that the world is divided into two regions – one of faith and another of infidelity (Hensmen, 2001). Such binary thinking has become part of the dominant geopolitical narrative, garnering support for both sides, and leaving little space in between for those who fail to identify with either side [. . .] Subject constitution and legitimacy require definition, something to define one's project against. Not only are such binaries logically questionable, they are also politically bankrupt, reproducing the dominant geopolitical narrative as the only political option [. . .] In the aftermath of the attacks on September 11th, the existence of an antagonist and an enemy territory (i.e. Afghanistan) has been crucial to the US government's response. Evidence that fifteen of the nineteen men on the four hijacked flights that crashed were Saudi nationals, and that Al Qaeda is a transnational network with operatives in 34 countries, including Germany, the US, and Canada, has been glossed over. A target was sorely needed and quickly identified.

Feminist political sociologist Cynthia Cockburn [. . .] contests the militarized either/or ultimatum with a logic of her own, noting that 'neither/nor' [is] an option [. . .] (Cockburn, 2000). The same feminist logic [. . .] applies to the events and aftermath of September 11th: neither is the killing of thousands of innocent civilians in New York, Washington, and Pennsylvania warranted, nor is the killing of thousands of innocent civilians in Afghanistan. The space to voice this kind of dissent has, however, been highly restricted in North America since the attacks, a point to which I will return.

FEW DEGREES OF SEPARATION

A brief political geography of the Taliban and Osama Bin Laden's early years also breaks down the convenient shorthand of 'either/or.' Certainly, an analysis of the history of all US foreign policy and its consequences is neither possible nor appropriate here, but a brief survey of US foreign policy as it relates to the formation and activities of the Taliban and Osama Bin Laden is surely justified. Mutually exclusive spaces of 'here' and 'there' and political dyads of 'us' and 'them' allow us to see the world more clearly, but less honestly. Michel Chossudovsky (2001) attempts to unsettle the dominant geopolitical narrative that emerged after the attacks of September 11th, highlighting instead the connections between the US, the Taliban, and Osama Bin Laden. He notes that the largest covert operation in the history of the CIA was launched in 1979 in response to the invasion of Afghanistan by the Soviet Union. Working together with the Pakistan Inter Services Intelligence (ISI), the CIA actively encouraged political instability in Afghanistan and beyond, with the idea that Muslim states could eventually defeat the Soviet Union. Between 1982 and 1992, some 35,000 radicals from 40 Islamic countries joined Afghanistan's fight. Tens of thousands more went to study in madrassas (Koranic schools) in Pakistan, as refugees fleeing the fighting in Afghanistan (Rashid, 1999).

Pakistan's ISI was the go-between between the CIA and the Mujahadeen (which included Osama Bin Laden) in Afghanistan. The CIA's support was covert and indirect, so as not to reveal its own geopolitical investments. While the Cold War began to fade and Soviet troops withdrew in 1989, the Islamic jihad based out of Pakistan did not, and the civil war in Afghanistan continued unabated [. . .] What is more disturbing is that the Clinton Administration appears to have known about links between Pakistani Intelligence (ISI) and Al Qaeda, including the former's use of Al Qaeda camps in Afghanistan to train covert operatives for use in a war of terror against India. The 'evildoers', as they have been called, were once US allies. One can speak more accurately, then, about degrees of separation between the US and the Taliban, than about historic enmity, longstanding hatreds, or the absence of political ties.

Just as Cold War geopolitics connected the US with Central Asia, the Taliban, and indirectly to Osama bin Laden, tracing the geopolitics of oil interests goes some distance in explaining the United States' hands-off approach to the Taliban and its treatment of women in Afghanistan, one which parallels US treatment of Saudi Arabia, its most important oil ally. When the Taliban won control of Kabul in 1996, Washington said nothing. In December 1997, Taliban leaders met with US State Department officials in Washington and visited Houston, Texas to meet with UNOCAL oil executives (Pilger, 2001). 'At that time the Taliban's taste for public executions and its treatment of Afghan women were not made out to be crimes against humanity' (Roy, 2001). Assured access for an oil pipeline from the Caspian Sea oil and gas reserves to the Indian Ocean would decrease US reliance on Middle Eastern sources. Turkmenistan, which borders on Afghanistan, holds the world's third largest gas reserves and an estimated six billion barrels of oil reserves. The desire for political stability in this region has a subtext.

The political economy of oil, underwriting an earlier war in the Persian Gulf, combined US efforts to destabilize the USSR not so long ago, illustrate the geopolitical designs of superpower on this region. Such designs shape what we see and hear in the mainstream media. Conventional state-centred and resource-driven geopolitics promote a dominant geopolitical narrative that ensconces the 'us'/'them' binary. Such politics obfuscate minor voices and non-militarized responses to the attacks, and muffle dissent where it finds expression. After September 11th, a dominant geopolitical narrative generated all-knowing maps of meaning that have been disseminated through the mainstream media. These god's-eye cartographies of peopleless places, mostly in Afghanistan, have mobilized consent for more violence in subtle ways. They enable military manoeuvres to proceed, despite

significant opposition to attacks that would harm innocent civilians.

Such maps construct particular sightlines that enable one to see 'enemy' positions and movements via remotely-sensed satellite data, but omit images of the Afghan civilians killed by the American and British attacks. Acts of omission are as much acts of commission in this context, and more accountable maps are in short supply. Matt Sparke (2001) has asked what other maps we might draw:

> maps that might trace where bin Laden's financial support has come from over the years; maps that might show how he has been 'harbored' by other states that the US would be much less inclined to bomb; and, maps that show how the dead terrorists themselves were once 'harbored' in states across America itself.

Maps that forge links and recognize extant networks among political actors resist 'either/or' reasoning and have the potential to enhance their accountability.

EMBODIED VISION AND VISIBLE BODIES

[. . .]
Since September 11th I have read the short, often moving, biographies of hundreds of the people killed, in *The New York Times*. Until the spring of 2002, an updated body count of the people lost in the World Trade Center, the Pentagon, and on the flights that never arrived at their destination was published every Sunday. As an audience, the human face of these horrific acts of violence in the US was everywhere apparent. It took a long time, however, before the same paper began to publish photos of civilians who had lost family members to the bombings in Afghanistan, and to cover controversial statistics about how many civilians had been killed in that country by US military planes equipped with smart and not-so-smart bombs (Bearak, 2002). The audible silence around the equally preposterous deaths of a people already ravaged by war and, in many regions, starvation was remarkable.
[. . .]
Reports of civilian deaths in Afghanistan took time to filter back to the US mainstream media, but once they did, an alarmingly visible landscape of death and destruction emerged. Bearak (2002) chronicles attacks on five towns and villages, 'mysteries' that remain

unresolved in which large numbers of civilians were killed. With information from other reporters, his article discusses the questionable use of cluster bombs, some of which fail to detonate on contact and are littered 'live' around the countryside of Afghanistan. While the tragedies at both of ends of this violence are not disproportionate, in terms of lives lost, the patriotic values placed on them (or not) vary tremendously. Body counts provide a reality check: violence kills civilians and is unwarranted wherever it occurs. Yet where is the space beyond retribution and the 'either/or' logic of militarization?

As Neil Smith (2002: 635) has argued, the 'need to nationalize September 11 arose from the need to justify war.' Smith argues that the World Trade Center catastrophe was a profoundly local and also global event, yet it was produced as a national tragedy. This politically strategic rescaling of September 11th serves at once to limit public and media expression that questions government tactics of combating terrorism and forces citizens to see these tactics as a matter of national interest and security.

[. . .]

WITHOUT CONCLUSION

The war on terrorism continues. In this context, feminist geopolitics represents a third space, beyond the binaries of either/or, here/there, us/them. As an ethnographic, rather than a strategic, perspective it does not promote an oppositional stance in relation to particular political principles or acts. Rather, as an analytic, it attempts to map the silences of the dominant geopolitical position[s] and undo these by invoking multiple scales of inquiry and knowledge production. Scrutinizing the prevailing nation-state-centred discourse of the war on terrorism is critical in recognizing the international and global dimensions of the terror perpetrated on September 11th and in seeing the terror invoked on Afghan civilians, in the name of justice.

In February 2002, Daniel Pearl, a *Wall Street Journal* reporter was abducted then killed in Pakistan. His tragic death was no doubt fuel for the fire in the war on terrorism, but the words of his wife upon his death are instructive:

Revenge would be easy, but it is far more valuable in my opinion to address this problem of terrorism with enough honesty to question our own responsibility as nations and as individuals for the rise of terrorism . . . [I hope] I will be able to tell our son that his father carried the flag to end terrorism, raising an unprecedented demand among people from all countries not for revenge but for the values we all share: love, compassion, friendship and citizenship, far transcending the so-called clash of civilizations (*The Globe and Mail*, 2002).

While the language of patriotism is clear in Ms. Pearl's commentary, her dismissal of Samuel Huntington's clash of civilizations thesis makes it clear that her husband's death should not become a justification for more violence, couched in the language of national security.

Dominant geopolitical narratives shape what we see and hear in profound ways. However, islands of opposition to the state-sponsored attacks on Afghanistan generate snippets of hope that dissident voices have not been silenced. On September 15th, the House of Representatives voted on a resolution permitting the President to use 'all necessary force against those nations, organization or persons he determines planned, authorized, committed, or aided the terrorist attacks' (Ibbitson, 2001). Defying unprecedented unity in Congress, all but one member voted for this resolution. The sole voice of dissent was lodged by Barbara Lee, congresswoman for the district of Berkeley and Oakland. She has since received numerous death threats and hate mail. Her dissent does not mean that women are categorically more peace-oriented than men. Many other US congresswomen supported the resolution. Rather her position has to do with a confluence between her politics and the geography of the district she represents. Berkeley City Council was the first municipal government in the US to pass a resolution criticizing the military campaign against Afghanistan. Its history of peace marches and anti-war activism are well-known. Lee has taken a courageous stand in a climate of patriotism that has been intolerant of criticism against the US government. Her stance has been applauded by feminists from countries whose experience of terrorism span decades for her ability to connect terrorism in the US with terrorism elsewhere (Cat's Eye, 2001). There are few degrees of separation between here and there, us and them, either/or. A feminist geopolitics aims to trace the connections between geographical and political locations, exposing investments in the dominant geopolitical rhetoric, in the pursuit of a more account-able and embodied geopolitics that contests the wisdom of violence targeted at innocent civilians, wherever they may be.

REFERENCES

Bearak, Barry. 2002. Unknown Toll in the Fog of War: Civilian Deaths in Afghanistan, *The New York Times*, February 10.

Cat's Eye Collective. 2001. America Attacked or America Attacks? *The Island*, Colombo, Sri Lanka, September 27.

Chossudovsky, Michel. 2001. Who Is Osama Bin Laden?, unpublished paper, Dept. of Economics, University of Ottawa/Centre for Research on Globalisation (CRG), Montréal; accessed September 16, 2001 at http://globalresearch.ca/articles/CHO109C.html.

Cockburn, Cynthia. 2000. Women in Black: Being Able to Say Neither/nor, *Canadian Woman Studies*, vol. 19, no. 4, pp. 7–11.

The Globe and Mail. 2002. Editorial: The Murder of Daniel Pearl, *The Globe and Mail*, February 23.

Hensmen, Rohini. 2001. The Only Alternative to Global Terror, South Asia Citizens Wire, Dispatch #2, October 24; reprinted in *Pravada*, October 2001.

Hyndman, Jennifer. 2001. Towards a Feminist Geopolitics, *The Canadian Geographer*, vol. 45, no. 2, pp. 210–222.

Ibbitson, John. 2001. The Defiant Dove of Capitol Hill, *The Globe and Mail*, November 5.

Pilger, John. 2001. Hidden Agenda behind War on Terror, *The Mirror* (UK), October 29.

Rashid, Ahmed. 1999. The Taliban: Exporting Extremism. *Foreign Affairs*, November–December.

Roy, Arundhuti. 2001. War is Peace. *Outlook Magazine*, October 29.

Smith, Neil. 2002. Editorial: Scales of Terror and the Resort to Geography: September 11, October 7, *Environment and Planning D: Society and Space*, vol. 19, pp. 631–637.

Sparke, Matthew. 2001. Maps, Massacres, and Meaning, unpublished paper, Department of Geography and Jackson School of International Studies, University of Washington.

36

Instant-Mix Imperial Democracy

(Buy One, Get One Free)

Arundhati Roy

from *The Ordinary Person's Guide to Empire* (2003)

[. . .]

As we lurch from crisis to crisis, beamed directly into our brains by satellite TV, we have to think on our feet. On the move. We enter histories through the rubble of war. Ruined cities, parched fields, shrinking forests, and dying rivers are our archives. Craters left by daisy cutters, our libraries.

So what can I offer you tonight? Some uncomfortable thoughts about money, war, empire, racism, and democracy.

[. . .]

Some of you will think it bad manners for a person like me, [. . .] an "Indian citizen," to [. . .] criticize the U.S. government. Speaking for myself, I'm no flag-waver, no patriot, and am fully aware that venality, brutality, and hypocrisy are imprinted on the leaden soul of every state. But when a country ceases to be merely a country and becomes an empire, then the scale of operations changes dramatically. So may I clarify that tonight I speak as a subject of the American Empire? I speak as a slave who presumes to criticize her king.

[. . .]

Public support in the U.S. for the war against Iraq was founded on a multi-tiered edifice of falsehood and deceit, coordinated by the U.S. government and faithfully amplified by the corporate media.

Apart from the invented links between Iraq and Al Qaida, we had the manufactured frenzy about Iraq's Weapons of Mass Destruction. George Bush the Lesser went to the extent of saying it would be "suicidal" for the U.S. not to attack Iraq. We once again witnessed the paranoia that a starved, bombed, besieged country was about to annihilate almighty America. (Iraq was only the latest in a succession of countries – earlier there was Cuba, Nicaragua, Libya, Grenada, and Panama.) [. . .] It ushered in an old doctrine in a new bottle: the Doctrine of Pre-emptive Strike, *a.k.a.* The United States Can Do Whatever The Hell It Wants, And That's Official.

The war against Iraq has been fought and won and no Weapons of Mass Destruction have been found. Not even a little one. Perhaps they'll have to be planted before they're discovered. And then, the more troublesome amongst us will need an explanation for why Saddam Hussein didn't use them when his country was being invaded.

[. . .]

Bush the Lesser has said Saddam Hussein was a "Homicidal Dictator." And so, the reasoning goes, Iraq needed a "regime change."

Never mind that forty years ago, the CIA, under President John F. Kennedy, orchestrated a regime change in Baghdad. In 1963, after a successful coup, the Ba'ath party came to power in Iraq. Using lists provided by the CIA, the new Ba'ath regime systematically eliminated hundreds of doctors, teachers, lawyers, and political figures known to be leftists. An entire intellectual community was slaughtered [. . .] The young Saddam Hussein was said to have had a hand in supervising the bloodbath. In 1979, after factional infighting within the Ba'ath Party, Saddam Hussein became the President of Iraq. In April 1980, while he was massacring Shias, the U.S. National Security Adviser Zbigniew Brzezinksi declared, "We see no fundamental incompatibility of interests between the United States and Iraq." Washington and London overtly and covertly supported Saddam Hussein. They financed him, equipped him, armed him, and provided him with dual-use materials to manufacture weapons of mass destruction. They supported his worst excesses financially, materially, and morally. They supported the eight-year war against Iran and the 1988 gassing of Kurdish people in Halabja, crimes which 14 years later were re-heated and served up as reasons to justify invading Iraq. [. . .] The point is, if Saddam Hussein was evil enough to merit the most elaborate, openly declared assassination attempt in history (the opening move of Operation Shock and Awe), then surely those who supported him ought at least to be tried for war crimes? Why aren't the faces of U.S. and U.K. government officials on the infamous pack of cards of wanted men and women?

Because when it comes to Empire, facts don't matter.

[. . .]

So here we are, the people of the world, confronted with an Empire [. . .] that has conferred upon itself the right to go to war at will, and the right to deliver people from corrupting ideologies, from religious fundamentalists, dictators, sexism, and poverty by the age-old, tried-and-tested practice of extermination. Empire is on the move, and Democracy is its sly new war cry. Democracy, home-delivered to your doorstep by daisy cutters. Death is a small price for people to pay for the privilege of sampling this new product: Instant-Mix Imperial Democracy (bring to a boil, add oil, then bomb).

[. . .]

At a media briefing before Operation Shock and Awe was unleashed, General Tommy Franks announced, "This campaign will be like no other in history." Maybe he's right.

I'm no military historian, but when was the last time a war was fought like this?

After using the "good offices" of UN diplomacy (economic sanctions and weapons inspections) to ensure that Iraq was brought to its knees, its people starved, half a million children dead, its infrastructure severely damaged, *after making sure that most of its weapons had been destroyed*, in an act of cowardice that must surely be unrivalled in history, the "Coalition of the Willing" [. . .] sent in an invading army!

Operation Iraqi Freedom? I don't think so. It was more like Operation Let's Run a Race, but First Let Me Break Your Knees.

As soon as the war began, the governments of France, Germany, and Russia, which refused to allow a final resolution legitimizing the war to be passed in the UN Security Council, fell over each other to say how much they wanted the United States to win. [. . .] Apart from hoping to share the spoils, they hoped Empire would honor their pre-war oil contracts with Iraq. [. . .] [A]t the moment of crisis, the unity of Western governments – despite the opposition from the majority of their people – was overwhelming.

[. . .] According to a Gallup International poll, in no European country was support for a war carried out 'unilaterally by America and its allies' higher than 11 percent. But the governments of England, Italy, Spain, Hungary, and other countries of Eastern Europe were praised for disregarding the views of the majority of their people and supporting the illegal invasion [. . .]

In stark contrast to the venality displayed by their governments, on the 15th of February, weeks before the invasion, in the most spectacular display of public morality the world has ever seen, more than 10 million people marched against the war on 5 continents. [. . .] They – we – were disregarded with utter disdain. [. . .]

Democracy, the modern world's holy cow, is in crisis. And the crisis is a profound one. Every kind of outrage is being committed in the name of democracy. It has become little more than a hollow word, a pretty shell, emptied of all content or meaning.

[. . .]

[M]odern democracies have been around for long enough for neo-liberal capitalists to learn how to subvert them. They have mastered the technique of infiltrating the instruments of democracy – the 'independent' judiciary, the 'free' press, the parliament – and molding them to their purpose. The project of corporate globalization has cracked the code. Free elections, a free press, and an independent judiciary mean little when the free market has reduced them to commodities on sale to the highest bidder.

[. . .]

Democracy has become Empire's euphemism for neo-liberal capitalism.

[. . .]

It is a cruel irony that the U.S., which has the most ardent, vociferous defenders of the idea of Free Speech, and (until recently) the most elaborate legislation to protect it, has so circumscribed the space in which that freedom can be expressed. In a strange, convoluted way, the sound and fury that accompanies the legal and *conceptual* defense of Free Speech in America serves to mask the process of the rapid erosion of the possibilities of actually *exercising* that freedom.

The news and entertainment industry in the U.S. is for the most part controlled by a few major corporations – AOL-Time Warner, Disney, Viacom, News Corporation. Each of these corporations owns and controls TV stations, film studios, record companies, and publishing ventures. Effectively, the exits are sealed.

[. . .]

So here it is – the World's Greatest Democracy, led by a man who was not legally elected. America's Supreme Court gifted him his job. What price have American people paid for this spurious presidency?

In the three years of George Bush the Lesser's term, the American economy has lost more than two million jobs. Outlandish military expenses, corporate welfare, and tax giveaways to the rich have created a financial crisis for the U.S. educational system. According to a survey by the National Council of State Legislatures, U.S. states cut 49 billion dollars in public services, health, welfare benefits, and education in 2002. They plan to cut another 25.7 billion dollars this year. That makes a total of 75 billion dollars. Bush's initial budget request to Congress to finance the war in Iraq was 80 billion dollars.

So who's paying for the war? America's poor. Its students, its unemployed, its single mothers, its hospital and home-care patients, its teachers, and health workers.

And who's actually fighting the war?

Once again, America's poor. [. . .] Only one of all the representatives in the House of Representatives and the Senate has a child fighting in Iraq. America's 'volunteer' army in fact depends on a poverty draft of poor whites, Blacks, Latinos, and Asians looking for a way to earn a living and get an education. Federal statistics show that African Americans make up 21 percent of the total armed forces and 29 percent of the U.S. army. They count for only 12 percent of the general population. [. . .]

[. . .] A study by the economist Amartya Sen shows that African Americans as a group have a lower life expectancy than people born in China, in the Indian State of Kerala (where I come from), Sri Lanka, or Costa Rica.

[. . .]

So we know who's paying for the war. We know

who's fighting it. But who will benefit from it? [. . .]

Operation Iraqi Freedom, George Bush assures us, is about returning Iraqi oil to the Iraqi people. That is, returning Iraqi oil to the Iraqi people via Corporate Multinationals. Like Bechtel, like Chevron, like Halliburton. Once again, it is a small, tight circle that connects corporate, military, and government leadership to one another. [. . .]

Consider this: the Defense Policy Board is a government-appointed group that advises the Pentagon. Its members are appointed by the under secretary of defense and approved by Donald Rumsfeld. Its meetings are classified. No information is available for public scrutiny.

The Washington-based Center for Public Integrity found that 9 out of the 30 members of the Defense Policy Board are connected to companies that were awarded defense contracts worth 76 billion dollars between the years 2001 and 2002. One of them, Jack Sheehan, a retired Marine Corps general, is a senior vice president at Bechtel, the giant international engineering outfit. Riley Bechtel, the company chairman, is on the President's Export Council. Former Secretary of State George Shultz, who is also on the Board of Directors of the Bechtel Group, is the chairman of the advisory board of the Committee for the Liberation of Iraq. [. . .]

Bechtel has been awarded a 680 million dollar reconstruction contract in Iraq. According to the Center for Responsive Politics, Bechtel contributed hundreds of thousands of dollars to Republican campaign efforts.

Arcing across this subterfuge, dwarfing it by the sheer magnitude of its malevolence, is America's anti-terrorism legislation. The U.S.A. Patriot Act, passed in October 2001, has become the blueprint for similar anti-terrorism bills in countries across the world. [. . .]

The Patriot Act ushers in an era of systemic automated surveillance. It gives the government the authority to monitor phones and computers and spy on people in ways that would have seemed completely unacceptable a few years ago. It gives the FBI the power to seize all of the circulation, purchasing, and other records of library users and bookstore customers on the suspicion that they are part of a terrorist network. It blurs the boundaries between speech and criminal activity creating the space to construe acts of civil disobedience as violating the law.

Already hundreds of people are being held indefinitely as 'unlawful combatants.' (In India, the number is in the thousands. In Israel, 5,000 Palestinians are now being detained.) Non-citizens, of course, have no rights at all. They can simply be 'disappeared' like the people of Chile under Washington's old ally, General Pinochet. More than 1,000 people, many of them Muslim or of Middle Eastern origin, have been detained, some without access to legal representatives.

Apart from paying the actual economic costs of war, American people are paying for these wars of 'liberation' with their own freedoms. [. . .]

So, as Lenin used to ask: What Is To Be Done? Well . . .

We might as well accept the fact that there is no conventional military force that can successfully challenge the American war machine. Terrorist strikes only give the U.S. Government an opportunity that it is eagerly awaiting to further tighten its strangle-hold. [. . .] The government's suppression of the Congressional committee report on September 11th, which found that there was intelligence warning of the strikes that was ignored, also attests to the fact that, for all their posturing, the terrorists and the Bush regime might as well be working as a team. They both hold people responsible for the actions of their governments. They both believe in the doctrine of collective guilt and collective punishment. Their actions benefit each other greatly.

The U.S. government has already displayed in no uncertain terms the range and extent of its capability for paranoid aggression. In human psychology, paranoid aggression is usually an indicator of nervous insecurity. It could be argued that it's no different in the case of the psychology of nations. Empire is paranoid because it has a soft underbelly.

Its 'homeland' may be defended by border patrols and nuclear weapons, but its economy is strung out across the globe. Its economic outposts are exposed and vulnerable. Already the Internet is buzzing with elaborate lists of American and British government products and companies that should be boycotted. Apart from the usual targets – Coke, Pepsi, McDonalds – government agencies like USAID, the British DFID, British and American banks, Arthur Andersen, Merrill Lynch, and American Express could find themselves under siege. These lists are being honed and refined by activists across the world. [. . .]

It would be naïve to imagine that we can directly confront Empire. Our strategy must be to isolate Empire's working parts and disable them one by one. No target is too small. No victory too insignificant. We could reverse the idea of the economic sanctions imposed on poor countries by Empire and its Allies. We could impose a regime of Peoples' Sanctions on every corporate house that has been awarded with a contract in postwar Iraq, just as activists in this country and around the world targeted institutions of apartheid. Each one of them should be named, exposed, and boycotted. Forced out of business. [. . .]

Another urgent challenge is to expose the corporate media for the boardroom bulletin that it really is. We need to create a universe of alternative information. We need to support independent media like Democracy Now!, Alternative Radio, and South End Press.

The battle to reclaim democracy is going to be a difficult one. Our freedoms were not granted to us by any governments. They were wrested from them by us. And once we surrender them, the battle to retrieve them is called a revolution. It is a battle that must range across continents and countries. It must not acknowledge national boundaries but, if it is to succeed, it has to begin here. In America. The only institution more powerful than the U.S. government is American civil society. The rest of us are subjects of slave nations. We are by no means powerless, but you have the power of proximity. [. . .] Empire's conquests are being carried out in your name, and you have the right to refuse. You could refuse to fight. Refuse to move those missiles from the warehouse to the dock. Refuse to wave that flag. Refuse the victory parade.

You have a rich tradition of resistance. You need only read Howard Zinn's *A People's History of the United States* to remind yourself of this.

Hundreds of thousands of you have survived the relentless propaganda you have been subjected to, and are actively fighting your own government. In the ultra-patriotic climate that prevails in the United States, that's as brave as any Iraqi or Afghan or Palestinian fighting for his or her homeland.

If you join the battle, not in your hundreds of thousands, but in your millions, you will be greeted joyously by the rest of the world. And you will see how beautiful it is to be gentle instead of brutal, safe instead of scared. Befriended instead of isolated. Loved instead of hated. [. . .]

History is giving you the chance.

Seize the time.

COPYRIGHT INFORMATION

PART THREE TWENTY-FIRST CENTURY GEOPOLITICS

PART FOUR THE GEOPOLITICS OF GLOBAL DANGERS

PART FIVE ANTI-GEOPOLITICS

Index